Springer Optimization and Its Applications

Volume 183

Aims and Scope

Optimization has continued to expand in all directions at an astonishing rate. New algorithmic and theoretical techniques are continually developing and the diffusion into other disciplines is proceeding at a rapid pace, with a spot light on machine learning, artificial intelligence, and quantum computing. Our knowledge of all aspects of the field has grown even more profound. At the same time, one of the most striking trends in optimization is the constantly increasing emphasis on the interdisciplinary nature of the field. Optimization has been a basic tool in areas not limited to applied mathematics, engineering, medicine, economics, computer science, operations research, and other sciences.

The series **Springer Optimization and Its Applications (SOIA)** aims to publish state-of-the-art expository works (monographs, contributed volumes, textbooks, handbooks) that focus on theory, methods, and applications of optimization. Topics covered include, but are not limited to, nonlinear optimization, combinatorial optimization, continuous optimization, stochastic optimization, Bayesian optimization, optimal control, discrete optimization, multi-objective optimization, and more. New to the series portfolio include Works at the intersection of optimization and machine learning, artificial intelligence, and quantum computing.

Volumes from this series are indexed by Web of Science, zbMATH, Mathematical Reviews, and SCOPUS.

More information about this series at http://www.springer.com/series/7393

Dionysis D. Bochtis • Dimitrios E. Moshou
Giorgos Vasileiadis • Athanasios Balafoutis
Panos M. Pardalos
Editors

Information and Communication Technologies for Agriculture—Theme II: Data

Springer

Editors
Dionysis D. Bochtis
Institute for Bio-economy and Agri-technology (iBO)
Centre for Research & Technology Hellas (CERTH)
Thessaloniki, Greece

Dimitrios E. Moshou
School of Agriculture
Aristotle Univeristy of Thessaloniki
Thessaloniki, Greece

Giorgos Vasileiadis
Institute for Bio-economy and Agri-technology (iBO)
Centre for Research & Technology Hellas (CERTH)
Thessaloniki, Greece

Athanasios Balafoutis
Institute for Bio-economy and Agri-technology (iBO)
Centre for Research & Technology Hellas (CERTH)
Thessaloniki, Greece

Panos M. Pardalos
Department of Industrial and Systems Engineering
University of Florida
Gainesville, FL, USA

ISSN 1931-6828 ISSN 1931-6836 (electronic)
Springer Optimization and Its Applications
ISBN 978-3-030-84147-8 ISBN 978-3-030-84148-5 (eBook)
https://doi.org/10.1007/978-3-030-84148-5

This Springer imprint is published by the registered company Springer Nature Switzerland AG
The registered company address is: Gewerbestrasse 11, 6330 Cham, Switzerland

Preface

In recent years, the rapid technological evolution has led to the development and application of a plethora of apparatuses for monitoring and recording or transmitting key variables within crop and livestock production. The adoption of information and communication technologies (ICT) in agricultural applications, has made possible the collection of a large amount of data. At the same time, the increasing computational capacity has increased the information management potential. Since the acquisition of data has become easier than ever, their effective management has become the biggest challenge to be tackled. The heterogeneity of the collected agri-data increases the difficulty degree of their management.

Towards that direction, the first part of this book focuses on data technologies and their relation to agriculture, and presents three key points in data management, namely data collection, data fusion, and their use in machine learning and artificial intelligent technologies. The second part of the book is devoted on the integration of these technologies in agricultural production processes by presenting specific applications in the domain. Finally, the last part of the book examines the added value of data management within agricultural products value chain.

Data collection technologies are continuously evolving, while new ones are being developed and applied widely in agriculture. However, the challenge still remains: How can this technological evolution be used for reforming agricultural processes and tasks? Towards that direction, the chapter "You Got Data... Now What: Building the Right Solution for the Problem" elaborates on the ways that the present technological revolution can be used to facilitate manual agricultural tasks and provide with the optimal management recommendations that can eventually result in their automation, since it is evident that technological developments lead to increased production and productivity. The chapter builds upon the realization that there are no "one fits all" solutions; thus, the comprehensive understanding of the potential of each option is a prerequisite for the development of suitable solutions for each problem. Additionally, the chapter assesses the potential of state-of-the-art sensing hardware and software, in offering credible solutions to current agri-food production challenges, acknowledging the importance of proper choices in their ergonomic and economic exploitation. In more detail, the work examines elements of the proper design of sensor networks, based on the required tasks and functions that need to be automated and optimized. Towards that direction, innovative solutions developed for a series of agri-food challenges are investigated.

The immediate availability and access to vast amounts of data at various temporal and spatial scales, where especially in the agricultural sector time and space diversifications are more intense, has created the need for suitable tools for the assessment and management of the different datasets that are collected. These tools call for effective techniques in order to aggregate the collected data. Data fusion can lead towards the optimal use of partial information, through the joint use of the diverse datasets and their effective combination, as a means of gaining the maximum benefit by obtaining a more comprehensive knowledge of the examined subjects.

The concept of data fusion involves all those methods that are used to merge information from various sources to create composite models of processes. The combined models can be considered as more accurate and exhaustive as they are generated with the utilization of information collected from more than one (and heterogenous) source. In the chapter "Data Fusion and Its Applications in Agriculture", the methodological tools available for the optimal fusion of multi-type datasets are presented. The chapter provides an overview of data fusion science along with wireless sensor networks (WSN) in order to examine the various applications in agricultural data processes. More specifically, the "whys" and "wherefores" of information fusion are presented along with the methods, techniques, and algorithms implemented for data fusion applications in agriculture. The chapter also provides a detailed overview of specific applications of data mining and artificial intelligence techniques for tackling agriculture-related issues such as weed disease detection.

Machine learning is a key technology for data analysis. In recent years, under the evolution and resonance of artificial intelligence, machine learning has been utilized in a wide range of applications covering almost all areas of human activity. Machine learning is a rapidly developing scientific field with the ultimate goal of transferring intelligence from human to machines. Computers practice through the provision of inputs and outputs and are trained by learning through their mistakes. In this manner, the computers/machines attempt to create rules that relate each input to a specific output. By doing this, machine learning algorithms can provide results in systems where there is not an explicit knowledge on the interconnection of the various processes (or at least in some of them). Agricultural production is such a system. It is a fact that agriculture is related to uncertainty, as it is a production domain affected by different and unstable time and space factors, such as varying climate conditions and fluctuations in the demand and supply of products. For now, the use of machine learning in the agri-food sector is mostly focusing on addressing specific issues, such as weed detection, plant species and disease detection, and yield forecasting. The evolution of its penetration in agriculture aims at using machine learning in the decision-making processes, along with the collection and analysis of data, in an integrated manner.

The chapter "Machine Learning Technology and Its Current Implementation in Agriculture" presents in detail the fundamentals of machine learning, starting from how it works, moving further to providing an overview of the basic machine learning methods and the relevant algorithms that are used for their realization. The chapter also thoroughly investigates the application of machine learning in agriculture by providing an extensive review of the recent applications.

For the optimization of agricultural processes in the digitized era of Industry 4.0, a series of components including Internet of Things (IoT), Big Data, robotics, simulation, horizontal and vertical integration, and cloud computing, among others, are becoming increasingly important. The second part of this book presents the application of the data technologies, described in the first part, on the agricultural domain. The application of IoT concepts, aiming at the acquisition of agricultural data, is related to the concept of Big Data, which concerns their management and utilization under the goal of decision-making. Several research, mainly on small-scale experiments, implemented several multifunctional, modular data acquisition and management systems, comprising of sensor networks, databases, application programming interfaces (API), and desktop or web-based management applications. Nevertheless, such research is usually limited to the data acquisition process not considering the entire data management workflow.

In the chapter "Application Possibilities of IoT-Based Management Systems in Agriculture", the issue of an integrated assessment of the entire workflow is revisited, expanding the research from data acquisition to data management and analytics. The chapter reviews the most recent advances on IoT and Big Data methods, to discover their development potential towards new advancements. The research presented firstly investigates the different methods through a quantitative bibliometric analysis and then examines the development of a data acquisition system and its application on a production system (specifically in greenhouse production systems). A practical experiment is also presented for the validation of the proposed system, which considers the use of agricultural data collected from various sensors.

The task of identifying plant species is a resource-intense task as it requires skills and experience and a significant number of workhours. Hence, the automatization of this tedious task by means of digital imagery processing is of a great interest. To cope with the requirements of the species identification task, an application would have to match the human performance (if not to outperform) requiring specialized knowledge and to be trained to assess and differentiate between complex identification features. A real-time recognition application is a valuable function both for experts and non-experts, aiding experts in their professional life while making plant identification less inaccessible to the non-expert farmers. Floral diversity is an index of ecological resilience and healthy functioning. Subsequently, by monitoring diversity, density, and plant health in an ecosystem, universal inferences about pollution, chemical contamination, and other negative impacting factors can be drawn.

In the chapter "Plant Species Detection Using Image Processing and Deep Learning: A Mobile-Based Application", the feasibility of the development of an automated system to identify plant species by means of digital imagery was proven.

Towards that scope, artificial intelligence and deep learning models can assist in achieving the high accuracy levels necessary for the task. It was concluded that convolutional neural networks perform efficiently and accurately for plant species identification, thus satisfying the key performance index set. Other deep learning models have been also used on the same training set to benchmark and evaluate performance on accuracy, reliability, and throughput levels. Throughout the process it became evident that a major challenge to be addressed is the volume of work related to pre-processing involved in compiling a high-quality training dataset.

The expected additional value of automation applications concerns increased production volume, reduced need of labor, processes' speed increase, and a controlled quality. This holds true in an agricultural context as characteristics of the sector enable it to benefit significantly from automation. Encompassing recent technological advances, cutting-edge technologies are making their presence noticeable in everyday operations. The range covers form robotics applications to artificial intelligence computer vision. This goes beyond the primary production up to the entire agri-food chain, with post-harvest processing being one of the links expected to benefit from such technologies. The chapter "Computer Vision–Based Detection and Tracking in the Olive Sorting Pipeline" presents an artificial intelligent-based automation paradigm in olive sorting process. Agri-food products sorting is a recurring task for each cultivating season. The workflow requires tedious manual labor that entails inspection and sorting of the harvested crop based on features such as color and ripeness level, among others. The task is suitable for automation, as speed of execution, accuracy, and repeatability are critical factors when dealing with large volumes of harvested crops. For introducing technological advances in harvested crops sorting, authors evaluated computer-vision technologies, to identify and register olive crops on a moving conveyor belt. The developed solution is based on the watershed transformation technique for the definition of the centroid of a fruit, along with Kalman filter for prediction and detection combined to a correction algorithm.

Image analysis applications are not exhausted at the level of identifying objects or their features but also cover areas of macroscopic interest. The objective of the work presented in chapter "Integrating Spatial with Qualitative Data to Monitor Land Use Intensity: Evidence from Arable Land – Animal Husbandry Systems" is to investigate changes in land use intensity of crop-livestock farming systems. Crop-livestock mixed faming systems have been always an object of study. Nonetheless, in a period between 1960 and 2000, such mixed farming systems have been decreasing in numbers, a trend that could be attributed to major structural landscape changes. Authors present an approach for monitoring land use intensity by means of remote sensing complemented by qualitative methods for a systematic classification of data, data analysis, data mapping, and finally information interpretation. To establish a robust methodological framework, authors implemented more than one method for recording the change of state in land-use intensity throughout the period examined. One of the methods used for data retrieval and processing was structure-from-motion, which utilized B&W aerial images. The outcome was a set of orthophoto maps, each depicting the state of land use in a certain period. Furthermore, object-based image analysis was implemented within the framework of remote sensing, with effective retrieval of useful information from panchromatic aerial imagery. The compiled information provided insight in land cover conversions, recording changes in land-use and agricultural coverage, reporting both expansion and abandonment of areas. Complimentary to horizontal approaches, monitoring cycles of crops to grasslands conversions are enriched with data regarding labor and capital investment. Qualitative methods also were used to document land-use management changes, for example, conversion of irrigated to rain-fed crops, as they were not traceable in remote sensing datasets. The implementation of multiple analysis frameworks results to a more information-dense and complete capture of state, which allows for the estimation of labor and capital investment in the period examined. These indicators are useful in elicitation of additional layers of information which can uncover underlying conditions, not being directly traceable. From this work, it becomes evident that one-dimensional approaches have limited applicability in macroscopic analyses such as land-use intensity changes.

Beyond changes in pure computational processes, Industry 4.0 has initiated a sequence of changes with a cascading effect to all technical sectors including agri-machinery industry. Machinery that copes with precision agriculture requirements has varying levels of integration in the so called "smart farming" concept, with respective challenges in operational efficiency. As, for example, in extensive agricultural systems, the machinery effectiveness is based on larger operating widths. However, large operating widths are linked to higher variability in the work quality. In the case of seeding operations, this translates to higher variability in plant spacing and higher rate of plant losses. Therefore, designing sub-systems to support the technological requirements of precision agriculture systems is a challenging endeavor. In building such a sub-system, simulation and data analysis are prerequisites.

The chapter entitled "Air Drill Seeder Distributor Head Evaluation: A Comparison Between Laboratory Tests and Computational Fluid Dynamics Simulations" is an example of such an implementation where for the design of a seeder distributor, in parallel to physical specifications, an additional layer of control is needed. In this layer, systems that monitor, control, and adjust operation of machinery are necessary to comply with real-time operational requirements. The presented approach is an example of a successful combination of numerical and experimental data in the development of a sub-system under precision requirements stemming from the Agriculture 4.0 framework.

Agricultural supply chains nowadays are more complicated than ever, involving a variety of different stakeholders in the different stages that go beyond the primary production, including the processing and distribution of the final products. Each element in an agricultural supply chain is linked to the generation of significant amounts of heterogenous data of different types and from different sources. For a successful management of agricultural supply chains, the effective assessment of these data flows in an integrated manner is imperative. However, the heterogeneity of the collected data leads to interoperability constraints between the involved agents in the chain. These problems are often deteriorated because of the inconsistency of the collected data, in terms of the format and the information resolution of the databases. Data collection, storage, and processing is performed in many ways; thus, the standardization of the information is especially complex and time consuming, while in many cases unprofitable. The last part of this book is dedicated to data management within agricultural value chains and the challenges involved.

Data management and business continuity in agriculture are two challenging and interconnected topics. Business continuity in agriculture is affected by the vulnerability of agricultural productivity due to imminent natural and man-made hazards. Research on business continuity management has highlighted the importance of using contemporary information and big data technologies in efficient strategies and management schemes towards the mitigation, or even the elimination, of the adverse impacts of such hazards. Data analysis for decision-making can be either descriptive or predictive based on the required outcome of the assessment. However, both scopes of analysis require considerable data volumes and intelligent systems for their successful collection and manipulation, while specific guidelines for the design of business continuity management policies should be followed, especially since inconsistent solutions lead to poor business continuity strategies. Towards that direction, business intelligence multidimensional models can promote the acquisition of knowledge for data-driven decision-making and strategy implementation ensuring business continuity.

The chapter "Data-Based Agricultural Business Continuity Management Policies" proposes a standardized data-based approach for the utilization of business continuity management data towards the creation of efficient multidimensional data models that promote business continuity and facilitate and improve decision-making. By utilizing data from two agricultural industries, authors propose two new multi-dimensional data warehouse schemas as tools for the analysis and storage of pattern-oriented agricultural business continuity management data. Data organized by these tools lead to dynamic data-based business continuity decisions that are represented with the use of business process modeling and notation diagrams. Lastly, decision trees are used to demonstrate an example of agricultural business continuity management classification, as evidence of the parallel predictive capability of the data used to create these schemas.

Predicting the price of agricultural commodities, in the short and in the long term, is an important issue that affects the revenues of all stakeholders across the value chain. Improving the predictability of prices can lead towards improved decision-making on a variety of aspects related to crop production, such as land cover allocation. In the short term, price prediction is related to the final revenue and the subsequent hedging that should be acquired. Additionally, speculation on the prices can be mitigated, by improving the predictability of prices. With respect to price time series, two main types of predictions can be distinguished: price prediction and price trend prediction, with the first being characterized as a regression and the second as a classification problem. In the price prediction problem, the focus is on the determination of the value for the next period, while in the price trend prediction problem, the focus is on predicting if the price will increase or decrease in the examined period.

The chapter entitled "Soybean Price Trend Forecast Using Deep Learning Techniques Based on Prices and Text Sentiments" delves into the issue of forecasting the price trend for the soybean value chain, which is considered one of the most important agricultural chains worldwide in terms of human food and animal feed production. Towards that direction, deep learning techniques are utilized to improve the predictability, compared to traditional models. More specifically, the chapter examines the use of deep learning techniques for the generation of a price signal and of a sentiment signal deriving from agricultural finance news along with the prediction of the price trend.

Data labelling is another issue in the assessment of agricultural supply chains that produce an enormous quantity of data, of different types and originations in terms of sources. To improve decision-making from the utilization of this inexhaustible source of data, their proper labelling with the information generated during their creation is imperative. The most important processes that generate data in agricultural supply chains are related to product identification and traceability, as well as to environmental and process monitoring. However, the manual processing of the collected data is not economically feasible. In an attempt to address this issue, the chapter "Use of Unsupervised Machine Learning for Agricultural Supply Chain Data Labeling" presents a dataset that represents a general agricultural supply chain, on which two unsupervised machine learning models (k-means and self-organizing maps – SOM) were implemented for the identification of patterns within data and the labeling of data points. The use of principle component analysis was also evaluated. Three groups of metrics were evaluated (traditional supervised learning, supervised clustering, and unsupervised clustering) and the SOM model demonstrated the best results in all of them, being the only one that managed to detect the smallest data points. Finally, the Gaussian neighborhood function demonstrated the best cluster separation.

This book aims at covering horizontally the "big issue" of data in agriculture. It provides the basic knowledge on the data technologies regarding data collection, data fusion, and data use by machine learning and artificial intelligence, alongside with narrative literature reviews on the implementation of these technologies on agricultural domain, and real-life applications covering topics from primary production up to the agri-food value chain. From the works presented here, it is evident that there are unlimited opportunities throughout the data-driven transformation of conventional agriculture to a digital one, but on the other hand, there are numerous limitations and challenges that must be faced.

Thessaloniki, Greece — Dionysis D. Bochtis
Thessaloniki, Greece — Dimitrios E. Moshou
Thessaloniki, Greece — Giorgos Vasileiadis
Thessaloniki, Greece — Athanasios Balafoutis
Gainesville, FL, USA — Panos M. Pardalos

Contents

Part I
Data Technologies

You Got Data.... Now What: Building the Right Solution for the Problem

Patrick Jackman

1 Introduction

Demands on agricultural and food resources are coming under ever-increasing pressure due to population increases which is now expected to be 8.5 billion by 2030 [1] in connection to the plateauing of arable land supply [2] and the broader decline of food production resources [3, 4]. Thus, feeding an increasingly hungry planet will require across-the-board optimization of production resources unless and until alternative food resources can be found. Pressure to optimize resources also comes from an increasingly globalized economy where the progressive lowering of trade barriers between nations and trading blocs has increased competition on suppliers to sell for the lowest possible price they can bear. An additional strain on agricultural production comes from an increasing labor crisis whereby tasks that were previously done by hand now need alternative methodologies [5–7].

In recent decades, technological solutions have been stepping up to the plate at all levels of the supply chain. At farm level, sensors and sensor networks have been used for arable farm production in multitude of scenarios [8]. A variety of counterpart software has been implemented to make the most of the acquired datasets, ranging from simply reporting sensor readings to implementing automatic controls based on the readings. The great challenge has been providing value for money for farmers and other production stakeholders and the historic high cost of specialized hardware has acted as a strong barrier to implementing technological solutions. Another limiting factor in some scenarios had been the unavailability of sufficient software power to optimally process the very large and nuanced datasets from which a decision must be divined.

P. Jackman (✉)
Technological University Dublin (formerly Dublin Institute of Technology), Dublin, Ireland
e-mail: patrick.jackman@tudublin.ie

D. D. Bochtis et al. (eds.), *Information and Communication Technologies for Agriculture—Theme II: Data*, Springer Optimization and Its Applications 183,
https://doi.org/10.1007/978-3-030-84148-5_1

We are now in the midst of two golden ages of information and communications technology and these revolutions have made data acquisition and processing that were previously considered infeasible in previous decades possible. More specifically, the opportunities have been created by the great fall in the cost of precision hardware components, leading to cheap and accurate sensors and the great expansion of computing capacity that have made the deep learning neural networks, dreamed of in the twentieth century, achievable.

This chapter considers how this technological revolution can ease the manual burden of agricultural and food production tasks and at least provide a suggested optimal course of action to such tasks and even go as far as automatically controlling such tasks. Simultaneously it is recognized that there are no "one size fits all" solutions and that the advantages and disadvantages of each option need to be fully understood and appreciated so the right solution is designed for the right problem. Hence, by embracing these golden ages in data the agri-food industry can increase production and productivity to meet the aforementioned ever-increasing needs of the world. Furthermore, this chapter evaluates the power of modern sensing hardware and analytical software to offer credible solutions to the current pressing agri-food production challenges, while also recognizing that the right choices need to be made if their potential is to be realized ergonomically and economically.

More specifically this chapter considers how an appropriate sensor network can be designed and engineered according to the envisaged tasks and functions that need to be automated and optimized; typical production scenarios that are in strong need of optimization and automation are both in-field and intensive crop production, as well as intensive animal production. A variety of modern technological solutions are examined, and the great diversity of solutions illustrate that almost any agri-food challenge can be met with some form of sensor network. Similarly, this chapter examines how the latest developments in artificial intelligence form the perfect counterpart to the sensor network datasets and in particular how Convolutional Neural Networks and Recurrent Neural Networks offer the ideal platform for processing the huge image and time-series datasets.

Various works have examined the "Big Data" question from a more fundamental viewpoint and provided a strong theoretical understanding of the principles of how to avail of the opportunities provided by the "Big Data" revolution. It is the intention of this chapter to provide an understanding of how a practical and ergonomic solution to current agri-food production challenges can be engineered to surf the wave of this revolution and use "Big Data" to provide "Big Answer".

2 Sensors and Their Readings

As mentioned, the cost of precision hardware has fallen tremendously in recent years leading to an abundance of affordable sensors entering the market that can measure a great variety of phenomena in many different ways. As such, we are often spoilt for choice when sourcing sensing hardware. However, great care is needed when choosing the right sensor for the right task as all sensors have limitations and restrictions on their correct use that must be understood and obeyed if trustworthy data is to be recorded.

It must be appreciated that a sensor is merely a device that produces a reaction when exposed to a stimulant; it only has utility if its reaction can be accurately converted into SI Units via a reliable calibration curve. If it cannot, then the sensor is of limited or even no practical value. While it is unlikely that a totally inappropriate sensor is chosen for a task, what is far more likely is that a sensor's limitations are not fully understood or appreciated when applying it to a task leading to plausible yet incorrect readings being recorded [9–11]; hence, leading to plausible yet flawed results and conclusions. An extreme example would be how flawed sensor design caused false yet plausible readings in the control room at the Three Mile Island Nuclear Power plant confusing the reactor engineers and contributing to the partial meltdown [12].

A sensor ontology is required to fully understand its limitations, and thus, ensure that it is only used appropriately for the tasks envisaged and its calibration curve is applied correctly. A comprehensive sensor ontology was produced by Compton et al. [13] that dissects each ontological element of a sensor and removes any assumptions that could lead to the sensor's performance being misinterpreted.

This ontology will reveal common phenomena that can skew a sensor's performance such as sensor drift [14] whereby a sensor's reaction to the same stimulation (including zero stimulation) gradually changes over time and often due to aging electronic components [15]. Detection limits [16] are the minimum stimulation to cause any reaction and the level of stimulation that causes the maximum possible reaction or "saturation" [17]. Sensor poisoning [18] is when a confounding substance or a very strong stimulation causes an irreversible change to the characteristic reaction to stimulation. Correctly anticipating these and other ontological phenomena will help lead to reliable conversion to SI Units.

In more practical terms a sensor's manufacturer will normally warrant a sensor's performance for a limited period of time on condition that the sensor is only used within specified limits. For example, this could be a temperature sensor is warranted for 12 months to provide accuracy within 0.1 °C, as long as the relative humidity of its environment is above 10% and below 80%; outside these limits the manufacturer will no longer warrant the sensor's performance. Thus, accurate knowledge of a sensor's limits will be essential to the acquisition of usable datasets as if a sensor is used outside its warranted limits then inadvertent collection of inaccurate data is likely and this will have a cascade effect throughout the entire data acquisition and processing chain. Furthermore, error identification will become very difficult if every sensor appears to be reporting consistent and plausible measurements.

A prime example of a limited yet highly useful sensor within its limits of operation is a thermal camera. Typically, thermal cameras can have excellent precision but poor bias [19]. As a result, they can be highly useful tools for measuring temperature contrast but poor tools for measuring absolute temperature. For example, if there were a leaking hot water pipe inside a wall of a building a thermal camera would be able to see the slightly higher wall temperature in the immediate vicinity of the leak and thus help identify the location of the leak. However, a thermal camera would not be able to estimate the absolute temperature of any part of the wall accurately.

There are numerous additional limitations to the implementation of sensors even if the issues above are properly addressed. A fundamental weakness of sensors is that, on their own, they are spatially and temporally blind. While Global Positioning System (GPS) detectors and internal computer chips that can count time are increasingly common, they are not available in all sensors and thus the location in space and time of the sensor reading must be inferred from the sensor lag and the installation and maintenance records. Sensor lag, whereby the sensor takes time to reach its equilibrium reading, can be characterized as the "rise time" and "fall time" [20] and should be specified by the manufacturer. For automatic control systems incorrect temporal assignment of sensor readings could lead to major process control errors especially where the process in question has a strong positive feedback. An extreme example of the dangers of positive feedback would be the old Soviet Union RBMK reactors [21].

If a sensor is attached to a moving piece of equipment, then incorrect temporal identification will also mean incorrect spatial identification; incorrect spatial identification could also come from errors in installation or maintenance where the technician incorrectly records where they mounted the sensor. In situations where the medium being sensed is well mixed this may not be a serious problem but where there is poor mixing then large errors can be propagated into process control decisions. A prime example would be a Continuous Stirred Tank Reactor or CSTR; generally, the region around the impeller has excellent mixing but the regions that are further away have poorer mixing and in some edge regions there may be little or no mixing in so-called "Dead-Zones" [22]. Thus, spatial placement of a sensor could be critical to the measurements recorded and their representivity of the true averages in the reactor.

As stated above increasingly "smart" sensors can be manufactured to make them spatially and temporally aware so it can be expected that the issues described will be less prominent in the coming years although there will always be some scenarios where it is impractical to add internal GPS and internal timestamping. Awareness of spatial and temporal location becomes increasingly important as sensing solutions move away from dependence upon a single sensor to provide a representative picture of the medium under investigation.

3 Networks of Sensors

In a large number of scenarios an array of sensors is required to capture the true spatial and temporal patterns of the medium of interest, and the fall in the price of accurate sensors and rapid telecommunications equipment has made sensor networks affordable for tackling many agri-food production problems. The parallel advancement of telecommunications capacity [23] has made networks of sensors feasible for an increasing number of tasks. Thus, where a single-point measurement cannot be trusted to provide a true picture of the medium of interest, a strategically placed network of sensors can provide the pseudo-spatially continuous monitoring of the medium. Of particular interest is how the recent advancement of wireless telecommunications has made the placement of sensor networks in very inaccessible and inhospitable locations possible [24].

It may be unavoidable to use a sensor that cannot detect its location in space and time, and this introduces a risk of spatial and/or temporal misplacement within the network of the recorded data. This can be mitigated by keeping accurate and diligent sensor installation and maintenance records. Where it is avoidable to use such sensors the emergence of low-cost microcomputers such as the Raspberry-Pi (Raspberry Pi Foundation, Cambridge, United Kingdom) and compatible low-cost GPS devices has made it possible to automatically add timestamps and GPS to sensor recordings by creating sensor "nodes" that know their place within the network.

The aforementioned low-cost microcomputers have allowed very efficient and effective network software to be implemented as the end nodes are "smart" in that they can communicate with the central server and accept commands and software updates from the central server as well as providing essentially "live" data and reports to the central server. Increased hardware capacity has made it possible to send much larger data files than previously through the network including "live" video feed [25] and thus a network is no longer limited to point sensor data.

Additionally, the emergence of low-cost wireless Wi-Fi has offered a realistic alternative to wired or Ethernet telecommunications. By eliminating the need for physical connection, it is possible to place sensors far from the central server and into difficult or hazardous locations where minimum maintenance is desired. However, as a general rule, wireless networks are less stable and secure than physically wired networks [26] and this is a factor that must be considered when designing a network of sensors.

Whichever options are chosen a robust buffering regime must be in place whereby if a sensor node loses communication with the central server its recordings are held in memory or storage until communication is restored and the batch of recordings can be transferred "en masse". This underscores the need for the nodes to be temporally aware with each individual recording having a timestamp, so there is no doubt as to what measurements were recorded at what time. While unlikely, it is possible that the sensor was moved during the out of communication period and in this case a local GPS stamp on the recordings would eliminate any spatial doubt.

A robust sensor network has the capability of providing comprehensive spatially rich and temporally rich accurate datasets that can reliably describe the medium of interest and thus lead to accurate decision making. Whether this takes the form of merely providing information to a human expert, recommending a decision to a human expert or fully automatic control such a network will lead to better decision making and thus better process outcomes. A host of pertinent applications is described below.

3.1 In-Field Crop Production

Sensor networks are an ideal solution to the in-field monitoring of crop production as they save a farmer or farm manager the time and effort of visiting each part of their crop-growing fields; for even the wealthiest farmer time, will always be in short supply. In countries like the USA crop fields can be enormous with most farms now at least 1100 acres in size [27] making it impossible to comprehensively survey a field by manual examination on most farms.

The advent of low-cost point sensor networks has made it possible to have a dense network of sensors providing comprehensive hyperlocal datasets of important features related to good crop production such as temperature, humidity, wind speed and direction, soil moisture, soil pH, wetness, etc. Armed with dense hyperlocal datasets problems that would otherwise be missed by a cursory manual examination can be identified. Strong Wi-Fi or mobile phone signals will provide a stable platform for the remote sensor nodes to communicate to the central server and there is a multitude of data transfer protocols such as Zigbee [28]. In recent years there have been more ambitious attempts to integrate biosensors and chemical sensors into in-field sensor networks such as that devised by crop giant Syngenta and partners [29] with the specific biosensor part described by West and Kimber [30]. This builds a comprehensive spatial and temporal map of common disease risks that can inform critical crop spraying strategies for farmers. While no longer cutting-edge technology, a point sensor network can still provide a very rich data resource from which reliable predictions can be derived.

More recent advancements in hardware have made it possible to use cameras as well as point sensors, so image processing can provide alternative comprehensive field assessments [31]. This is particularly useful when infrared images are captured along with Red, Green, and Blue, as crop indices can be calculated as convoluted vegetation index images. These images show at a hyper-local level where a multitude of problems may or may not exist such as poor nutrition, poor growth, poor moisture, etc. [32].

There are three main options in this regard; firstly, cameras can be mounted on towers to act much like "security cameras" with a line of sight to all parts of the field [33]. This has the advantage of simplicity and the facility to mount a lot of hardware together such as a strong Wi-Fi transmitter. Although for the same reason they have the disadvantage of cumbersome installation and maintenance, they also have the disadvantage that some regions of the field will be viewed at a steep angle and some at a shallow angle leading to inconsistency in the image data resolution.

Secondly, the cameras could be mounted onto agricultural machinery that would frequently visit the field so proximal image data is collected in the same manner that existing technology collects proximal spot average data [34–36]. This eliminates the aforementioned problems of cumbersome installation and maintenance and the inconsistency in the line of sight of the camera. The essential advantage of proximal sensing is the effective image resolution and granularity is greatly increased meaning more hyperlocal features can be extracted and can lead to better hyper-local predictions of key crop characteristics [37]. These can in turn lead to more precisely targeted remedies and mitigations that are increasingly important as restrictions on the use of pesticides and other crop chemicals are growing [38]. The key disadvantage of this approach is that data can only be collected on occasions when the machinery visits that field; however, limited temporal granularity may be acceptable.

A third option is to effectively integrate Unmanned Aerial Vehicle Imagery or Satellite Imagery into the sensor network by consolidating these images with all other datasets on the central server. The use of Aerial Vehicles, be it drones, or piloted small planes have the key advantage of being able to survey a very large area in a short time and the ability to cross otherwise inhospitable and inaccessible terrain. The disadvantage being a lot of effort is required to fly the Aerial Vehicle and data collections can only take place when such resources are available. The integration of satellite images requires no such efforts only the fees associated with their purchase; however, satellite images have limited resolution and granularity which may be insufficient for the desired applications [39].

In summary, a spectrum of options exists for building a sensor network for in-field crop monitoring depending upon what limitations are acceptable and what the intended applications are. Low resolution and granularity datasets might be adequate if non-targeted applications need to be better informed, such as when is the optimal moment to harvest an entire field. In other applications, high resolution and high granularity datasets might be essential such as for targeted application of herbicides for weeds [40]. Either way, some kind of sensor network solution is feasible and ultimately desirable for in-field crop production.

3.2 Intensive Crop Production

For a variety of reasons crops will be grown intensively in an enclosed environment but primarily this will be done to simulate the climate under which the crop is most efficiently grown. It might also be done for increased biosecurity [41]. The most critical climate simulation controls will be manipulating ambient temperature and humidity to replicate peak growing conditions where the natural environment is incapable of or unable to sustain for sufficient timeframes [42]. The creation and maintenance of an optimal growing climate will require a network of temperature and humidity sensors feeding data to an automatic control system and this will be a relatively simple task in terms of identifying a problem and implementing remedial action, which will usually take the form of increasing or decreasing power to the heating systems or adjusting the water injected and withdrawn from air circulating through a humidifier.

The emergence of affordable camera-based systems has created a new dimension in process controls, as image data analysis is possible and problems that would otherwise be “lost in the average” can be rapidly identified. This would allow a locus of poor performance to be quickly discovered so an investigation could take place. Such poor growth may be due to a hyperlocal lack of nutrients, a defect of the climate control systems in that locality, an ingress of the outside environment due to damage to the polytunnel or greenhouse or a local biosecurity breach. The type of camera systems that could be effective in these scenarios would be multispectral with near infrared to indicate biomass via the Normalized Difference Vegetation Index [43], thermal imaging to scan the housing for crop stress [44] or color imaging to identify the plant phenotype [45]. Image processing algorithms can then mine such image data to generate process control decisions and this will be discussed in-depth in the following sections.

3.3 Intensive Animal Production

Intensive animal production faces similar challenges to intensive crop production in that point sensor networks and image-based networks will be useful in tackling similar process control challenges. Furthermore, the capability of animals for rapid movement means that video imaging rather than static imaging is likely to be required to characterize the performance of the animal crop or herd. The importance of imaging solutions is underscored by the fact that some intensive animal production does not involve a completely enclosed building such as cattle farming which complicates any attempt at the estimation of environmental parameters as external ambient air can freely enter the building.

A prime example would be intensive poultry production, where production takes place within an enclosed building but strong local variations are possible within the building and a single sensor could paint a misleading picture of the true state of production, and large process control errors could go unnoticed. The key process control parameters for optimal bird growth are air temperature, humidity, velocity, carbon dioxide level, and ammonia level [46–48].

A multitude of sensors exist to measure these features, and these can be easily mounted in a grid across the building to identify transient and chronic variations from the set parameters [48, 49]. More advanced and ambitious sensors will attempt a direct measurement of biohazards inside the growing house [50]; however, this is a very difficult challenge especially due to the dusty environment of the house as dust and dirt can obscure the sensor assay. These types of biosensors can be integrated into an automatic control system just the same as a conventional sensor with the correct electronic engineering.

When a transient deviation occurs, the automatic control systems can take mitigating action such as switching on a burner to raise local temperature or opening a venting board if humidity or gas concentration is too high. When a chronic variation is occurring, an investigation can take place as to the cause such as a draught caused by a failing seal on a wall. A multitude of such automatic control systems is already available [51–53].

Advances in telecommunications capacity have now made rapid video data capture and collation possible which can provide much more complete reporting than point sensors alone and a farmer can directly observe the behavior of their flock remotely [54]. Furthermore, bird behavior algorithms can be devised from such video datasets [54, 55]. In this approach bird behavior is determined to be stereotypically normal or abnormal according to the trained algorithm and once abnormal behavior is identified then the farmer can be alerted.

Intensive pig production largely faces very similar challenges to poultry production in that a comfortable internal environment is required for pigs to optimally gain weight and a pig-growing house will be mostly isolated from the external atmosphere. A key factor in maximizing weight gain like poultry production is avoiding heat stress [56]. Thus, very similar sensor networks and automatic control systems can be used to optimize process control for pigs.

As mentioned above, other intensive animal production such as cattle production takes place in growing houses that are not insulated from the external environment and herds will spend limited time inside the houses. Thus, it is futile to attempt to use environmental measurements for process control as it is impossible to control the weather. In scenarios such as this, it will be analyzing animal behavior and development that will be critical and this will involve at least imaging but most likely video-based data acquisition and processing [57, 58]. The determination of normal and abnormal behavior will be based on motion and position and thus it will follow the same principles as for poultry and pig production.

4 Using Machine Learning

The previous sections have discussed the opportunities for acquiring, collating, and consolidating point sensor, image and video datasets and how they can be used to inform process control systems and decision and how this has been made possible by the "Golden Age" of low-cost electronics. The parallel "Golden Age" in computing power has allowed the state of the art in machine learning and artificial intelligence to be applied to these datasets to extract the optimal value from them so better process control decisions are possible.

Previously, an automatic control system would be based upon point sensor data and use approaches such as "Proportional, Integral and Differential" or PID control to determine the appropriate response to the deviation from optimal conditions. This is where the response considered the acute deviation, the chronic deviation, and the instantaneous trend in the deviation from optimal [59]. For many applications this was and still is adequate and systems as important as nuclear reactions were successfully controlled via PID (proportional–integral–derivative) [60].

The emergence of two state-of-the-art advancements in Neural Networks has opened up opportunities for enhanced automatic control systems. The first is Convolutional Neural Networks [61] which allow for intelligent and rapid image processing; whereby images or sequences of images (i.e., videos) can be analyzed in their entirety without the need for predetermined explicit feature extraction; thus the classification label is applied to the entire image. There would be at least a normal or abnormal classification or more ambitiously a classification of the mis-performance type. This would be particularly useful in intensive animal production whereby very difficult problems such as classifying animal gaits and postures can be tackled [57, 62] or in crop production whereby the crop morphology could be rapidly classified in the field [63].

The second state-of-the-art advancement is in Recurrent Neural Networks that can absorb any time delays, lags, and shifts once they are exposed to enough training data [64] thus solving a key weakness of the classic PID method or any other predictive model that has to assume or predict a temporal lag or shift in the data. For example, in intensive poultry production a loss of control that would upset the birds from optimal eating and drinking may not manifest itself for a number of days in the readings from the automatic weighing scales [49] and thus would be an ideal candidate for a Recurrent Neural Network. A strength of these types of neural networks is that they can handle both point and image-based datasets.

As computing power continues to expand, these state-of-the-art neural networks will be able to cope with progressively larger datasets leading to increasingly precise and accurate control systems. The pending quantum computing revolution [65] thus holds out the opportunity for a whole new generation of automatic control systems.

5 Remaining Challenges and Opportunities

The technological advancements described above offer solutions to many of the greatest challenges of agri-food production; namely, the agricultural labor crisis and the chronic inefficiency and ineffectiveness of some current farming methods. However, there can often be too many options for developing technological solutions for optimal agri-food productivity due to the great variety of hardware and software commercially available. Hence, reducing to a small number of credible options deserving of comprehensive examination is the first and ever-present challenge of building any solution.

That step underscores the most fundamental challenge of all; the need for an accurate and precise problem definition, i.e., without the right questions asked the answer will most likely be wrong and expensively wrong. A problem definition will determine what measurements could plausibly improve decision making, how frequently would the measurements need to be taken? What spatial granularity in the measurements would be required? How much missing data and downtime can be tolerated? Would a point sensor be adequate, or would imagery be required at each measurement node? What software and processing power is needed to analyze the measured data in sufficient time? What contingencies are needed if the system misperforms or fails? What other systems would the proposed system depend upon? How quickly does a prediction or recommendation has to be returned, i.e., is real time essential or can results be reviewed at the end of a working day? Can maintenance be easily sourced, etc.? In short, unless it is clearly defined what needs to be measured and delivered a functional and adequate sensor or sensor network and control system simply cannot be defined. Similarly, failures to define the problem accurately at the initial stages can make reverse engineering of hardware and patching of software very difficult and costly.

Following on from defining the problem and thus determining what is possible or what could work, a determination of what is likely to work has to be performed to reduce down the available options resulting from the initial problem definition phase. For agri-food production the first and most important screening factor is interference with normal operating procedures; this is important as even the wealthiest farmers will inevitably be time poor and be dependent on a relatively small workforce due to the aforementioned agricultural labor crisis. Any solution that requires substantial additional human input is not feasible and hence solving the ergonomics of otherwise brilliant technological solutions will continue to be a great challenge of the future.

The second most important screening factor will be cost as even though the cost of electronics has greatly fallen a finished product will be what is ultimately purchased or rented by a farmer or an agri-food producer. Thus, the total costs of the production of the technology (e.g., marketing, research and development, labor, asset depreciation, etc.) will still be integrated into the final sale or lease price and thus costs might still remain prohibitive, particularly for a small-scale farmer. Thus, a great challenge for these kinds of technological solutions remains how to reduce the entire cost base not just the individual hardware components. If these main challenges are met, then the two "Golden Ages" of technology can be availed of to improve agri-food production systems.

However, it should be noted that a very exciting opportunity in the future of agri-food production lies in autonomous robotics which could pave the way for a third technological revolution as some vital farming procedures such as fruit picking still require wholescale manual tasks. Furthermore, autonomous robots could be assigned to highly dangerous tasks such as emptying grain silos that have led to many fatal accidents. If robot production costs can be reduced and comprehensive artificial intelligence algorithms can be programmed into them, then it is possible that a fleet of autonomous robots could perform many of the tedious, physically demanding, and hazardous tasks typical of agri-food production. This would free up the farmer or farm manager so they can devote their efforts and time into more productive tasks and business decisions leading to more productive and profitable supply chains.

In summary, now more than ever before it is possible to acquire very large and very comprehensive datasets in real time or very close to real time, these datasets can measure automatically, objectively, and more cost-effectively that what was previously measured manually or not measured at all. Similarly, it is possible to analyze datasets that were previously too complex to process in a feasible timeframe. It is still vitally important not to lose sight of the production context and these opportunities must be harnessed into a workable solution that reflects the realities "on the factory floor" so even though we have data we can actually use it efficiently and effectively and build a profitable solution for agri-food producers.

The fusion of modern technological advancements used to solve the major agri-food production problems discussed in this chapter has been made possible by the more fundamental technological advancements in data acquisition, storage, and processing that are described in the other chapters in this book. Computer scientists and electronic engineers lay the foundations into which the necessary levels of certainty and clarity in datasets can be assured; and it is on these foundations that agri-food engineers can build the required practical, cost-effective, and ergonomic solutions to current and future production challenges.

References

1. United Nations. (2020a). *Global issues: Our growing population*. United Nations Neutral Zone.
2. United Nations. (2020b). *Looking ahead in world food and agriculture*. United Nations Neutral Zone.
3. Gerland, P., Raftery, A. E., Ševcíková, H., Li, N., Gu, D., Spoorenberg, T., Alkema, L., Fosdick, B. K., Chunn, J., Lalic, N., Bay, G., Buettner, T., Heilig, G. K., & Wilmoth, J. (2014). World population stabilization unlikely this century. *Science (New York, N.Y.), 346*(6206), 234–237.
4. Kummu, M., Guillaume, J. H. A., de Moel, H., Eisner, S., Flörke, M., Porkka, M., Siebert, S., Veldkamp, T. I., & Ward, P. J. (2016). The world's road to water scarcity: Shortage and stress in the 20th century and pathways towards sustainability. *Scientific reports, 6*, 38495.
5. Agriland. (2020). *Labour availability is now a critical issue within agri-food*. Agriland Media.
6. Food Manufacture. (2020). *Labour shortage reaching crisis point for agricultural sector*. William Reed Business Media.
7. Food Processing Technology. (2020). *UK food industry suffers from labour shortage*. Global Data.
8. Henriksen, A. V., Edwards, G. T. C., Pesonen, L. A., Green, O., & Sorensen, C. A. G. (2020). Internet of Things in arable farming: Implementation, applications, challenges and potential. *Biosystems Engineering, 191*(1), 60–84.
9. Jamieson, J. A. (1976). Passive Infrared Sensors: Limitations on Performance. *Journal of Applied Optics, 15*(4), 891–909.
10. Maxbotix. (2019). *Ultrasonic sensors: Advantages and limitations*. Maxbotix Inc..
11. National Safety Council. (2020). *The pros and cons of electrochemical sensors*. National Safety Council Congress & Expo.
12. Donald, N. (1988). *The design of everyday things*. Basic Books.
13. Compton, M., Barnaghi, P., Bermudez, L., Garcia-Castro, R., Corcho, O., Cox, S., Graybeal, J., Hauswirth, M., Henson, C., Herzog, A., Huang, V., Janowich, K., Kelsey, W. D., Le Phouc, D., LeFort, L., Leggieri, M., Neuhaus, H., Nikolov, A., Page, K., … Taylor, K. (2012). The SSN ontology of the W3C semantic sensor network incubator group. *Journal of Web Semantics, 17*(1), 25–32.
14. Liu, H., & Tang, Z. (2013). Metal oxide gas sensor drift compensation using a dynamic classifier ensemble based on fitting. *Sensors, 13*(7), 9160–9173.
15. Irish, J. (2005). *Ocean instrumentation – Instrumentation specifications*. Massachusetts Institute of Technology.
16. Loock, H. P., & Wentzell, P. D. (2012). Detection limits of chemical sensors: Applications and misapplications. *Sensors and Actuators B: Chemical, 173*(2), 157–163.
17. Dang, Q. K., & Suh, Y. S. (2014). Sensor saturation compensated smoothing algorithm for inertial sensor based motion tracking. *Sensors, 14*(5), 8167–8188.
18. Palmisano, V., Weidner, E., Boon-Brett, L., Bonato, C., Harskamp, F., Moretto, P., Post, M. B., Burgess, R., Rivkin, C., & Buttner, W. J. (2015). Selectivity and resistance to poisons of commercial hydrogen sensors. *International Journal of Hydrogen Energy, 40*(35), 11740–11747.
19. Sparkfun Electronics. (2020). *FLIR Radiometric Lepton Dev Kit V2*. Sparkfun Electronics.

20. Ward, W. K., Engle, J. M., Branigan, D., El Youssef, J., Massoud, R. G., & Castle, J. R. (2012). The effect of rising vs. falling glucose level on amperometric glucose sensor lag and accuracy in type 1 diabetes. *Journal of Diabetic Medicine, 29*(8), 1067–1073.
21. World Nuclear Association. (2019). *RBMK reactors – Appendix to nuclear power reactors*. World Nuclear Association.
22. Corrigan, T. E., & Beavers, W. O. (1968). Dead space interaction in continuous stirred tank reactors. *Chemical Engineering Science, 23*(9), 1003–1006.
23. Hilbert, M., & Lopez, P. (2011). The World's technological capacity to store, communicate, and compute information. *Science, 332*(2), 60–65.
24. Fidanova, S., Shindarov, M., & Marinov, P. (2017). Wireless sensor positioning using ACO algorithm. In *Recent contributions in intelligent systems* (pp. 33–44). Springer.
25. Abbas, N., Yu, F., & Fan, Y. (2018). Intelligent video surveillance platform for wireless multimedia sensor networks. *Journal of Applied Sciences, 348*(8), 1–14.
26. Cisco Systems. (2020). *What is a Wi-Fi or wireless network vs. a wired network?* Cisco Systems.
27. MacDonald, J. M., Korb, P., & Hoppe, R. A. (2016). *Farm size and the organization of U.S* (Crop Farming). United States Department of Agriculture Economic Research Service.
28. Zigbee Alliance. (2020). *What is Zigbee?* Zigbee Alliance.
29. Jackman, P., Gray, A. J. G., Brass, A., Stevens, R., Shi, M., Scuffell, D., Hammersley, S., & Grieve, B. (2012). Processing online crop disease warning information via sensor networks using ISA ontologies. *CIGR Journal, 15*(3), 243–251.
30. West, J., & Kimber, R. B. E. (2015). Innovations in air sampling to detect plant pathogens. *Annals of Applied Biology, 166*(1), 4–17.
31. He, Y., Peng, J., Liu, F., Zhang, C., & Kong, W. (2015). Critical review of fast detection of crop nutrient and physiological information with spectral and imaging technology. *Transactions of the Chinese Society of Agricultural Engineering, 31*(3), 174–189.
32. Henrich, V., Krauss, G., Gotze, C., & Sandow, C. (2020). *Index database: A database for remote sensing indices*. University of Bonn.
33. Ahamed, T., Tian, L., Jiang, Y., Zhao, B., Liu, H., & Ting, K. C. (2012). Tower remote-sensing system for monitoring energy crops; image acquisition and geometric corrections. *Biosystems Engineering, 112*(2), 93–107.
34. CLAAS. (2020). *Forage harvesters – Jaguar*. CLAAS Harsewinkel.
35. John Deere. (2020). *HarvestLab 3000*. John Deere.
36. YARA. (2020). *N-Sensor ALS – to variably apply nitrogen*. YARA.
37. Oerke, E. C., Mahlein, A. K., & Steiner, U. (2014). Proximal sensing of plant diseases. In *Detection and diagnostics of plant pathogens*. Springer.
38. European Parliament. (2020). Chemicals and pesticides, factsheets on the European Union. .
39. European Space Imaging. (2020). *Our satellites: Earths most advanced constellation*. European Space Imaging.
40. Partel, V., Kakarla, S. C., & Ampatzidis, Y. (2019). Development and evaluation of a low-cost and smart technology for precision weed management utilizing artificial intelligence. *Computers and Electronics in Agriculture., 157*(3), 339–350.
41. Benke, K., & Tompkins, B. (2017). Future food-production systems: vertical farming and controlled-environment agriculture. *Journal of Sustainability: Science, Practice & Policy., 13*(1), 13–26.
42. Jha, M. K., Pakira, S. S., & Sahu, M. R. (2019). *Protected cultivation of horticulture crops*. Educreation Publishing.
43. Rouse, J. W., Haas, R. H., Scheel, J. A., & Deering, D. W. (1974). *Monitoring vegetation systems in the great plains with ERTS*. In: Proceedings, 3rd earth resource technology satellite (ERTS) symposium, vol. 1, p. 48–62.
44. Ryu, K. H., Kim, G. Y., & Chae, H. Y. (2000). Monitoring greenhouse plants using thermal imaging. *IFAC Proceedings Volumes, 33*(29), 181–186.

45. Li, L., Zhang, Q., & Huang, D. (2014). A Review of Imaging Techniques for Plant Phenotyping. *Journal of Sensors, 14*(11), 20078–20111.
46. Corkery, G., Ward, S., Kenny, C., & Hemmingway, P. (2013). Incorporating smart sensing technologies into the poultry industry. *World Poultry Research, 3*(4), 106–128.
47. Jackman, P., Penya, H., & Ross, R. (2020). The role of information and communication technology in poultry broiler production process control: A review. *Agricultural Engineering International (CIGR Journal), 22*(3), 284–299
48. Ward, S. (2012). *BOSCA – A smart networked sensing system in agriculture: A poultry industry focus*. Science Foundation Ireland.
49. Jackman, P., Ward, S., Brennan, L., Corkery, G., & McCarthy, U. (2015). Application of wireless technologies to forward predict crop yields in the poultry production chain. *CIGR Journal, 17*(2), 287–295.
50. Astill, J., Dara, R. A., Fraser, E. D. G., & Sharif, S. (2018). Detecting and predicting emerging disease in poultry with the implementation of new technologies and big data: A focus on avian influenza virus. *Frontiers in Veterinary Science, 5*(1), 1–12.
51. Agrologic. (2017). *Poultry products*. Agrologic Online Service.
52. Fancom. (2017). *Broiler climate controllers*. Fancom Online Service.
53. Rotem. (2014). *Platinum plus controller manual, rotem control and management online service*. Petach-Tikva.
54. Ross, R. J. (2015). *Precise poultry: Analytics supported decision systems in poultry farming*. Enterprise Ireland.
55. Neves, D. P., Mehdizadeh, S. A., Tscharke, M., deAlancar-Naas, I., & Banhazi, T. M. (2015). Detection of flock movement and behaviour of broiler chickens at different feeders using image analysis. *Information Processing in Agriculture, 2*(2), 177–182.
56. Ross, J. W., Hale, B. J., Gabler, N., & Rhoads, R. P. (2015). Physiological consequences of heat stress in pigs. *Animal Production Science, 55*(11), 1381–1390.
57. Ter-Sarkisov, A., Ross, R., & Kelleher, J. (2017). *Bootstrapping labelled dataset construction for cow tracking and behavior analysis*. In: 14th Conference on computer and robot vision. Edmonton, AL, Canada. May 17–19, 2017.
58. Yukun, S., Pengju, H., Yujie, W., Ziqi, C., Yang, L., Baisheng, D., Runze, L., & Yonggen, Z. (2019). Automatic monitoring system for individual dairy cows based on a deep learning framework that provides identification via body parts and estimation of body condition score. *Journal of Dairy Science, 102*(11), 10140–10151.
59. Bennett, S. (1993). Development of the PID controller. *IEEE Control Systems Magazine, 13*(6), 58–62.
60. Liu, C., Peng, J.-F., Zhao, F.-Y., & Li, C. (2009). Design and optimization of fuzzy-PID controller for the nuclear reactor power control. *Nuclear Engineering and Design, 239*(11), 2311–2316.
61. Lu, X., Duan, X., Mao, X., Li, Y., & Zhang, X. (2017). Feature extraction and fusion using deep convolutional neural networks for face detection. *Mathematical Problems in Engineering, 1*(1), 1–9.
62. Pereira, D. T., Aldarondo, D. E., Willmore, L., Kislin, M., Wang, S. S.-H., Murthy, M., & Shaevitz, J. W. (2019). Fast animal pose estimation using deep neural networks. *Nature Methods, 16*(1), 117–125.
63. Shakoor, N., Lee, S., & Mockler, T. C. (2017). High throughput phenotyping to accelerate crop breeding and monitoring of diseases in the field. *Current Opinion in Plant Biology, 38*(1), 184–192.
64. Graves, A. (2012). *Supervised sequence labelling with recurrent neural networks*. Springer Press.
65. Trabesinger, A. (2017). Quantum computing: towards reality. *Nature Outline, 543*(1).

Data Fusion and Its Applications in Agriculture

Dimitrios E. Moshou and Xanthoula Eirini Pantazi

1 Introduction

The data volume and complexity are growing day-by-day and this is where Artificial Intelligence (AI) needs to play an important role. The major reason why AI is having the potential to boost Internet of Things (IoT)-driven initiatives is its ability to pick insights from a huge amount of data. AI provides:

- Recommendations, insights, predictions, and solutions: AI can analyze huge chunks of data in minutes and generate meaningful results.
- Real-time and rapid response: In order to decide future course-of-action, AI enables real-time processing and response when connected with IoT devices.

AI and IoT can create more adaptive learning and analytical systems that help businesses to collect insights. It facilitates synchronization, communication, and integration in a better way. It informs businesses about proactive actions needed to be taken in order to keep up with the changes. The AI-enabled IoT systems are intelligent, self-learning, and capable of establishing highly cognitive enterprises while they automate the processes across. They also boost the productivity, performance, and maintenance activities. Large-scale organizations across the world are already embracing this synergy to take their businesses a notch higher. Additionally, in the near-term future, AI is expected to have a profound impact on how people are hired at work and what skills they possess in order to yield a better result in this machine-driven automated work environment. The AI becomes a prerequisite for IoT as we "can do nothing" only with machine-generated data. There should be a

D. E. Moshou (✉) · X. E. Pantazi
Aristotle University of Thessaloniki, School of Agriculture, Thessaloniki, Greece
e-mail: dmoshou@auth.gr

D. D. Bochtis et al. (eds.), *Information and Communication Technologies for Agriculture—Theme II: Data*, Springer Optimization and Its Applications 183,
https://doi.org/10.1007/978-3-030-84148-5_2

mechanism for analyzing, collecting, manipulating, and generating insights from an enormous amount of data at a high speed.

The phrasing identified with frameworks, models, applications, techniques, and hypotheses about the combination of information originated from numerous sources expands to distinctive terms. For instance, Sensor/Multi-sensors fusion is usually used to determine that sensors provide the information being melded. Regardless of the philosophical issues about the distinction among information and data, the terms Information fusion and Data fusion are generally acknowledged as identical terms.

Numerous meanings of information combination have been given as the years progressed, a large portion of them got from military and remote detecting fields. In 1991, the information combination work gathering of the Joint Chiefs of Research facilities (JDL) sorted out a push to define a vocabulary with a few terms of reference for information combination. They characterize information fusion as a:

> staggered, multifaceted process managing the programmed identification, affiliation, connection, estimation, and mix of information and data from various sources.

Klein [1] sums up this definition, expressing that information can be given by a solitary source or by numerous sources. The two definitions are general and can be applied in various fields, including remote detecting.

Hall and Llinas [2] characterize information combination as:

> the blend of information from different sensors and data given by related databases, to accomplish enhanced exactness and more explicit inductions than the one that could be accomplished by the utilization of a solitary sensor alone.

Here, information combination is performed with a goal: precision enhancement. Notwithstanding, this definition is limited to information given by sensors; it doesn't predict the utilization of information from a solitary source.

Guaranteeing that every single past definition is centered around strategies, means, and sensors, Wald [3] changes the concentration to the system used to meld information. Wald states that:

> information combination is a formal structure in which there are communicated means and devices for the partnership of information starting from various sources. It goes for acquiring data of more noteworthy quality; the correct meaning of 'more prominent quality' will rely on the application.

Furthermore, Wald considers information taken at various moments from particular sources, as synergistic. "Quality" is a free term deliberately received to mean that the combined information is in some way more fitting to the application than the first information. Specifically, for WSNs, information can be intertwined with somewhere around two targets: precision enhancement and bandwidth sparing.

Even though Wald's definition and wording are entirely acknowledged by the Geoscience and Remote Detecting Society [4], and authoritatively received by the Data Fusion Server [5], the term Multisensor fusion has been utilized with a similar significance by different authors, for example, Hall and Llinas [2].

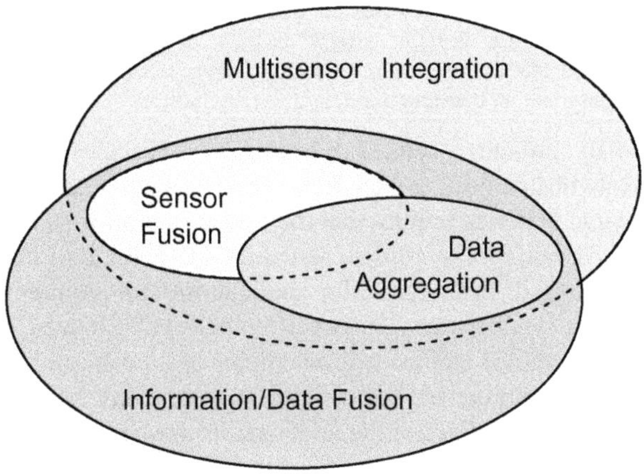

Fig. 1 The relationship among the fusion terms: multisensor/sensor fusion, multisensor integration, data aggregation, data fusion, and information fusion

Multisensor fusion is a term utilized in mechanical technology and/or computer vision [6], and modern mechanization [7]. As indicated by Luo et al. [8], multisensor integration is:

> the synergistic utilization of data given by numerous agents to aid the achievement of an undertaking by a framework; and multisensor combination manages the blend of various wellsprings of tangible data into one authentic arrangement amid any phase in the coordination procedure.

Multisensor integration is a more extensive term than multisensor fusion (Fig. 1). It expresses how the fused information is utilized by the entire framework to communicate with the earth. However, it might suggest that only sensory data are used in the fusion and integration processes.

This disarray of terms is featured by Dasarathy [9] who introduced the term Data Combination [10] expressing that:

> with regards to its utilization in the general public, it envelops the hypothesis, systems and instruments made and connected to use the cooperative energy in the data obtained from various sources (sensor, databases, data accumulated by people, and so on.), so that the subsequent choice or activity is in some sense better (subjectively or quantitatively, as far as exactness, integrity, and so forth.) than the one that would be conceivable if any of these sources were utilized separately without such collaboration in place.

Potentially, this is the broadest definition grasping any kind of source, learning, and asset used to intertwine distinctive snippets of data. The term Data Combination and Dasarathy's definition are likewise accepted by the International Society of Information Fusion [5]. Kokar et al. [11] additionally utilize this term in a system of formal rationale and classification hypothesis where the structures, speaking to the importance of data (speculations and models) are really combined, while information is simply handled and sifted through such structures.

The term Information Accumulation has turned out to be prevalent in the remote sensor arrange community as an equivalent word for data combination [12, 13]. As for Cohen et al. [14]:

information total involves the gathering of crude information from inescapable information sources, the adaptable, programmable structure of the crude information into less voluminous refined information, and the convenient conveyance of the refined information to information consumers.

By utilizing "refined information", exactness enhancement is recommended. Notwithstanding, as van Renesse [13] states: *"conglomeration is the capacity to outline"*, which implies that the measure of information is diminished. For example, by methods for synopsis capacities, the volume of information being controlled is diminished. In any case, for applications that require unique and exact estimations, such a synopsis may be out of context [15]. Indeed, albeit numerous applications may benefit from information combination, one can't generally reach the conclusion that the summarized information is more exact that the initially obtained one.

Figure 1 delineates the relationship among the ideas of multisensor/sensor fusion, multisensor integration, data aggregation, data fusion, and information fusion. Here, the two terms, information combination and data combination, can be utilized with a similar significance. Multisensor/sensor combination is the subset that works with physical sources. Information total characterizes another subset of data combination that plans to decrease the information volume (regularly, rundown), which can control any kind of information/data, including tangible information. Then again, multisensor incorporation is a somewhat unique term as it applies data combination to deduce inferences utilizing tangible gadgets and related data (e.g., from database frameworks) to communicate with nature. Along these lines, multisensor/sensor combination is completely contained in the crossing point of multisensor coordination and data/information combination.

Here, data combination was utilized as the general term with the goal that sensor and multisensor combination can be considered as the subset of data combination that handles information gained by tangible gadgets. In any case, as information combination is additionally acknowledged as a general term, we fortify Elmenreich's proposal [16], which expresses that combination of crude (or low dimension) information ought to be unequivocally expressed as low dimension information fusion to avoid confusion with the information utilized by the Geoscience and Remote Sensing Society [4].

In this chapter, we provide an overview of data fusion and Wireless Sensor Networks (WSN) with the aim of providing a review of different approaches to data fusion in agricultural applications. Section 2 also includes an introduction to Data fusion in agriculture.

2 Data Fusion

2.1 Introduction

At present, land managers have available enormous amounts of data at a wide variety of spatial and temporal scales and often they do not have appropriate tools to analyze and manage many disparate datasets. For example, remote sensing methods such as Light Detection and Ranging (LIDAR) do not generally require repeated measurements, and data are usually provided at a resolution of less than 1 m. Yield is measured once a growing season and for a 12-row combine, traveling at 5 km h^{-1}, harvesting corn planted on 0.76 m centers, yield data have a resolution of approximately 2.1 m in the direction of travel and 9.1 m between passes. Soil bulk electrical conductivity (ECa) generally needs only to be measured once and this occurs in the spring or fall when the soil water content is close to field capacity. The spatial resolution of the ECa survey data depends on ground speed and distance between passes (generally 10–20 m). Soil samples are often collected in the fall or spring, every 2–4 years with a grid sampling (e.g., 100 m spacing) or with a regular or irregular zone sampling approach. Sentinel 2 imagery has bands with a resolution variable of 10 m in the visible and near infrared (NIR) range and 20 m and 60 m in the NIR and Middle infrared (MIR) ranges. Soil survey data are available for many areas within the United States at scales between 1:12,000 and 1:26,000, but the intensity of ground soil profile observations varies substantially at different sites and depends on many factors.

An information revolution is currently occurring in agriculture resulting in the production of massive datasets at different spatial and temporal scales; therefore, efficient techniques for processing and summarizing data will be crucial for effective management. With the profusion and wide diversification of data sources provided by modern technology, such as remote and proximal sensing, sensor datasets could be used as auxiliary information to supplement sparsely sampled target variables. Remote and proximal sensing data are often massive, taken on different spatial and temporal scales, and subject to measurement error biases. Moreover, differences between the instruments are always present; nevertheless, a data fusion approach could take advantage of their complementary features by combining the sensor datasets in a manner that is statistically robust.

It would then be ideal to jointly use (fuse) partial information from the diverse today-available sources so efficiently to achieve a more comprehensive view and knowledge of the processes under study. However, spatial data often turn out to be "incompatible" [17], owing to their heterogeneities in terms of nature (continuous or categorical), quality (soft or hard data), spatial and temporal scales and resolutions, dimensions, and sample locations. Further, the occurrence of complex spatial dependence and inter-dependence structures among spatial variables [18] contributes in making the combination (fusing) of "incompatible" spatial data, a more difficult and a rather challenging problem.

Many definitions for data fusion exist in the literature. Hall [19] defines data fusion as:

> the process of combining information from heterogeneous sources into a single composite picture of the relevant process, such that the composite picture is generally more accurate and complete than that derived from any single source alone.

Although the principle seems simple and data fusion has been widely applied in many disciplines, nevertheless, the objectives and the approaches used to achieve them differ greatly in the diverse fields and strongly depend on the specific applications [20]. Moreover, depending on context, data fusion may assume different meanings such as information fusion, sensor fusion, or image fusion. Information fusion is the process of merging information from different sources; sensor fusion is the combination of data from different sensors and image fusion is the fusion of two or more images into one, which should be a more useful image. In the statistical context, data fusion is considered as an inference problem; in other words, the problem consists in, given diverse heterogeneous datasets, developing a methodology that allows to optimally estimate the variable of primary interest and obtain uncertainty measures associated with inference [21].

Fusion processes are usually categorized in a triad of levels of modes depending on the processing level that the data have undergone, low-, intermediate-, and high-level fusion, as follows:

1. Low-level data fusion combines various sources of typically the same type of raw minimally processed data to construct a new raw dataset that is expected to be more informative and useful than the inputs. To comply with the further processing, the data should be transformed in the required engineering units, or might it be raw unprocessed data. This can be expanded to include the fusion of estimated states of a system or object by using data from several, often dissimilar, sensors and is called state vector fusion. The entire process is called kinematic data fusion, which can include kinematic states as well as dynamic states.
2. Intermediate-level, mid-level fusion, or feature-level fusion combines various features such as edges, lines, corners, textures, or positions into a feature map. This map is used for segmentation of images, detection of objects, etc. This process of fusion is called pixel-, feature-, or image-level fusion.
3. High-level fusion, or decision fusion, combines decisions from several experts. Methods of decision fusion are voting, fuzzy logic, and statistical methods.

2.2 The "Whys" and "Wherefores" of Information Fusion

The data fusion approach joins information from numerous sensors (and related databases if fitting) to accomplish enhanced accuracies and more specific inferences that couldn't be accomplished by the utilization of just a solitary sensor [2]. This idea is not new: living beings have the capacity to utilize various faculties to find out about the earth. The cerebrum melds accessible data to play out a choice undertaking. One of the first definitions of information combination came from the North American Joint Directors of Laboratories (JDL) [22, 23], who defined information combination as:

> a staggered, multifaceted process managing the programmed identification, affiliation, correlation, estimation and blend of information from single and various sources.

A Wireless Sensor Network (WSN) [24, 25] is a unique sort of specially appointed system, made from an extensive number of local nodes furnished with various sensor units. This system is upheld by innovative advances in low-power remote correspondences alongside various functionalities, for example, detecting, correspondence, and handling. WSNs are developing as an imperative information device class dependent on another processing stage and system administration structure that will empower novel applications that are identified with various zones, for example, natural checking, oriented industry and production mechanization, health care, and military. Ordinarily, remote sensor systems have strict limits with respect to power consumption and computational limit.

A WSN might be structured with various destinations. It might be intended to accumulate and process information from nature to have a superior comprehension of the conduct of the observed element. It might likewise be intended to screen a domain for the event of a lot of conceivable occasions, with the goal that the best possible move might be made at whatever point important. A crucial issue in WSNs is the way the gathered information is prepared. In this specific situation, data combination emerges as a control that deals with how information accumulated by sensors can be handled to expand the pertinence of such a mass of information. In any case, data combination can be characterized as the blend of various sources to get enhanced data (less expensive, or more noteworthy quality, or more prominent pertinence).

Information fusion is ordinarily utilized in location and order undertakings in different application in agriculture, for example, applied autonomy [26]. Of late, these systems have been utilized in new applications, for example, yield prediction [27]. Inside the WSN space, basic states (e.g., maximum, least, and normal) have been utilized to decrease the general information traffic to spare bandwidth [28, 29]. Moreover, data combination procedures have been connected to WSNs to enhance location appraisals of sensor hubs [30], identify steering failures [31], and gather connect insights for steering conventions [32].

Given the significance of data combination for WSNs, this section studies the best in class identified with data combination and how it has been utilized in WSNs and sensor-based frameworks. A few unique terms (e.g., data fusion, sensor fusion, and information fusion) have been utilized to depict parts of the fusion subject (including theories, processes, systems, frameworks, tools, and methods). Therefore, there is a wording perplexity.

WSNs are proposed to be conveyed in situations where sensors can be presented to condition that they may meddle with their estimations. Such conditions incorporate solid fluctuations in temperature and pressure, electromagnetic interference, and radiation. Consequently, sensors' estimations might be uncertain (or even pointless) in such situations. Notwithstanding, when ecological conditions are perfect, sensors may not give immaculate measurements. Basically, a sensor is an estimation device, and an imprecision estimation is generally connected with its perception. Such imprecision is attributed to the sensor inaccuracy and signal processing strategies used by the sensor to quantify a physical property.

Failures are not an exemption in WSNs. For example, consider a WSN that screens a timberland forest to recognize an occasion, for example, fire or the nearness of a creature. Sensor hubs can be annihilated by flame, creatures, or even people; they may introduce producing issues; and they may quit working because of an absence of vitality. Every hub, that winds up as inoperable, may trade off the general recognition and additionally the correspondence capacity of the system. Here, recognition ability is identical to the introduction idea [33, 34].

Both spatial and lifetime inclusion present constraints to WSNs. The detecting capacity of a hub is confined to a constrained locale. For instance, a thermometer in a room reports the temperature close to the gadget, however it may not decently speak to the general temperature inside the room. Spatial inclusion in WSNs [35] has been investigated in various situations, for example, target following [36], note scheduling [37], and sensor's condition [38]. Lifetime inclusion can be comprehended as the capacity to satisfy the system during its lifetime. For example, in a WSN for occasion recognition, lifetime inclusion goes for guaranteeing that no significant occasion will be remembered favorably, in the occasion that there was no sensor seeing the area at the explicit time the occasion happened. Along these lines, transient inclusion relies upon the sensor's examining rate, correspondence delays, and the hub's obligation cycle (time when it is on alert or sleeping).

To deal with sensor failures, mechanical constraints, spatial and transient inclusion issues, three properties must be guaranteed: participation, repetition, and complementarity [8, 39]. As a rule, an area of intrigue must be completely secured by the utilization of a few sensor hubs, each collaborating with a fractional perspective of the scene; data combination can be utilized to create the total view from the pieces given by every hub. Repetition makes the WSN helpless against disappointment of a solitary hub and covering estimations can be intertwined to get progressively exact information; Rao [40] indicates how data combination can perform in any event and the best sensor. Complementarity can be accomplished by utilizing sensors that see distinctive properties of the earth; data combination can be utilized to join reciprocal information so the resultant information permits deductions that may be unrealistic to be gotten from the individual estimations (e.g., point and separation of an inevitable danger can be melded to acquire its position).

Because of repetition and collaboration properties, WSNs are frequently made of an extensive number of sensor hubs representing an adaptability challenge caused by potential impacts and transmissions of excess information. As to vitality limitations, correspondence ought to be decreased to build the lifetime of the sensor hubs. In this way, data combination is additionally vital to decrease the general correspondence stack in the system, by evading the transmission of repetitive messages. What's more, any undertaking in the network that handles flags or needs to influence deductions can possibly utilize data combination.

2.3 *Information Fusion: Methods, Techniques, and Algorithms*

This section introduces models of data fusion, architectures of data fusion and performance aspects and data alignment in order to perform fusion of attributes.

2.4 *Models of Data Fusion*

Data fusion depends on the control of different estimations, where classifiers work on highlights extricated from these present reality estimations (Fig. 2).

An audit of the courses for consolidating classifiers can be found in [41]. Creators recognize the two combination classes:

- Data combination, where the classifier works on either the crude information or highlights separated specifically from the estimations;
- Decision combination, where the choices from the individual classifiers for various information channels are consolidated.

The decision relies upon the factual connection between the information channels, common entropy, or joint Gaussianity [42]. The primary issues are flag nonlinearity (with related non-Gaussianity), nonstationary, discontinuous information natures, and commotions. This makes it exceptionally difficult to perform an estimation with standard techniques since no presumption on the information model and conveyance can be discovered. In a few applications, for example, useful Magnetic Reverberation Imaging (fMRI), there is even no "ground truth", to depend upon. Multisensor down to earth frameworks consequently go for giving higher exactness and enhanced vigor against vulnerability and sensor breakdown [43], and furthermore for the data removed from different sources to be incorporated into a solitary flag or amount.

Flag handling calculations for "sensor" or "information combination" can be founded on [44]:

- Probabilistic models: Bayesian thinking, proof hypothesis, strong insights.
- Least squares: Kalman filtering, regularization, set enrollment.
- Intelligent fusion: Fuzzy logic, neural systems, genetic algorithms.

Fig. 2 General data fusion concept

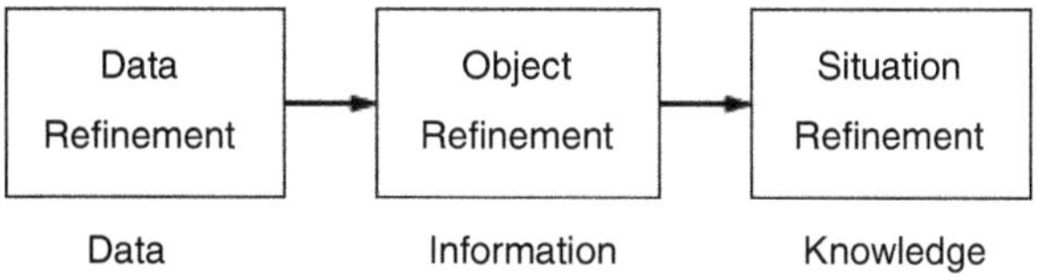

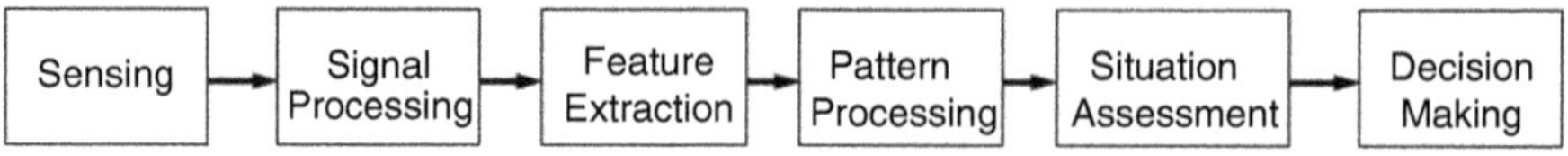

Fig. 3 The waterfall model

Intelligent fusion refers to the use of artificial intelligence and machine learning techniques for weighing the importance of each sensor modality so that automatic inferences are produced.

Neural systems have been connected to data fusion with application to Automatic Target Recognition (ATR) through utilizing numerous heterogeneous sensors [45, 46]. Their effective application lays on their ability to provide parallel techniques for handling bit datasets, through their robustness despite various uprising issues, as noise [47]. Baran [48] proposed a data fusion approach for ATR that utilizes a neural system functioning as an associative memory that directs the example coordinating procedure for target recognition [49]. Cain et al. [50] utilized neural systems for classifying targets dependent on data acquired by a multispectral NIR sensor and a bright laser radar.

A special case of neural systems is Deep Learning. Deep Artificial Neural Networks (ANNs) are most generally alluded to as Deep Learning (DL) or Deep Neural Networks (DNNs) [51]. They are a generally new territory of Machine Learning (ML) focusing on permitting computational models that are made of numerous handling layers to learn complex information portrayals, utilizing different dimensions of deliberation. One of the principle points of interest of DL is that now and again, the progression of highlight extraction is performed by the model itself. DL models have drastically enhanced the best in class in a wide range of divisions and ventures, including agribusiness. DNNs are just an ANN with various concealed layers between the information and yield layers and can be either managed, somewhat regulated, or even unsupervised. A typical DL example is the Convolutional Neural Networks (CNNs), in which include maps are extricated by performing convolutions in the picture area. A thorough presentation on CNNs is given in Goodfellow et al. [52].

Genetic algorithms provide an optimization framework for sensor fusion. The data fusion problem can be described as a function of measured data and weight coefficients. Then we can use the optimization technique to calculate the weight coefficient values adjusted to measured data. Gene expression programming (GP) suits such an optimization task, and it can be used to derive a more advanced function [53]. Such methods require complex computation process. Yet, the best fitting of the chromosome algorithm can be used in order to decrease this complexity [54]. The GP is used in order to derive the function that best fits the setup conditions.

One of the first proposed information combination models was the "cascade show" (Fig. 3), created for the UK Safeguard Assessment Exploration Office (DERA) [23]. The sensors produce raw data, which must be submitted for a preprocessing to reduce the noise and extract useful features. Then pattern recognition algorithms are employed to extract semantic information in order to label the data samples and subsequently characterize the situation, which in turn leads to decision making by the decision support system.

Architectures and Performance Aspects

Consolidating multisensor information in the information combination structure has the capability of quicker and less expensive preparing and new interfaces, together with decreasing of, generally speaking, vulnerability. Such information can be consolidated in different routes, for example, by: i) straight combiner, ii) mix of rear ends (loads, display significance), and iii) result of rear ends (free data). Considering the different methods for joining data and different semantic dimensions, we differentiate between the accompanying information combination models, as follows:

- Centralized: basic calculations, however inflexible to sensor changes
- Hierarchical: communitarian preparing, two-way correspondence
- Decentralized: powerful to sensor changes and failures, complex calculations

The synergy of data sources through fusion [55] offers a few benefits compared to individual sources, for example,

- Improved confidence because of reciprocal and excess data
- Robustness and dependability in noisy conditions (smoke, commotion, impediment)
- Better discrimination between classes due to fusing more data and approaching the total information content
- Forming robust systems, operating efficiently regardless of possible few sensor failure

The worldview of ideal combination in this sense is to limit the likelihood of inadmissible blunder.

In light of the scientific categorization, contingent upon the phase at which combination happens, information combination is frequently sorted as the low-level fusion (LLF), intermediate level fusion (ILF) or high-level fusion (HLF) combination, where:

- LLF (information combination) joins crude information sources to give better data
- ILF (include combination) joins includes that originate from heterogeneous or homogeneous raw information
- HLF (choice combination), consolidates choices or confidence levels originating from a few specialists (hard and delicate combination)

Data Alignment and Fusion of Attributes

Depending upon when the combination procedure takes place, open literature differentiates between the temporal, spatial, and area change combination. It should be noticed that the last two can be considered as instances of the low- or intermediate-level fusion. Temporal fusion is different, because it can be performed at any level, since the data acquired from one sensor taken at different time steps are fused.

The data entering a fusion procedure ought to be adjusted, a difficult problem for which there is no broad supporting hypothesis. Arrangement ought to be connected to both homogeneous (equivalent) and heterogeneous (incomparable) data, which may require change or change of perceptions.

The idea of arrangement accepts "standard dialect" between the contributions, for example:

- Standardization of estimation units
- Sensor alignment or
- Corrections for different illuminants and shading [56]

Arrangement may work at any of the three semantic dimensions: estimations, properties, and principles, with conceivable intersections between levels [56]. For example, for adjusted and related sources of data, fusion characteristics link traits of a similar article, got from different snapshots of the item. Fusion of snapshots performs meta-tasks and subsequently can be performed with different kinds of fusion algorithms.

Information combination likewise applies to the internet, where interruption identification (ID) systems meld information from heterogeneous appropriated organized sensors to make "situational mindfulness" [57], for example, the discovery of system inconsistencies and infection assault.

Execution parts of a fusion framework [55] are domain-specific:

- Detection performance quality (false detection rate)
- Spatial/transient goals and capacity to recognize signals
- Spatial and temporal inclusion (the sensor's range)
- Detection/following mode (examining, following, different target following)
- Measurement precision and dimensionality

2.5 *Applications of Information Fusion in Agriculture*

Neural systems for data combination can likewise be found in different applications other than ATR. Lewis and Powers [58] utilized neural systems to intertwine various media data for broad media discourse acknowledgment. Cimander et al. [59] utilize a two-organize combination technique that works on signs from bioreactors (e.g., temperature, pH, and oxygen) to control the yogurt maturation process. Yiyao

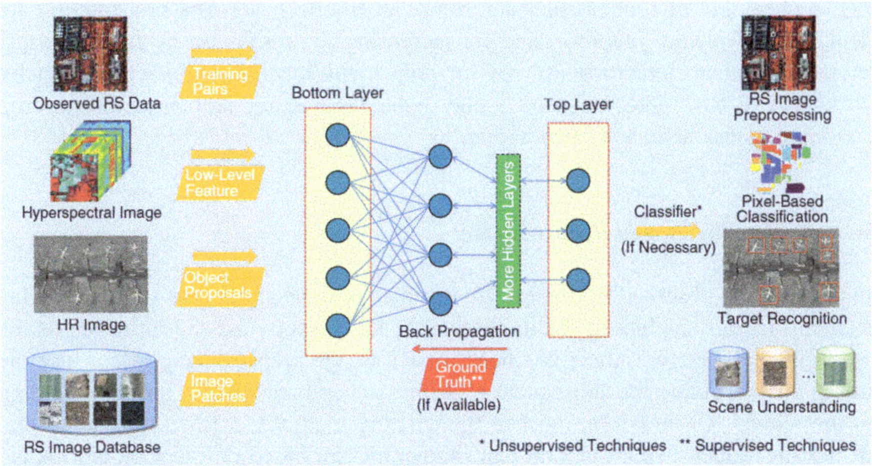

Fig. 4 Sensor fusion structure by an ANN and DNN

et al. [60] proposed a combination plot named Knowledge-based Neural Network Fusion (KBNNF) to intertwine edge maps from multi-unearthly sensor pictures procured from radars, optical sensors, and infrared sensors.

Regardless of the complex various leveled structures, the majority of the DL-based strategies can be fused into a general system. Figure 4 outlines a general system of DL for Remote Sensing (RS data). The flowchart incorporates three fundamental segments: the input information, the processing layers, and the expected output data. Practically speaking, the input–output datasets depend on the specific application. For instance, for RS picture preprocessing, they are the High-Resolution (HR) and Low-Resolution (LR) picture patches from the panchromatic (PAN) pictures [61]; for pixel-based classification, they are the spectral–spatial highlights and their feature arrangement (unsupervised form) or class information (supervised version) [62]; while, for undertakings of target recognition [63] and scene understanding [61], the inputs are the features extracted from the data samples, and also the raw pixel digital numbers acquired from the HR pictures and RS image databases individually, and the output data are the same as in the case of pixel-labeling.

At the point when the input–output datasets have been appropriately categorized, the inherent and natural connection between the input and output datasets forms a deep network structure comprising of numerous dimensions of nonlinear activities, where each dimension is demonstrated by a shallow module. It ought to be noticed that, if an adequate preparing test set is accessible, such a deep system ends up being a supervised approach. It may be further calibrated by the utilization of the label data, and the network's top layer output is the so-called labeled metadata instead of the conceptual feature representation learned by an unsupervised deep network. At the point when the core deep network has been trained, it can be utilized to foresee the expected output of a test data sample. Alongside with the general system in

Fig. 4, a couple of fundamental algorithms in the deep network construction are depicted: RS picture preprocessing for increasing the resolution of the RS image, pixel-based characterization utilized for crop identification, target recognition for weed and disease detection, and scene understanding for crop monitoring, crop condition assessment, and phenotyping.

Remote Sensing Image Preprocessing

In practice, the observed RS images are not always as satisfactory as we demand due to many factors, including the limitations of the sensors and the influence of the atmosphere. Therefore, there is a need for RS image preprocessing to enhance the image quality before the subsequent classification and recognition tasks. According to the related RS literature, most of the existing methods in RS image denoising, deblurring, super-resolution, and pan sharpening are based on the standard image-processing techniques in the signal processing society, while there are very few machine-learning-based techniques. In fact, if we can effectively model the intrinsic correlation between the input (observed data) and output (ideal data) by a set of training samples, then the observed RS image could be enhanced by the same model. According to the basic techniques in the previous section, such an intrinsic correlation can be effectively explored by DL.

Restoration and Denoising

For RS image restoration and denoising, the original image is the input to a certain network that is trained with the clean image to obtain the restored and denoised image. Pan sharpening can be realized by introducing deep neural networks, There is a key hypothesis that the HR and LR multispectral (MS) image patches have the same relationship as that between the HR and LR PAN image patches; thus, it is a learning-based method that requires a set of HR–LR image pairs for training. Since the HR PAN is already available, we have designed an approach to obtain its corresponding LR PAN. Therefore, we can use the fully trained DL network to reconstruct the HR MS image from the observed LR MS image. The experimental results demonstrated that the DL-based pan sharpening method outperforms the other traditional and state-of-the-art methods. The aforementioned methods are just two aspects of DL-based RS image preprocessing. In fact, we can use the general framework to generate more DL algorithms for RS image-quality improvement for different applications.

Pixel-based Classification

Pixel-based classification is one of the most popular topics in the geoscience and RS community. However, from the DL point of view, most of the existing methods can extract only shallow features of the original data (the classification step can also be treated as the top level of the network), which is not robust enough for the classification task. DL-based pixel classification for RS images involves constructing a DL architecture for the pixel-wise data representation and classification. By adopting DL techniques, it is possible to extract more robust and abstract feature representations and thus improve the classification accuracy. The scheme of DL for RS image pixel-based classification consists of three main steps: (1) data input, (2) hierarchical DL model training, and (3) classification. In the first steps, the input vector could be the spectral feature, the spatial feature, or the spectral–spatial feature, as we will discuss later. Then, for the hidden layers, a deep network structure is designed to learn the expected feature representation of the input data.

Spectral Feature Classification

The spectral information usually contains abundant discriminative information. A frequently used and direct approach for RS image classification is spectral feature-based classification, i.e., image classification with only the spectral feature. Most of the existing common approaches for RS image classification are shallow in their architecture, such as Support Vector Machines (SVMs) and k-nearest neighbor (KNN). Instead, DL adopts a deep architecture to deal with the complicated relationships between the original data and the specific class label. For spectral feature classification, the spectral feature of the original image data is directly deployed as the input vector. The input pixel vector is trained in the network part to obtain the robust deep feature representation, which is used as the input for the subsequent classification step.

Classification with Spatial Information

Land covers are known to be continuous in the spatial domain, and adjacent pixels in an RS image are likely to belong to the same class. For a certain pixel in the original RS image, it is natural to consider its neighboring pixels to extract the spatial feature representation. However, due to the hundreds of channels along the spectral dimension of a hyperspectral image, the region-stacked feature vector will result in too large an input dimension. As a result, it is necessary to reduce the spectral feature dimensionality before the spatial feature representation. Principle Components Analysis (PCA) is commonly executed in the first step to map the data to an acceptable scale with a low information loss. Then, in the second step, the spatial information is collected using a w # w (w is the size of window) neighboring region of every certain pixel in the original image [64].

Target Recognition

Target recognition in large HR RS images, such as ship, aircraft, and vehicle detection, is a challenging task due to the small size and large numbers of targets and the complex neighboring environments, which can cause the recognition algorithms to mistake irrelevant ground objects for target objects. However, objects in natural images are relatively large, and the environments in the local fields are not that complex compared to RS images, making the targets easier to recognize. This is one of the main differences between detecting RS targets and natural targets. Although many studies have been undertaken, we are still lacking an efficient location method and robust classifier for target recognition in complex environments. In the literature, Cai et al. [65] showed how difficult it is to segment aircraft from the background. The performance of target recognition in such a complex context relies on the features extracted from the objects. DL methods are well suited for this task, as this type of algorithm can extract low-level features with a high frequency, such as edges, contours, and outlines of objects, whatever the shape, size, color, or rotation angle of the targets. This type of algorithm can also learn hierarchical representations from the input images or patches, such as the parts of the objects that are compounded by the lower-level features, making recognition of RS targets discriminative and robust.

Scene Understanding

Satellite imaging sensors can now acquire images with a spatial resolution of up to 0.41 m. These images, which are usually called very high-resolution (VHR) images, have abundant spatial and structural patterns. However, due to the huge volume of the image data, it is difficult to directly access the VHR data containing the scenes of interest. Due to the complex composition and large number of land-cover types, efficient representation and understanding of the scenes from VHR data have become a challenging problem, which has drawn great interest in the RS field. Good internal feature representations are hierarchical. In an image, pixels are assembled into edgelets, edgelets into motifs, motifs into parts, and parts into objects. This suggests that recognizing and analyzing scenes from VHR images should have multiple trainable feature-extraction stages stacked on top of each other, and we should learn the hierarchical internal feature representations from the image.

2.6 Data Mining and Artificial Intelligence in Agriculture

Yield Prediction

Yield prediction is highly significant among the recent applications in precision farming. In literature, there are numerous ML applications, targeted to this field of research, for example, the study by Ramos et al. [66], which demonstrated a classification approach for determining the harvesting crop readiness in coffee. The technique evaluated the growth and the maturity level of coffee crops aiming to assist coffee growers to raise their profits and make a more efficient crop management. Another research targeted on yield prediction is presented by Amatya et al. [67]. In this study, a machine vision framework for automating vibrational harvesting, crop identification in cherry orchards has been utilized, by identifying blocked cherry branches. The primary aim of the study focused on alleviating work burden with tasks. A similar approach has been presented by Sengupta and Lee [68], where an early yield mapping framework has been proposed for recognizing juvenile green citrus in citrus orchards.

Ali et al. [69] built up a model for the estimation of field biomass (kg dry matter issue per ha per day) with the help of ANNs and multi-temporal remote detecting sensing. An approach for predicting yield productivity in wheat crop has been presented by Pantazi et al. [27]. The study focused mainly on the effective fusion of satellite vegetation index features and associated attributes with soil properties for predicting yield productivity. Senthilnath et al. [70] demonstrated an approach for identifying the location of tomatoes dependent on Expectation Maximization technique applied on RGB pictures, acquired by an Unmanned Aeronautical Vehicle (UAV). Moreover, Su et al. [71] demonstrated a technique for the rice growth prediction through SVM approach and essential geographic data acquired from climate stations in China.

Disease Detection

Disease detection and yield prediction are the subjects with the higher number of articles in literature, an important issue concerning the most critical risks in agribusiness stress and disease control in field crops (arable cultivations) and nursery conditions [72]. The most generally utilized practice in the case of crop infection control is to uniformly apply pesticides over the canopy. ML techniques have been targeted on offering solutions to precision agriculture cases. In Pantazi et al. [73], a device is displayed for the recognition and separation of healthy Silybum marianum plants and those infected by *Microbotyum silybum* during growth. Ebrahimi et al. [74] built up a technique based on picture preprocessing system for grouping parasites and identifying thrips in strawberry crops. Chung et al. [75] introduced a technique for recognition and monitoring of Bakanae illness in rice seedlings. The primer aim of the study was the precise identification of the pathogen *Fusarium fujikuroi* infecting two rice cultivars. The automated approach of recognizing the infected plants expanded grain yield and was less tedious compared to the expert's eye examination.

Wheat stands out among the most financially noteworthy crops around the world. Pantazi et al. [76] introduced an approach for the identification of nitrogen stress, yellow rust infection, and healthy condition in winter wheat plots based on hierarchical SOMs and hyperspectral reflectance imaging data. The study focused on the effective classification of the above-presented crop health conditions, aiming to the targeted fungicides and fertilizers application customized to the plant's needs. Moshou et al. [77] demonstrated a classification method focusing on the recognition of *Septoria tritici* infection and healthy wheat crops. The methodology utilized a Least Square Support Vector Machines (LS-SVM) classifier with optical multisensor fusion. Least-squares support-vector machines (LS-SVM) are least-squares versions of support-vector machines (SVM), which are a set of related supervised learning methods that analyze data and recognize patterns, and which are used for classification and regression analysis.

In this version, one finds the solution by solving a set of linear equations instead of a convex quadratic programming (QP) problem for classical SVMs. Least-squares SVM classifiers were proposed by Suykens and Vandewalle [78]. LS-SVMs are a class of kernel-based learning methods. Moshou et al. [79] exhibited a technique to discriminate yellow rust infected from healthy wheat crops and further separate it from drought stress as presented in Moshou et al. [80] with the help of ANN models and spectral reflectance features. The precise and effective discrimination of the two above-mentioned crop heath status encourages the targeted pesticides application. In Moshou et al. [81], a remote sensing framework is introduced for recognizing yellow rust infected from healthy wheat crops. The presented work used a Self-Organizing Map (SOM) neural network together with information fusion of hyperspectral reflectance and multispectral fluorescence imaging. The objective of the study was the early-stage identification. An approach has been introduced by Ferentinos [82] using a CNN-based strategy for the identification of several crop infections on leaf pictures with high accuracy.

Weed Detection

Weed detection is another important task in agriculture. The majority of growers consider weeds as the most imperative danger for yield productivity, since their presence is often hard to distinguish and discriminated from the cultivated crop. There have been several ML applications combined effectively with different types of sensory systems that provided solutions for the accurate identification and discrimination of weeds from the cultivated crop in the field non-destructively and environmentally friendly. These ML-based solutions encourage the evolution of sensory systems and platforms oriented to weed management. There have been two recent ML applications for weed identification issues in the field of precision agriculture. Pantazi et al. [83] introduced an approach based on Counter-propagation Artificial Neural Network (CPANN) and multispectral UAS images for the identification of Silybum marianum, a noxious weed that its presence results to low yield productivity, causing subsequently significant financial losses to cultivators. Moreover, Pantazi et al. [84] used another ML approach combined with hyperspectral imaging for different weed species recognition and discrimination from the main crop. More precisely, an operational active learning framework is presented aiming to the identification of Zea mays, as the main crop cultivated in the field and its discrimination from the following different weed species: *Ranunculus repens, Cirsium arvense, Sinapis arvensis, Stellaria media, Tarraxacum officinale, Poa annua, Polygonum persicaria, Urtica dioica, Oxalis europaea*, and *Medicago lupulina*.

Species Recognition

An important sub-task of yield prediction is the crop type recognition. The primary objective is the autonomous discrimination and the grouping of plant species in order to avoid involving human experts into this laborious task and additionally to decrease the characterization time. Grinblat et al. [85] presented an approach for the discrimination and the sorting of three vegetable species, like white beans, red beans, and soybean, based on leaf vein patterns.

3 Conclusions and Future Challenges

The purpose of the chapter was to propose methodological tools for optimally fusing multi-type datasets. Data fusion is generally defined as the process of combining information from heterogeneous sources into a single composite picture of the relevant process, such that the composite picture is generally more accurate and complete than that derived from any single source alone. Data fusion can be approached from several different points of view; in particular, from statistical perspective it aims to infer the true value of an underlying process of interest at a specific location by using all multi-source information available and taking into account differences in support and measurement error characteristics. It is based on the belief that data fusion can improve the knowledge of the process by making optimal inferences derived from multiple complementary data sources. However, an effective methodology must be capable of addressing all the difficulties of data fusion: scalability, change of support, non-stationarity and non-isotropy, production of confidence measures, and bias correction.

Perhaps the most vital benefit of sensor fusion for the future is the context that it provides, taking disparate IoT inputs and interpreting their meaning within the environment. One example here is lidar and radar systems in vertical applications, such as agricultural drone systems, where pressure sensors are a vital tool for flight control and positioning in situations where GPS signals are unreliable. This sensor fusion-derived context is particularly important in the world of IoT and smart farming development, where thousands of “dumb” sensors, once networked, can be built into field-scale responsive systems once successfully fused. There are of course privacy and security concerns here if in-field data are used in an identifiable way, and wider public security concerns.

There are many challenges here, of course, not least the difficulties in reliably responding to variations in environmental conditions, on top of the challenge of extracting meaningful data from a wide range of sensors in varying implementations, any of which potentially add device error, noise, and flaws in the data gathering process. Smoothing out these errors was a near-impossible task just a few years ago, but with the rise of relatively affordable machine learning and artificial intelligence toolsets, the potential to deliver real benefits from sensor fusion has grown exponentially. The promise of artificial intelligence technology goes much further, of course, generating new use cases and therefore new markets for sensor suppliers and designers. In the shorter term, artificial intelligence and sensor fusion can minimize security risks by providing enhanced local processing of data, significantly reducing the requirements to securely transmit, process, and store personal data offsite. This may well become a vital value proposition, reducing business risk and overhead costs as well as providing a clear benefit to the end user.

References

1. Klein, L. A. (1993). *Sensor and data fusion concepts and applications*. Vol. TT14. SPIE Optical Engineering Press.
2. Hall, D. L., & Llinas, J. (1997). An introduction to multisensor data fusion. *Proceedings of the IEEE, 85*, 6–23.
3. Wald, L. (1999). Some terms of reference in data fusion. *IEEE Transactions on Geoscience and Remote Sensing, 13*(3), 1190–1193.
4. Geoscience and Remote Sensing Society. (2004). (Online) Available: http://www.dfc-grss.org.
5. International Society of Information Fusion. (2004). (Online) Available: http://www.inforfusion.org.
6. Luo, R. C., Kay, M. G., Eds. (1995). *Multisensor Integration and Fusion for Intelligent Machines and Systems*, Reissue edition Computer Engineering and Computer Science. Ablex Publishing, New Jersey, USA.
7. Brokmann, G., March, B., Romhild, D., & Steinke, A. (2001). *Integrated multisensors for industrial humidity measurement*. In Proceedings of the IEEE international conference on multisensor fusion and integration for intelligent systems. IEEE, Baden-Baden, Germany, 201–203.
8. Luo, R. C., Yih, C.-C., & Su, K. L. (2002). Multisensor fusion and integration: Approaches, applications, and future research directions. *IEEE Sensors Journal, 2*(2), 107–119.
9. Dasarathy, B. V. (1997). Sensor fusion potential exploitation-innovative architectures and illustrative applications. *Proceedings of the IEEE, 85*(1), 24–38.
10. Dasarathy, B. V. (2001). What, where, why, when, and how? *Information Fusion, 2*(2), 75–76. Editorial.
11. Kokar, M. M., Tomasik, J. A., Weyman, J. 1999. *A formal approach to information fusion*. In Proceedings of the 2nd international conference on information fusion (Fusion'99). Vol. 1. ISIF, Sunnyvale, 133–140.
12. Kalpakis, K., Dasgupta, K., & Namjoshi, P. (2003). Efficient algorithms for maximum lifetime data gathering and aggregation in wireless sensor networks. *Computer Networks, 42*(6), 697–716.
13. Van Renesse, R. (2003). The importance of aggregation. In A. Schiper, A. A. Shvartsman, H. Weatherspoon, & B. Y. Zhao (Eds.), *Future directions in distributed computing: Research and position papers* (Lecture notes in computer science) (Vol. 2584, pp. 87–92). Springer.

14. Cohen, N. H., Purakayastha, A., Turek, J., Wong, L., & Yeh, D. (2001). *Challenges in flexible aggregation of pervasive data*. IBM research report RC 21942 (98646), IBM Research Division, Yorktown Heights, NY (January).
15. Boulis, A., Ganeriwal, S., & Srivastava, M. B. (2003). Aggregation in sensor networks: An energy-accuracy trade-off. *Ad Hoc Networks, 1*(2–3), 317–331. Special Issue on Sensor Network Protocols and Applications.
16. Elmenreich, W. (2002). Sensor fusion in time-triggered systems. Ph.D. thesis, Institut f ¨ ur Technische Informatik, Vienna University of Technology, Vienna, Austria.
17. Gotway, C. A., & Young, L. J. (2002). Combining incompatible spatial data. *Journal of the American*.
18. Cao, G., Yoo, E., & Wang, S. (2014). A statistical framework of data fusion for spatial prediction.
19. Hall, D. L. (2004). *Mathematical techniques in multisensor data fusion*. Artech House.
20. Bogaert, P., & Fasbender, D. (2007). Bayesian data fusion in a spatial prediction context: A general formulation. *Stochastic Environmental Research and Risk Assessment, 21*, 695–709.
21. Nguyen, H., Cressie, N., & Braverman, A. (2012). Spatial statistical data fusion for remote sensing applications. *Journal of the American Statistical Association, 107*(499), 1004–1018.
22. White, Jr., F. E. (1990). *Joint directors of laboratories data fusion subpanel report*. In: Proceedings of the joint service data fusion symposium, DFS–90, 496–484.
23. Worden, K., & Dulieu-Barton, J. M. (2004). An overview of intelligent fault detection in systems and structures. *Structural Health Monitoring, 3*, 85–98.
24. Akyildiz, I. F., Su, W., Sankarasubramaniam, Y., & Cyirci, E. (2002). Wireless sensor networks: A survey. *Computer Networks, 38*(4), 393–422.
25. Pottie, G. J., & Kaiser, W. J. (2000). Wireless integrated network sensors. Comm. *ACM, 43*(5), 51–58.
26. Brooks, R. R., & Iyengar, S. (1998). *Multi-sensor fusion: Fundamentals and applications with software*. Prentice Hall PTR.
27. Pantazi, X.-E., Moshou, D., Alexandridis, T. K., Whetton, R. L., & Mouazen, A. M. (2016). Wheat yield prediction using machine learning and advanced sensing techniques. *Computers and Electronics in Agriculture, 121*, 57–65. https://doi.org/10.1016/j.compag.2015.11.018
28. Intanagonwiwat, C., Govindan, R., & Estrin, D. (2000). *Directed diffusion: A scalable and robust communication paradigm for sensor networks*. In Proceedings of the 6th annual international conference on mobile computing and networking (MobiCom'00). ACM Press, Boston, MA, 56–67.
29. Krishnamachari, B., Estrin, D., Wicker, S. (2002). *The impact of data aggregation in wireless sensor networks*. In International workshop of distributed event based systems (DEBS). IEEE, Vienna, Austria, 575–578.
30. Savvides, A., Han, C., & Strivastava, M. B. (2003). The n-hop multilateration primitive for node localization. *Mobile Networks and Applications, 8*(4), 443–451.
31. Nakamura, E. F., Nakamura, F. G., Figueiredo, C. M., & Loureiro, A. A. (2005). Using information fusion to assist data dissemination in wireless sensor networks. *Telecommunication Systems, 30*(1–3), 237–254.
32. Woo, A., Tong, T., & Culler, D. (2003). *Taming the underlying challenges of reliable multihop routing in sensor networks*. In Proceedings of the 1st international conference on embedded network sensor systems (SenSys'03), 14–27.
33. Megerian, S., Koushanfar, F., Qu, G., Veltri, G., & Potkonjak, M. (2002). Exposure in wireless sensor networks: Theory and practical solutions. *Wireless Networks, 8*(5), 443–454.
34. Meguerdichian, S.,Koushanfar, F., Potkonjak, M., & Srivastava, M. (2001). *Coverage problems in wireless ad-hoc sensor networks*. In Proceedings of IEEE infocom 2001. Vol. 3. IEEE, Anchorage, AK, 1380–1387.
35. Meguerdichian, S., Slijepcevic, S., Karayan, V., & Potkonjak, M. (2001). *Localized algorithms in wireless ad-hoc networks: Location discovery and sensor exposure*. In Proceedings of the

2001 ACM international symposium on mobile ad hoc networking & computing. ACM Press, Long Beach, CA, 106–116.
36. Chakrabarty, K., Iyengar, S. S., Qi, H., & Cho, E. (2002). Grid coverage for surveillance and target location in distributed sensor networks. *IEEE Transactions on Computers, 51*(12), 1448–1453.
37. Tian, D., Georganas, N. D. (2002). *A coverage-preserving node scheduling scheme for large wireless sensor networks*. In Proceedings of the 1st ACM international workshop on wireless sensor networks and applications (WSNA'02). ACM Press, Atlanta, GA, 32–41.
38. Dhillon, S. S.,Chakrabarty, K., & Iyengar, S. S. (2002). *Sensor placement for grid coverage under imprecise detections*. In Proceedings of the 5th international conference on information fusion (Fusion 2002).Vol. 2. IEEE, Annapolis, Maryland, 1581–1587.
39. Durrant-Whyte, H. F. (1988). Sensor models and multisensor integration. *The International Journal of Robotics Research, 7*(6), 97–113.
40. Rao, N. S. V. (2001). On fusers that perform better than the best sensor. *IEEE Transactions on Pattern Analysis and Machine Intelligence, 23*(8), 904–909.
41. Tax, D. M., et al. (2000). Combining multiple classifiers. *Pat. Rec., 33*, 1475–1485.
42. Brooks, R. R., Ramanathan, P., & Sayeed, A. M. (2003). Distributed target classification and tracking in sensor networks. *Proceedings of the IEEE, 91*, 1162–1171.
43. Zhao, F., et al. (2003). Collaborative signal and information processing: An information–Directed approach. *Proceedings of the IEEE, 91*, 1199–1209.
44. Sasiadek, J. Z. (2002). Sensor fusion. *Annual Reviews in Control, 26*, 203–228.
45. Filippidis, A., Jain, L. C., & Martin, N. (2000). Fusion of intelligent agents for the detection of aircraft in SAR images. *IEEE Transactions on Pattern Analysis and Machine Intelligence, 22*(4), 378–384.
46. Luo, R. C., & Kay, M. G. (1992). Data fusion and sensor integration: State-of-the-art 1990s. In M. A. Abidi & R. C. Gonzalez (Eds.), *Data fusion in robotics and machine intelligence* (pp. 7–135). Academic Press, , Chapter 3.
47. Castelaz, P. F. (1988). *Neural networks in defense applications*. In Proceedings of the IEEE international conference on neural networks. Vol. II. IEEE, San Diego, CA, 473–480.
48. Baran, R. H. (1989). *A collective computation approach to automatic target recognition*. In Proceedings of the international joint conference on neural networks. Vol. I. IEEE, Washington, D.C., 39–44.
49. Kohonen, T. (1997). *Self-organizing maps*. Springer-Verlag.
50. Cain, M. P., Stewart, S. A., & Morse, J. B. 1989. *Object classification using multispectral sensor data fusion*. In Proceedings of SPIE sensor fusion II. Vol. 1100. SPIE, Orlando, FL, 53–61.
51. LeCun, Y., Bengio, Y., & Hinton, G. (2015). Deep learning. *Nature, 521*, 436–444. https://doi.org/10.1038/nature14539
52. Goodfellow, I., Bengio, Y., & Courville, A. (2016). *Deep learning* (pp. 216–261). MIT Press.
53. Gandomi, A. H., Alavi, A. H., Arjmandi, P., et al. (2010). Genetic programming and orthogonal least squares: A hybrid approach to modeling the compressive strength of CFRP-conned concrete cylinders. *Journal of Mechanics of Materials and Structures, 5*, 735–753.
54. Li, H., Jiao, Y. C., Zhang, L., & Gu, Z. W. (2006). Genetic algorithm based on the orthogonal design for multidimensional knapsack problems. In L. Jiao (Ed.), *Adv. Nat. Comput. Lecture notes in computer science*. Advances in Natural Computation: Proceedings Second International Conferernce, ICNC 2006, Xi'an, China, September 24–28, (Vol. 4221, pp. 696–705). Springer.
55. Waltz, E., & Llinas, J. (1990). Multisensor data fusion. *Artech House*.
56. Pau, L. F. (1988). Sensor Data Fusion. *Journal of Intelligent & Robotic Systems, 1*, 103–116.
57. Bass, T. (2000). Intrusion detection systems and multisensor data fusion. *Communications of the ACM, 43*(4), 99–105. ACM Press.

58. Lewis, T. W. Powers, D. M. W. 2002. *Audio-visual speech recognition using red exclusion and neural networks*. In Proceedings of the 25th Australasian conference on computer science. Australian Computer Society, Inc., Melbourne, Victoria, Australia, 149–156.
59. Cimander, C., Carlsson, M., & Mandenius, C. (2002). Sensor fusion for on-line monitoring of yoghurt fermentation. *Journal of Biotechnology, 99*(3), 237–248.
60. Yiyao, L., Venkatesh, Y. V., & Ko, C. C. (2001). A knowledge-based neural network for fusing edge maps of multi-sensor images. Inform. *Fusion 2, 2*(June), 121–133.
61. Zhang, F., Du, B., & Zhang, L. (2015). Saliency-guided unsupervised feature learning for scene classification. *IEEE Transactions on Geoscience and Remote Sensing, 53*(4), 2175–2184.
62. Chen, Y., Zhao, X., & Jia, X. (2015). Spectral-spatial classification of hyperspectral data based on deep belief network. *IEEE Journal of Selected Topics in Applied Earth Observations and Remote Sensing, 8*(6), 2381–2392.
63. Chen, X., Xiang, S., Liu, C. L., & Pan, C. H. (2014). Vehicle detection in satellite images by hybrid deep convolutional neural networks. *IEEE Geoscience and Remote Sensing Letters, 11*(10), 1797–1801.
64. Zhang, F., Du, B., Zhang, L., & Zhang, L. (2016). Hierarchical feature learning with dropout k-means for hyperspectral image classification. *Neurocomputing, 187*, 75–82.
65. Cai, K., Shao, W., Yin, X., & Liu, G.. (2012). *Co-segmentation of aircrafts from high-resolution satellite images*. In Proc. IEEE international conference on signal processing, Beijing, pp. 993–996.
66. Ramos, P. J., Prieto, F. A., Montoya, E. C., & Oliveros, C. E. (2017). Automatic fruit count on coffee branches using computer vision. *Computers and Electronics in Agriculture, 137*, 9–22. https://doi.org/10.1016/j.compag.2017.03.010
67. Amatya, S., Karkee, M., Gongal, A., Zhang, Q., & Whiting, M. D. (2015). Detection of cherry tree branches with full foliage in planar architecture for automated sweet-cherry harvesting. *Biosystems Engineering, 146*, 3–15. https://doi.org/10.1016/j.biosystemseng.2015.10.003
68. Sengupta, S., & Lee, W. S. (2014). Identification and determination of the number of immature green citrus fruit in a canopy under different ambient light conditions. *Biosystems Engineering, 117*, 51–61. https://doi.org/10.1016/j.biosystemseng.2013.07.007
69. Ali, I., Cawkwell, F., Dwyer, E., & Green, S. (2016). Modeling managed grassland biomass estimation by using multitemporal remote sensing data—A machine learning approach. *IEEE Journal of Selected Topics in Applied Earth Observations and Remote Sensing, 10*, 3254–3264.
70. Senthilnath, J., Dokania, A., Kandukuri, M., Ramesh, K. N., Anand, G., & Omkar, S. N. (2016). Detection of tomatoes using spectral-spatial methods in remotely sensed RGB images captured by UAV. *Biosystems Engineering, 146*, 16–32. https://doi.org/10.1016/j.biosystemseng.2015.12.003
71. Su, Y., Xu, H., & Yan, L. (2017). Support vector machine-based open crop model (SBOCM): Case of rice production in China. *Saudi Journal of Biological Sciences, 24*, 537–547. https://doi.org/10.1016/j.sjbs.2017.01.024
72. Anagnostis, A., Tagarakis, A. C., Asiminari, G., Papageorgiou, E., Kateris, D., Moshou, D., & Bochtis, D. (2021). A deep learning approach for anthracnose infected trees classification in walnut orchards. *Computers and Electronics in Agriculture*. https://doi.org/10.1016/j.compag.2021.105998
73. Pantazi, X. E., Tamouridou, A. A., Alexandridis, T. K., Lagopodi, A. L., Kontouris, G., & Moshou, D. (2017b). Detection of Silybum marianum infection with Microbotryum silybum using VNIR field spectroscopy. *Computers and Electronics in Agriculture, 137*, 130–137. https://doi.org/10.1016/j.compag.2017.03.017
74. Ebrahimi, M. A., Khoshtaghaza, M. H., Minaei, S., & Jamshidi, B. (2017). Vision-based pest detection based on SVM classification method. *Computers and Electronics in Agriculture, 137*, 52–58. https://doi.org/10.1016/j.compag.2017.03.016
75. Chung, C. L., Huang, K. J., Chen, S. Y., Lai, M. H., Chen, Y. C., & Kuo, Y. F. (2016). Detecting Bakanae disease in rice seedlings by machine vision. *Computers and Electronics in Agriculture, 121*, 404–411. https://doi.org/10.1016/j.compag.2016.01.008

76. Pantazi, X. E., Moshou, D., Oberti, R., West, J., Mouazen, A. M., & Bochtis, D. (2017). Detection of biotic and abiotic stresses in crops by using hierarchical self-organizing classifiers. *Precision Agriculture, 18*, 383–393. https://doi.org/10.1007/s11119-017-9507-8
77. Moshou, D., Pantazi, X.-E., Kateris, D., & Gravalos, I. (2014). Water stress detection based on optical multisensor fusion with a least squares support vector machine classifier. *Biosystems Engineering, 117*, 15–22. https://doi.org/10.1016/j.biosystemseng.2013.07.008
78. Suykens, J. A. K., & Vandewalle, J. (1999). Least squares support vector machine classifiers. *Neural Processing Letters, 9*(3), 293–300.
79. Moshou, D., Bravo, C., West, J., Wahlen, S., McCartney, A., & Ramon, H. (2004). Automatic detection of "yellow rust" in wheat using reflectance measurements and neural networks. *Computers and Electronics in Agriculture, 44*, 173–188. https://doi.org/10.1016/j.compag.2004.04.003
80. Moshou, D., Bravo, C., Wahlen, S., West, J., McCartney, A., De Baerdemaeker, J., & Ramon, H. (2006). Simultaneous identification of plant stresses and diseases in arable crops using proximal optical sensing and self-organising maps. *Precision Agriculture, 7*, 149–164. https://doi.org/10.1007/s11119-006-9002-0
81. Moshou, D., Bravo, C., Oberti, R., West, J., Bodria, L., McCartney, A., & Ramon, H. (2005). Plant disease detection based on data fusion of hyper-spectral and multi-spectral fluorescence imaging using Kohonen maps. *Real-Time Imaging, 11*, 75–83. https://doi.org/10.1016/j.rti.2005.03.003
82. Ferentinos, K. P. (2018). Deep learning models for plant disease detection and diagnosis. *Computers and Electronics in Agriculture, 145*, 311–318. https://doi.org/10.1016/J.COMPAG.2018.01.009
83. Pantazi, X. E., Tamouridou, A. A., Alexandridis, T. K., Lagopodi, A. L., Kashefi, J., & Moshou, D. (2017a). Evaluation of hierarchical self-organising maps for weed mapping using UAS multispectral imagery. *Computers and Electronics in Agriculture, 139*, 224–230. https://doi.org/10.1016/j.compag.2017.05.026
84. Pantazi, X.-E., Moshou, D., & Bravo, C. (2016). Active learning system for weed species recognition based on hyperspectral sensing. *Biosystems Engineering, 146*, 1–10. https://doi.org/10.1016/j.biosystemseng.2016.01.014
85. Grinblat, G. L., Uzal, L. C., Larese, M. G., & Granitto, P. M. (2016). Deep learning for plant identification using vein morphological patterns. *Computers and Electronics in Agriculture, 127*, 418–424. https://doi.org/10.1016/j.compag.2016.07.003

Machine Learning Technology and Its Current Implementation in Agriculture

Athanasios Anagnostis, Gabriela Asiminari, Lefteris Benos, and Dionysis D. Bochtis

1 Introduction

Artificial intelligence has been around in computer science since the 1950s [1, 2]. Till then, computers were considered to be "dumb", because they only did what they were "told" by the programmers in a pre-determined manner. Scientists, however, always wanted to make computers being able to "learn" from data, without being explicitly told on how to do so. The ability of a computer to learn from its errors and try to correct them is the foundation of artificial intelligence However, this learning capability is found only in living beings [3, 4]. Specifically in human beings, the learning ability of their brain has been studied extensively and in some cases, this ability is even implemented in computer science [5, 6]. Seventy years later, artificial intelligence is probably the most influential area of computer science of the modern world [7–9]. Its applications cover almost all sectors including agriculture [10], medicine [11], energy [12], manufacturing [13], and so on. Its infiltration to our everyday lives is very deep, even though most times goes unrecognizable [14]. Artificial intelligence is applied in our smartphones, smartwatches, autonomous cars, searching engines, shopping carts, music players, video platforms, and so on. We are becoming more and more accustomed to the idea that machines can understand more than they used to [15].

A. Anagnostis (✉) · G. Asiminari · L. Benos
Institute for Bio-economy and Agri-technology (iBO), Centre for Research and Technology Hellas (CERTH), Thessaloniki, Greece
e-mail: a.anagnostis@certh.gr

D. D. Bochtis
Institute for Bio-economy and Agri-technology (iBO), Centre for Research and Technology Hellas (CERTH), Thessaloniki, Greece

farmB Digital Agriculture P.C, Thessaloniki, Greece

D. D. Bochtis et al. (eds.), *Information and Communication Technologies for Agriculture—Theme II: Data*, Springer Optimization and Its Applications 183,
https://doi.org/10.1007/978-3-030-84148-5_3

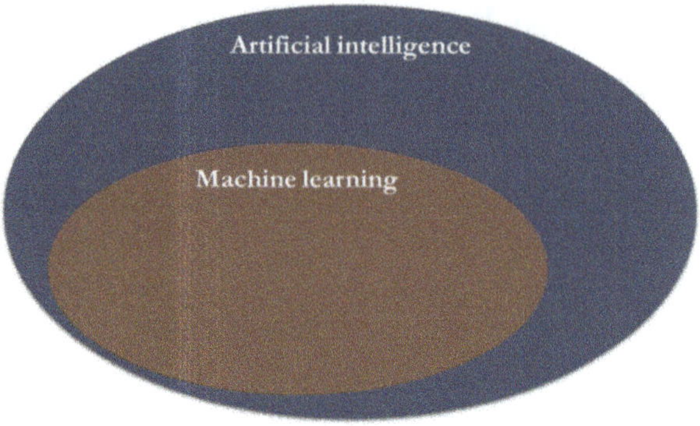

Fig. 1 Machine learning as part of artificial intelligence

Artificial intelligence is built upon various mathematical concepts that are being used in ways and combinations so that a computer, or "machine" in general, is able to learn from data [16]. Different mathematical and algorithmic approaches have found fertile ground on this area and were able to make machines learn. The set of all of these algorithms, methods, and mathematical concepts is called machine learning [17]. Machine learning is the base on which the machines are programmed to achieve intelligence. In fact, machine learning is only a part of the general idea of artificial intelligence [18], visually shown in Fig. 1, however it is possibly the most important one and definitely one that is of high interest.

2 Machine Learning Versus Conventional Programming

Machine learning and conventional programming are both parts of computer science. However, they differ in one major aspect [19]. In conventional programming algorithms are structured so that they "explain" the computer in each step exactly what to do. So, given an input, there is a set of rules that computer must follow in order to achieve a desired outcome. In machine learning, the premise is exactly the opposite. In particular, the computer is provided with sets of input-output and then, the computer tries to find and build the rules that lead from the particular input to the particular output. This diversion can be illustrated as in Fig. 2.

This set of rules can be either clear or obscure to the programmer, depending on which algorithm is implemented. However, this is of minor practical importance to the application of machine learning, when it can ultimately be proven that it works with extreme efficiency in the application that is tested.

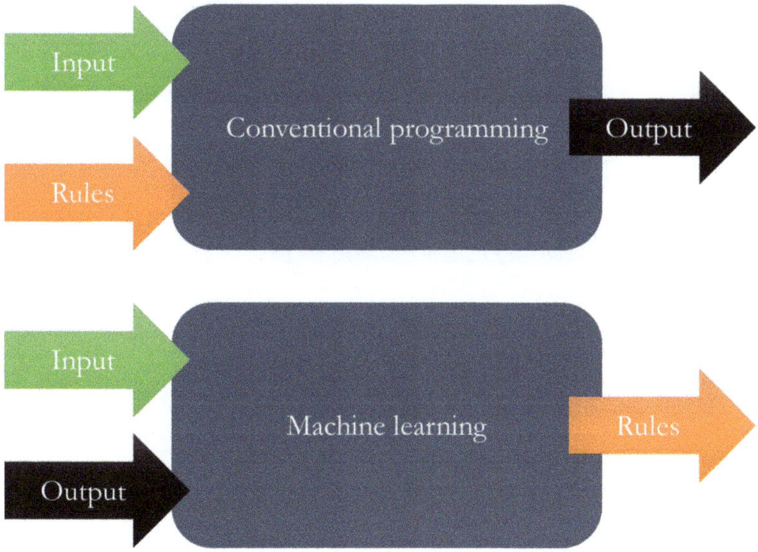

Fig. 2 Conventional programming versus machine learning

3 Fundamental Features of Machine Learning

Machine learning strongly depends on data. In most cases, the more data it uses, the better it works. This make sense, given that machine learning is built on "experience". As far as living creatures are concerned, we learn from examples; the first human that touched fire probably never did it a second time. The more experiences someone has, the better understanding is accomplished towards a given subject or task. Similarly with physical abilities, where someone must throw a rock many times to learn how to throw it far, or to build strength for it. The same applies also to the machines. In an attempt for a machine to build a set of rules that describe the function that turns input onto output, it needs a lot of input and output data to be used for tryouts. Again, tryouts stand for the testing of as many possibilities that will lead to the desired outcome, in the same manner as babies try to find the right shape that goes to the right hole. The mathematics behind this thought process is a bit more complicated than this example, however the idea is simple; more data means more experience, leading to more tryouts, causing deeper learning, and consequently, better understanding.

At the time being, machine learning allows for achieving something called narrow artificial intelligence [20]. This is the intelligence that is confined within some specific limits. This means that we can build models, where models are the rules that a machine learning algorithm produces given explicit input/output that can be applied only to specific tasks, e.g., identification of faces in images. The greater picture is general artificial intelligence [21]. This is the hypothetical, for the moment,

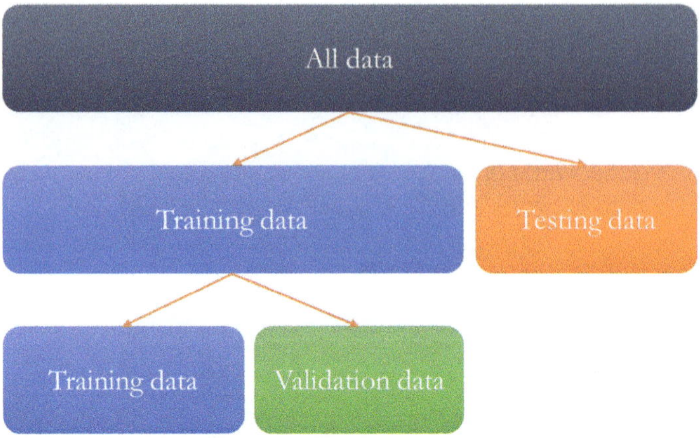

Fig. 3 Splitting the data for machine learning

intelligence that gives a machine the capacity to perform intellectual tasks like an adult human.

A simplified version of machine learning working pipeline reads as follows:

- Input and output data are obtained
- The desired algorithm is selected and fed the data
- The algorithm makes an attempt to solve the problem
- After the attempt, results are evaluated
- The algorithm proceeds to correct its parameters and attempts again
- Repeat until a performance condition is met

The final condition is highly relative to the type of problem, the desired outcome and the algorithm that has been chosen, but simplified again, aim is to build a model that covers most cases successfully or generalizes well. This means that the model should not only be able to predict the provided examples, but also examples that are completely unknown to it. This is achieved by using the data that are in possession, in a smart way. Thus, in most of the cases of a finite amount of data, data are split into three categories. Before starting anything, an amount of data is concealed from the rest of the process. This set is called "testing data" and will be used in the final step again. The rest of the data, that is usually the larger amount compared to the testing set, will be used for the learning process. Out of the training data however, a small portion will be used for the validation of the learning process itself. This validation data helps the model to learn from its mistakes and improve its predictions after each iteration, until it reaches a condition that is being set. After the model has completed training, it is evaluated against the testing data. Given on how well it performs on the testing data, its accuracy and general efficiency is evaluated. The splitting of the data is visualized in Fig. 3.

The above analysis is a description on how supervised machine learning works [22]. However, supervised learning is not the only type of machine learning. There are more which will be elaborated next.

4 Types of Machine Learning Methods

As it has already been mentioned, machine learning incorporates a large variety of algorithms and mathematical methods that allow machines to conduct operations without being explicitly told on how to perform them, as long as the desired result is reached. There are cases where there is a knowledge on how these results are supposed to be; this is called supervised learning and has been described in detail above. However, there are cases where the knowledge on the expected results is just an intuition. For example, if we have a pile of t-shirts of various colors and sizes and ask someone to separate them into categories, this person would have to choose to separate them based on color or on size. In this simplified example, we would have to wait and see the outcome of the separation first and decide afterwards if we are satisfied with the result. This is an example of unsupervised learning [23], where as the name suggests, we have no clear view of the outcome. Therefore, the algorithm must come up with an outcome on its own. There are also semi-supervised learning methods [24] which can be considered as a combination of supervised and unsupervised learning.

There are also algorithms that "reward" desired actions and "punish" undesired ones as they happen. This type of learning is called reinforcement learning and has received great acceptance, especially because of its application on digital games. The most famous is AlphaGo [25], a machine learning program that has been trained to play the game "Go", rather famous in Asia. AlphaGo was set to play games with human players online, without being programmed on any of the moves. It slowly started to learn patterns and in time started to win its human opponents and reached a high position in the world rank. It was at that time that it was disclosed that AlphaGo was a machine learning program and not a human player. It ended up facing the world champion of Go and defeating him with "moves that made no sense, but somehow worked", paraphrasing the former champion's words. AlphaGo went on and beat more world-class players, establishing its superiority and bringing forth the advantages of machine learning. This fame engaged more people into trying similar applications, resulting in funny animations of simulated human models trying to walk/run/jump [26]. Furthermore, some extreme Super Mario levels [27] that no human could ever cleared were ultimately cleared by machine learning models.

Another particular category of machine learning class is called active learning and is widely used for building recommender systems. As the name suggests, they are designed to provide with recommendations that fit a pattern. This is probably the most day-to-day interaction we have with machine learning. Services like Spotify [28] and Netflix [29] use it for recommending songs and movies that fit the taste of the user. Amazon also uses machine learning for recommending items that will complement what you are buying or what are highly possible that you like and buy them nevertheless [30]. Finally, Google [31] uses machine learning (and develops tools and services associated with it) to almost every aspect, such as in its search engine, mail service, maps, and most importantly, advertising. Next, follows an analysis of how each type of machine learning works.

4.1 *Supervised Learning*

Supervised learning [22] is the method where the input and the output are known, and the machine should find the best way to reach an output given an input. This is based on the premise that the entire output is known. Especially for classification, which will be described below, it means that all data are labeled, i.e., they have a label that describes what is expected in terms of the characteristic at hand. There are two model categories that fall under supervised learning, namely regression and classification.

Regression

In a regression model, the desired output is a continuous variable, which means that it can be any number depending on what it predicts. Some common examples of regression are the prediction of real-estate values given as input their size and location, the forecasting of stock market based on its past values, and the prediction of words while typing a text. In all these examples, a common denominator is that the prediction can be almost any value. The visual representation of regression on a plane can be seen in Fig. 4, where the blue dots are the actual data, and the green line is the prediction curve.

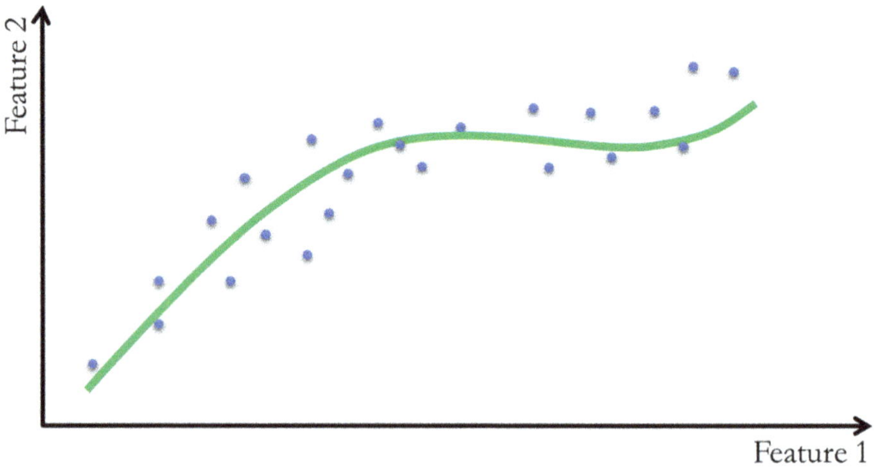

Fig. 4 Qualitative representation of regression

Classification

In classification model, the target set is specific, as the input is assigned to a predefined class. This is used widely in imaging applications, such as medical or agricultural, where the aim is to classify an image based on its content and reach a conclusion. Therefore, in medical applications for example, machine learning algorithms are used to identify cancers and metastases in whole body scans or other imaging methods. All data are labeled in this case, which means that each image is appointed to a class prior to the learning process. Classification can be considered like regression except from the final step, where the outcome runs through an activation function. The only purpose of it is to assign new values to the outcome so that it will fall under an assigned class. Of course, this is not exactly how it works with all algorithms. However, it is a good guide to follow when thinking of classification. A visual representation of classification is shown in Fig. 5 where the blue line separates the data into two categories.

4.2 *Unsupervised Learning*

In unsupervised learning [23] the outcome is unclear to the user. Consequently, machine must provide with "intuition" on how to handle a problem and provide an output. In fact, there is a general sense on what to do, but it is reached without prior knowledge, set by examples. There are three categories under unsupervised learning: clustering, dimensionality reduction, and association.

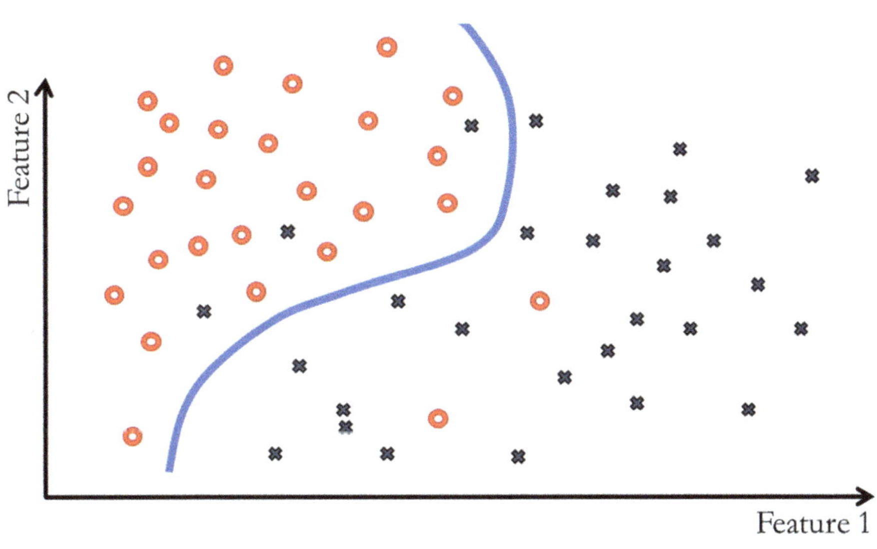

Fig. 5 Qualitative representation of *classification*

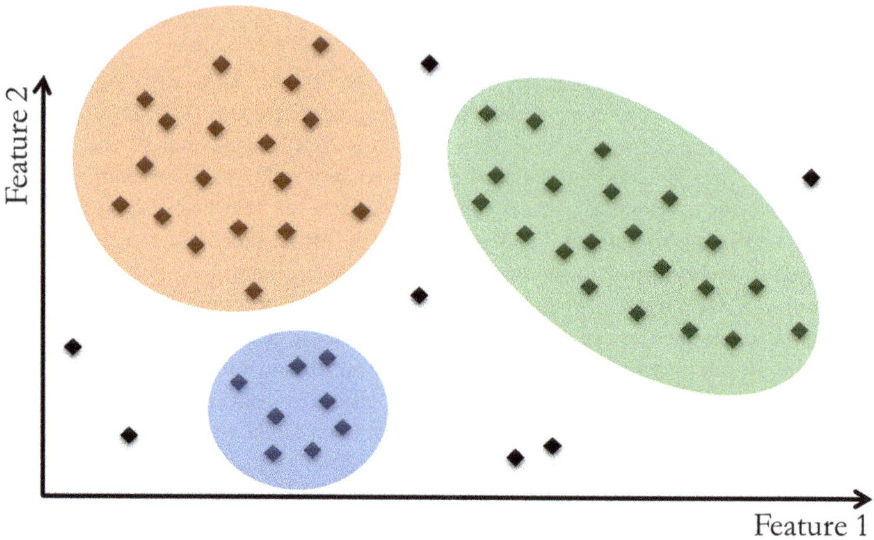

Fig. 6 Qualitative representation of *clustering*

Clustering

To put it simply, this is the task of grouping (or dividing) data based on their similarities. It can be considered as a collection or clusters of instances that have similar characteristics. It is different than classification, where the classes are predefined, because clustering tries to find the characteristics of instances and assign them into non-predefined groups. It makes more sense when shown visually, like in Fig. 6 where the three clusters are shown with data of similar characteristics (and some that belong to none of the clusters).

Dimensionality Reduction

In this methodology, the aim is to reduce the amount of information that is available. Large amounts of data do not necessarily mean plenty of valuable information. Therefore, dimensionality reduction techniques try to find relations between the data and remove features that do not offer any value. Thus, the amount of data can be significantly reduced, while the amount of information diminishes as little as possible. A simple example of how we go from 2-dimension space to 1-dimension space is depicted in Fig. 7.

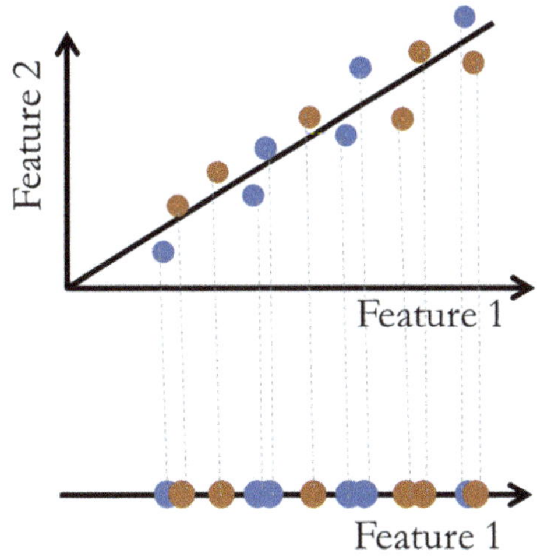

Fig. 7 Dimensionality reduction schematic illustration

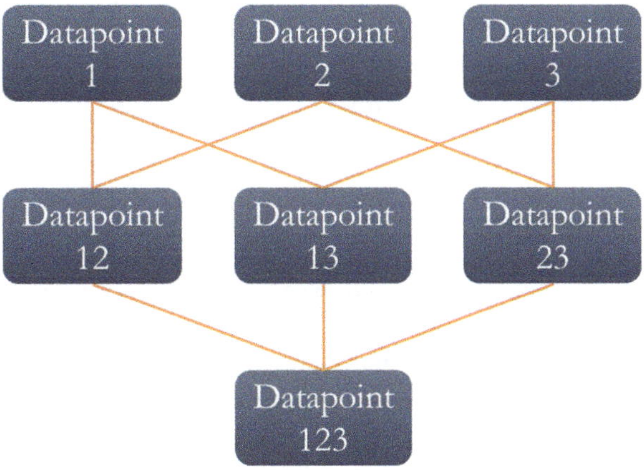

Fig. 8 Association rules in data sets

Association

As the name states, association aims to find association rules between large amounts of data. This cognitive function is relatively easy for human; however, computers can also perform this with the help of machine learning, and sometimes even better, especially when there is a huge amount of data with thousands of attributes each. A simplified visualization is shown in Fig. 8.

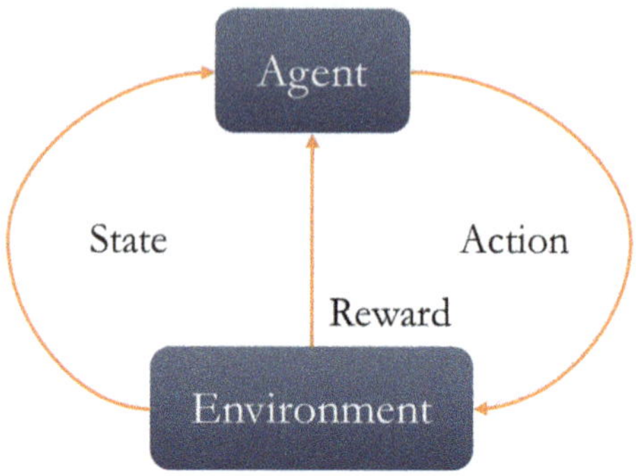

Fig. 9 Reinforcement learning

4.3 *Reinforcement Learning*

This type of learning basically rewards "good" actions and punishes "bad" actions. It maybe closer to the way that most living creatures learn compared to the rest of the methods. The way reinforcement learning works can be illustrated in Fig. 9, where the relationship is shown between an agent that acts and its actions are rewarded on the basis of the outcome they produce.

In machine learning it is used mainly for the following described two practices: classification and control.

Classification

Classification has already been described and in essence is not much different in reinforcement learning. The main difference is that the assigned class can change over time as the outcome of an action. An example is the optimized marketing where someone who browses an online shop one day browses sound-related equipment and is classified as an audiophile, and another day searches for car parts and is assigned to the car-enthusiast class. Real-life applications are more complicated than this simple instance, however the main idea remains still.

Control

This is probably the most famous application of reinforcement learning because its applications are widely marketable. In this task, the learning part aims at controlling driverless cars and robotic arms without explicit programming. A set of actions related to the hardware that needs to be controlled is given. Subsequently, based on the input at any given time and the outcome, the machine learning model must choose which action is more appropriate to reach the goal. Mistakes can be made but are "punished" so that they will not happen again and behaviors that lead to the target are "rewarded".

4.4 *Recommender Systems (Active Learning)*

Active learning differs from the rest of the types, because in this case the learning algorithm can actively request feedback from the information source in order to label previously unlabeled data. It can also be described as iterative supervised learning. This is more clearly shown in Fig. 10 for the example of a user choosing an item.

There are two main categories for recommender systems, namely content-based and collaborative filtering, that are briefly described next.

Content-based

In content-based systems, the algorithm constantly tries to add information about a user with the intention of improving its predictions. In most cases, the algorithms invoke the user into providing with more data such as ratings or comments. This is an iterative process and provides adaptability for changes in behavior from the side of the user.

Collaborative Filtering

In contrast to content-based systems, collaborative filtering systems are built explicitly on past interaction of a user and their targets. They are built on the premise that historical information is enough to make a prediction about the user.

To sum up, all machine learning methods tackle different problems in different ways. Machine learning as a science is constantly expanding, with more algorithms being developed day by day. The main pillars however, as described above, remain the same and are presented in Fig. 11.

5 Families of Machine Learning Algorithms

In this section, the families of machine learning algorithms are presented. Each algorithm learns in different ways, however there are some similarities on their basic functions that allow us to apply this categorization.

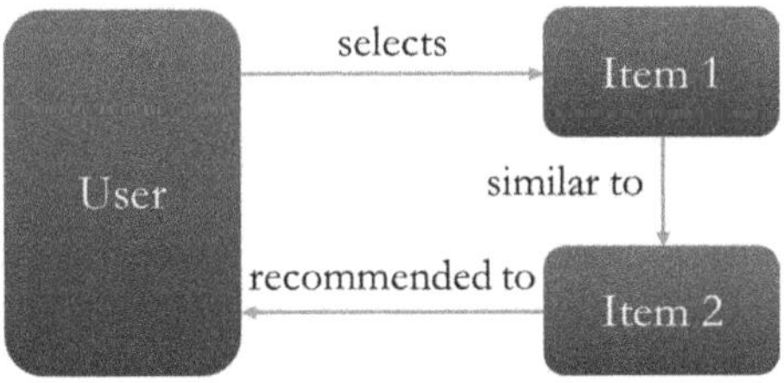

Fig. 10 Recommender system

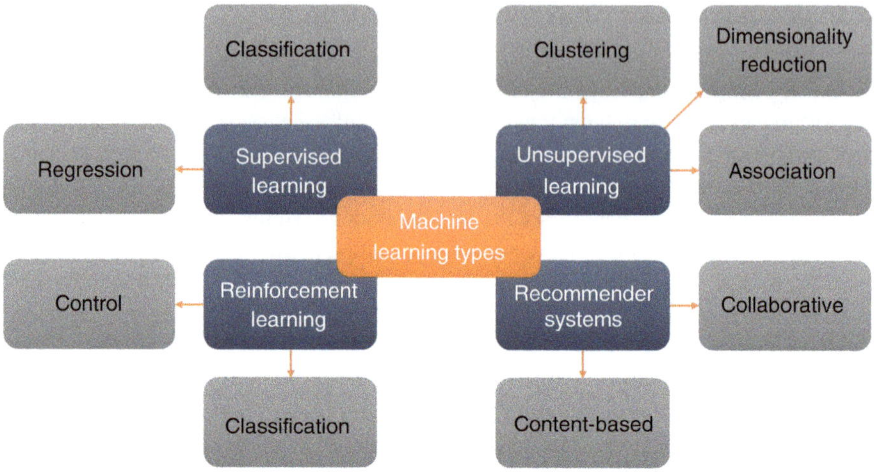

Fig. 11 Machine learning types and their respective categories

5.1 Regression

Regression is one of the most basic machine learning families and is set up on trying to find relations between variables. The most common one is linear regression [32] which is a line that describes the relationship of two-dimensional data. Other types of regression are multiple [33], logistic [34], stepwise [35], ordinary least squares [36], multivariate adaptive splines [37], and locally estimated scatterplot smoothing [38].

5.2 Regularization

This family of algorithms could be merged with the previous since they are basically regression, however they have a significant difference. All algorithms here apply regularization terms during their learning process, meaning that they penalize complex models and favoring simpler ones in order to achieve generalization. Some of the most famous algorithms are least angle [39] and ridge [40] regression, elastic-net [41] and LASSO [42] which stands for Least Absolute Shrinkage and Selection Operator.

5.3 Bayesian

The Bayesian family of algorithms are probabilistic algorithms that base their learning methods on Bayes' theorem [43]. Simply put, this theory defines the probability of something that will happen, based on the things that have already happened. It is widely used in statistics [44] and it is probably the oldest mathematical concept here since it was published by Thomas Bayes in 1761. The most famous Bayesian machine learning algorithms are naïve [45] and Gaussian naïve [46], Bayesian network [47], and belief networks [48].

5.4 *Instance-based*

The algorithms in this family base their learning by treating data as examples, or instances in a space. When a new instance (testing example or unknown data) is introduced, it is compared with all (or some) previous instances, and based on a similarity measure, a prediction takes place. The representation of data in space is the key feature here especially in algorithms such as support vector machines [49], which is the most famous in this category. Instance-based methods are also known as memory-based methods and some other noteworthy examples are k-nearest neighbors [50], self-organizing maps [51], learning vector quantization [52], and locally weighted learning [53].

5.5 *Decision Tree*

Decision tree methods are algorithms that build tree-like structures of decisions (or outcomes) based on if-else conditions which are applied on the features of the examples. When the desired output is reached, the best branch is selected. They are usually quite fast and accurate but can suffer from overfitting (not generalizing) quite easily. The most famous are classification and regression tree [54], conditional trees [55], iterative dichotomizer [56], and C4.5-C5.0 [57].

5.6 *Ensemble*

This algorithmic family applies the concept that the combination of weaker models can result in a strong model that can offer generalization. This is exactly the case with random forest [58], where, as the name suggests, it takes multiple decision trees and aggregates their outcome. This way the problem of overfitting disappears, and the ensemble model generalizes well. Boosting algorithms occupy the majority of implementations in this family with most prominent examples being adaptive [59] and gradient boosting [60], bootstrapped aggregation [61], and stacked generalization [62].

5.7 Clustering

In the clustering family, a category of algorithms is set up to build models by being given some reference points, also known as centroids, and trying to assign the data based on these centroids. Most famous algorithms in this category are the k-means [63] and k-medians [64]. However, there is another category where the algorithms build the model and appoint data on hierarchical pre-existing structures.

5.8 Dimensionality Reduction

All algorithms in this family aim to reduce the number of variables, in order to minimize the data that will be used later on. The minimization of data and variables is done with focus on having as little impact as possible on the retained information. Prominent examples of dimensionality reduction algorithms are principal component [65], quadratic [66], mixture [67] and flexible discriminant analysis [68], partial least squares [69] and principal component regression [70], multidimensional scaling [71], and projection pursuit [72].

5.9 Association Rule

The association rule algorithms try to find associations between data variables. For very large datasets with large number of variables, the a priori [73] algorithm is the most prominent example. However, it is computationally complex and quite slow. For small and medium datasets, the eclat [74] algorithm is the most suitable and faster solution.

5.10 Artificial Neural Networks

Artificial neural network is probably the most famous algorithmic family, and the one that is more associated with machine learning at the present time. Artificial neural networks were designed as the equivalent of the human brain, where neurons are represented by nodes with units and synapses as mathematical operations [75]. The nodes are stacked together and form layers which are positioned so that the data will run through each layer consecutively. In the perceptron [76], the most basic implementation of neural networks, there is an input layer, a hidden layer, and an output layer. Adding more hidden layers gives us the multilayer perceptron [77]. The more complicated the neural networks became, the more difficult was to optimize their variables. Optimization methods such as stochastic gradient descent

[78] have evolved in neural network algorithms themselves. A groundbreaking algorithm, called backpropagation [79], was developed in order to be able to compute the gradients of variables comparing predicted with the real output, and reiterate operations backward in order to correct them toward a better prediction. Other mentionable algorithms in this family are the radial basis function [80] and the Hopfield [81] networks.

5.11 Deep Neural Networks

Deep neural networks are the byproduct of artificial neural networks and were introduced on more complex problems that simple artificial neural networks could not handle well [82]. They are the newest addition to the machine learning science and have seen immense growth in the past 10 years due to vast availability of huge amounts of data, and the abundant cheap computational power from central, graphical, and tensor processing units (CPUs, GPUs, and TPUs). In essence, they are deeper and more complex architectures of the aforementioned artificial neural networks. This allows the algorithms to uncover hidden features from large amounts of data, sometimes without prior feature engineering [83]. The type of learning based on deep neural networks is also called deep learning [84]. There are notable mentions in this family that radicalized computer science and technology and continue to do so.

Recurrent neural networks were developed for forecasting purposes such as time-series predictions [85]. Values of previous time steps of a variable are iteratively used as input, stored in memory cells and then the network is making future predictions. This has applications to weather prediction, energy, stock market value predictions, but also to other applications such as word prediction when typing texts. Recurrent neural networks can handle a large amount of data and are able to store in memory mostly the valuable information in order to learn the best trends that fit the forecasting they make. Recurrent networks are best represented by the long short-term memory [86] and the gated recurrent unit [87] algorithms.

Convolutional neural networks have added immensely to the popularity of deep neural networks because of their wide applicability [88]. They apply filters, also known as convolutions in mathematics, that help bring out characteristics, such as edges or corners in images for example, that will help with the solution of the problem. They have become famous as they outperformed artificial neural networks by a large margin in problems of image classification, such as the identification of handwritten digits [89]. Besides image classification, convolutional neural networks are widely used for object detection problems, where objects are located within the image and categorized appropriately [90]. A more advanced implementation of this is the instance aware semantic segmentation where each pixel of the image is assigned with an object class, therefore whatever is seen in the image is given a label [91]. Autonomous cars and surveillance systems run based on convolutional neural networks.

Based on these networks and their performance superiority on image-related tasks, more complicated architectures and concepts have been developed. Variational autoencoders [92] were developed where a network encodes input images based on their features and then another network decodes images based on these encodings, resulting in similar but definitely not identical images. This algorithm has found application in counterfeiting problems. Similar to autoencoders, generative adversarial networks [93] use two networks: one for the generation of images and one for the discrimination of the generated images compared to real images. This way both networks are trained on opposing tasks, one is trying to generate images that are so close to real that can deceive the discriminator, and the other is trying to distinguish counterfeits in an increasingly improved generator. Other notable mentions are deep belief networks [94] that are generative models, and deep Boltzmann machines [95] that are recurrent models, where stochastic principles are applied.

6 Machine Learning in Agriculture

Agriculture plays a vital role in the economic development of a country. It is the main source of food for all people, raw materials, and income, as well as employment for a significant percentage of population worldwide. Nowadays, agriculture faces considerable challenges as the global population increases [96, 97] and in the meantime, agricultural land becomes smaller and poorer over the years [98, 99]. Furthermore, it is of great importance that the global food system provides healthy and nutritious food, while being able to minimize the environmental impact during production. In order to overcome these challenges, agricultural systems need to become more comprehensible which means monitoring and analyzing repeatedly a big amount of physical aspects and phenomena [100–102].

Machine learning is a powerful tool that has been recently introduced in agriculture, as it has the capacity to process a big amount of input data and cope with non-linear tasks [103, 104]. Machine learning models have been applied in many agricultural applications, especially for crop management [105, 106]. In particular, crop management provides information to farmers about harvesting, planting, spraying, and other field operations. Additionally, it contributes to crucial decision-making based on this information. In the case of crop management, machine learning predictions are mainly obtained without using a vast amount of data from other resources. Other implementations of machine learning in agriculture involve livestock management, water management, and soil management. However, these applications are less in number as compared to those of crop management, because many data recordings are needed along with a huge effort for the data analysis task.

For the purpose of capturing the recent progress in crop management, a preliminary scholarly literature survey was conducted, focused on articles that make references of machine learning algorithms in the agricultural domain. To this end, different combinations of the keywords "precision agriculture", "machine learning",

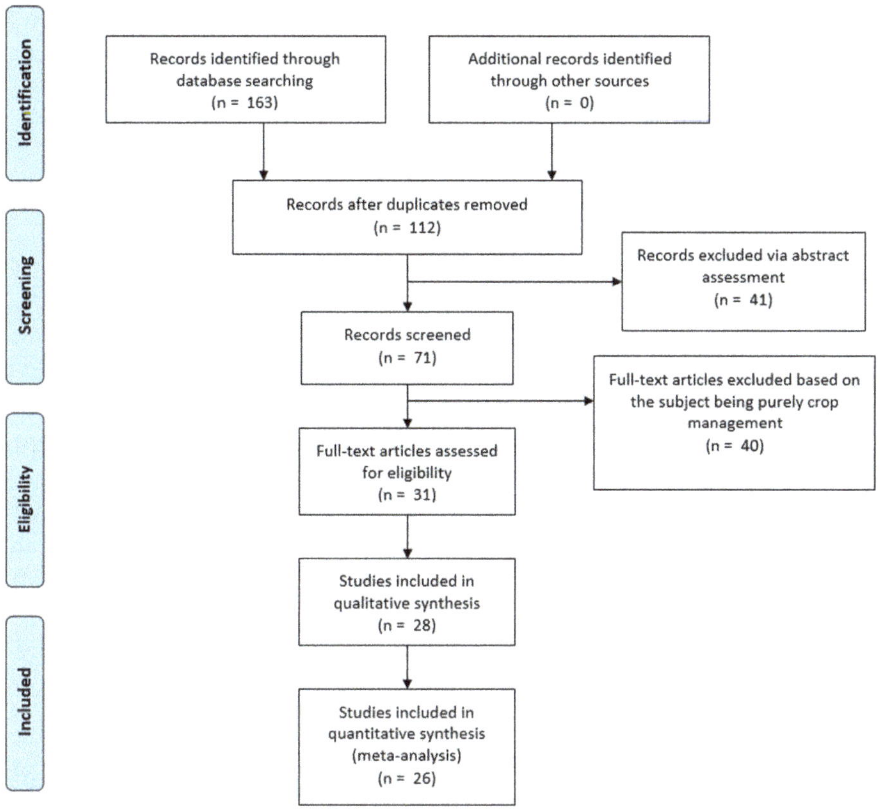

Fig. 12 Flowchart of the present survey methodology based on [107]

and "deep learning" were used in the common search engines of Scopus and Google Scholar. All reviewed papers were published in Scientific Journals and, hence, conference articles, Master and PhD theses as well as non-English studies were excluded. In addition, only articles published between 2018 and 2020 were considered valid for the present literature survey, since a recent review study already exists in the relative literature capturing the progress in this field up to 2018 [105]. Figure 12 depicts a flowchart of the present methodology according to the PRISMA guidelines [107].

In order to get a first idea of the subject and main methodologies followed by the reviewed articles, a keyword information clustering was created. To that end, about 130 keywords were utilized from the 26 selected publications, which are listed in the corresponding part of the journals. Out of all keywords, the 10 most frequently appearing ones were selected which are summarized in Fig. 13. The font size expresses the frequency of the appearance of each keyword and keywords with the same colors indicate that they have the same frequency.

Fig. 13 Keyword information clustering of the 26 reviewed articles

According to Fig. 13, "precision agriculture" was the most common keyword, as it was the dominant research topic. "Deep Learning" followed, as most of the applied algorithms belonged to this family of machine learning methods. Keywords "Convolutional Neural Network" (proving that this is the most prevailing algorithm) and "Disease" (indicating that most studies focused on disease detection) came next with the same frequency. "Image Processing", "Machine Learning", and "Unnamed Aerial Vehicles" (UAV) were the keywords that followed. UAVs were used in many works in order to acquire high-resolution images and this reason justifies its frequency appearance. Furthermore, "Image Processing" techniques were combined with machine learning algorithms to succeed better results. Keyword "Yield prediction" that came next demonstrated that there were many works associated with this sub-category. Finally, the keywords "Artificial Neural Network" (ANN) and "Feature Selection" presented with the same frequency. The former indicated that ANN is an algorithm that is applied in precision agriculture, whereas the latter proves that a subset of features is selected for use in machine learning algorithms.

Finally, the applications are classified into four identified categories according to their purpose, namely yield prediction, crop disease detection, weed detection, and quality assessment, which are described next.

6.1 Yield Prediction

Yield prediction is one of the most significant sub-categories of crop management. It provides yield estimation, yield mapping, and helps farmers to make essential financial decisions. Moreover, based on the information that are provided, seed companies can predict the performances of new hybrids in different environments. To that end, many studies have been conducted upon examining the performance of machine learning models in yield prediction.

For example, in [108] real-world data were used to predict the performance of corn hybrids in 2017 in locations across the United States and Canada. Two deep neural networks were trained to predict yield based on forecasted weather data, crop genotype, and older yield performance. It was deduced that prediction accuracy was better than other popular methods with a root-mean-square error reaching 12% of the average yield and 50% of the validation dataset's standard deviation. Additionally, convolutional neural networks (CNN) were applied in the studies of [109, 110]. In the former, CNN was used based on NDVI (Normalized Difference Vegetation Index) and RGB (Red, Green, Blue) data acquired by unnamed aerial vehicles from nine crop fields to predict the yield of these fields. An 8.8% mean absolute percentage error (MAPE) was achieved for data collected during the early period of the growth season. In the latter work, CNN was tested for extracting important features for rice grain yield prediction on the basis of high-resolution UAV captured images. It was proven that CNNs trained by RGB images reached the lowest MAPE (26.61%) in the middle of the ripening stage.

Region-based convolutional neural network (R-CNN) was implemented in [111] to automatically detect and count the number of flowers as well as mature and immature strawberries. This information is necessary for predicting the upcoming strawberry yield and create distribution maps. The mean average precision (mAP) reached 0.83 for objects at 2 m heights and 0.72 for objects at 3 m height. In another work [112], the well-known regression method (LASSO) was compared with three machine learning methods (support vector machine, random forest, and neural network) for modeling crop yield in Australia. It was inferred that machine learning algorithms exceeded the regression method. The results demonstrated that combining satellite and climate data, high performance can be reached at the statistical division level ($R^2 \sim 0.75$).

Furthermore, eight machine learning models were tested in [113] for their capability to predict winter wheat yield in China. Models achieved accurate yield prediction for 1–2 months before harvesting dates. It was proven that support vector machine, Gaussian process regression, and random forest algorithms had better performance among all the algorithms with higher R^2 ($0.79 \sim 0.81$). Finally, in [114] machine learning algorithms and more specifically gradient boosting and random forests were applied to create crop meta-models trained on the global gridded crop models, to predict crop model outputs at fine spatial resolutions. The results indicated high accuracy with $R^2 > 0.96$ in case of predictions of maize yields.

6.2 Crop Disease Detection

Crop diseases are a crucial problem in agriculture, as they can destroy crop production and have catastrophic economic impact. For that reason, many works are focused on the creation of machine learning models that automatically detect and classify plant diseases. Most studies use images that contain the diseased parts of the plant such as leaves or seeds for training machine learning models. In particular, leaf disease detection can be seen in [115], where a CNN model was used to classify images of walnut leaves into two categories, namely healthy and infected by anthracnose, reaching 98.7% accuracy. Similarly, in the study of [116] a one-class support vector machine (SVM) model for each plant healthy condition (healthy, downy mildew, powdery mildew, and black rot) was trained on images depicting vine leaves in the above-mentioned four different health conditions. The proposed algorithm accomplished a high generalization behavior in other crops as well. Remarkably, 95% total success rate was achieved, meaning that 44 of the 46 tested plant-condition combinations were successfully classified.

In [117], CNN was combined with color information to detect disease in vine yards based on images taken from UAVs. Many combinations of color spaces and vegetation indices were examined in order to improve the performance of the model. It was proven that the best combinations reached accuracy more than 95.8%. An effort to improve the identification accuracy of maize leaf diseases was conducted in [118]. Two improved models, which were GoogLeNet and Cifar10, were used to train and test images containing nine kinds of maize leaf images. These two models achieved accuracy of 98.9% and 98.8%, respectively, when identifying eight kinds of maize leaf diseases. In another study [119], a hybrid method was introduced so as to detect and classify diseases in citrus plants. Initially, lesion spots on leaves and fruits were detected and then a Multi-Class Support Vector Machine was used to classify citrus diseases including anthracnose, black spot, canker, scab, greening, and melanose. This technique reached 97% classification accuracy on image gallery dataset showing citrus disease images, 90.4% on a local dataset and 89% on combined dataset.

SVM in conjunction with k-means clustering algorithm was also applied in [120] for classifying papaya diseases. At the beginning, k-means clustering algorithm segmented out the diseased region from images taken from mobile or handheld devices. Then, support vector machine extracted features to classify the diseased succeeding 90% classification accuracy. In [121], k nearest-neighbor method was utilized to detect canker in leaves achieving detection accuracy 96% for late disease stage in indoor conditions. In [122], a pre-trained CNN model, VGG16, was implemented for identification of mildew disease in pearl millet, demonstrating a 95% accuracy. Finally, in [123], deep residual neural network (DRNN), a powerful category of neural networks, was implemented for the early detection of many plant diseases using wheat images. The algorithm was applied in a smartphone application and reached 0.87 accuracy under exhaustive testing and 0.96 accuracy on a pilot test conducted in Germany.

6.3 *Weed Detection*

Weed is an undesirable plant, growing along with useful agriculture products. It can decrease the growth of the crop and cause yield losses. In the USA, the cost of weeds is considered to be over $ 26 billion [124]. Plenty of methods were developed for weed management in agriculture for controlling or eliminating weeds. These methods include the use of herbicides and weed removal. However, herbicides are expensive and harmful to the environment and human health. The best way to control weed is to prevent their spreading to other fields. The first step to succeed, that is by detecting weed across the field.

Machine learning and deep learning techniques are an important tool for weed detection and many studies have been conducted to examine their effectiveness. Based on the study of [125], several deep CNNs were reported for their performance in weed detection in bermudagrass. VGGNet, a pre-trained CNN model, reached F1 score values higher than 0.95 for detecting dollar weed, old world diamond-flower, and Florida pusley. Moreover, DetectNet outperformed the other networks concerning annual bluegrass growing with broadleaf weeds reaching an F1 score higher than 0.99. Moreover, ANN and SVM were implemented in [126] in order to integrate many shape features and create a pattern for every variety of the plants. More specifically, four species of weeds in sugar beet fields were examined. The results indicated that ANN and SVM correctly classified 92.5% and 93.33% of weed, respectively.

In [127], a fully automatic learning method was proposed using CNN. UAV captured images were collected and used as an unsupervised training dataset for the network. The results indicated that the performance was close to the performances of traditional supervised training data labeling, whereas differences in accuracy were 1.5% in the spinach field and 6% in the beam field. In [128, 129], object-based image analysis (OBIA) algorithm combined with machine learning techniques, Random Forest (RF), were implemented. The aim of [128] was to classify soil, weeds, and maize. Five-fold cross-validation was used to evaluate classifier, reaching an overall accuracy of 94.5. In the work of [129], an innovative OBIA algorithm was created based on UAV images. Then, RF classifier used OBIA-based plant heights as a feature in the automatic sample selection. Finally, prescription maps were created according to weed maps. Most of the weeds were accurately identified and spatially located reaching 84% of weeds in the cotton field and 81.1% in the sunflower field. Fully convolutional neural networks were applied in [130, 131]. In the first case, a fully convolutional neural network which integrates sequential information is applied in order to provide a crop-weed classification system. The system uses the crop arrangement information taken from the image sequences and provides a pixel-wise classification of the images into weed and crop. The obtaining precision was 98.3%, 99.1%, and 85.5% for crop weed and intra weed, respectively. In the second case, UAV images were collected to create a precise weed cover map. For that purpose, a fully convolutional network in combination with transfer learning was applied. The overall accuracy was exceeded 93.5% and the accuracy for weed recognition reached 88.3%.

6.4 Quality Assessment

Classifying crops regarding their quality can assist farmers in determining their market goals and reduce waste. Machine learning techniques can help to detect crop characteristics that indicate the quality grade of the crop. In [132], pepper seeds were classified as high-quality and low-quality. A multilayer perceptron neural network (MLP) used 15 physical traits of seeds as variables and succeeded 99.4% stability rate (ratio of accuracies of the test and training sets). In addition, [133] suggested principal component analysis (PCA) and two neural network algorithms (genetic algorithm optimized BP neural network and T-S fuzzy neural network) to categorize soybean seed varieties. In this way, best quality seeds were identified quickly reaching 96% accuracy of the training set and 84% accuracy of the test set. Study [134] aimed at the development of a machine learning predictive model, which evaluates the intensity levels of 10 beer sensory descriptors. An artificial neural network (ANN) was proved to be accurate and effective to evaluate quality of beer concerning color, foamability, and related sensory descriptors.

7 Summary of the Basic Aspects of the Reviewed Studies

The total number of articles included in our survey for all sub-categories of crop management was 26, while Table 1 summarizes all the information of these articles. More specifically, it presents the year of publication, the type of crop, the purpose of the study, the algorithm that was applied, and the results that emerged for each study.

Out of the total 26 reviewed articles, nine of them (which constitute the largest group) were found to be involved in crop disease detection (34.62%). Furthermore, seven articles were related to yield prediction (26.92%) and seven articles focused on weed detection (26.92%). The sub-category with the less articles was quality assessment, as only three articles were found (11.54%). The distribution of the four categories of crop management is depicted in Fig. 14.

From the investigation of these articles, it was found that in total 12 different machine learning algorithms were applied. In particular, as can be deducted from Fig. 15, the most highly implemented algorithm was CNN, which was applied in nine studies (28.13%). This can be explained as most of the studies relied on features extracted from high-resolution images. These images have high dimensionality since each pixel is a different feature. CNNs can decrease the number of parameters without losing much information. Hence, they are suitable for image classification and image preprocessing. The second most applied algorithm is SVM, which was applied in six studies (15.63%). RF follows, as it was used in five works (15.63%). ANN, FCN, and NN came next, since two of the reviewed articles utilized them (6.25%). Finally, DNN, GPR, KNN, MLP, RNN, and G Boost were found in one study (3.13%).

Table 1 List of the reviewed publications along with the year of publishing, category of crop management, crop type, purpose of the study, implemented algorithm, and main results

Ref	Year	Cat.	Crop	Purpose	Algor.	Results
[108]	2019	YP	Corn	Predictions of yields for new hybrids planted	DNN	RMSE = 12% av. yield RMSE = 50% STD for validation dataset
[109]	2019	YP	Wheat Malting barley	Crop YP from NDVI and RGB data	CNN	Early period: MAE = 484.3 kg/ha MAPE = 8.8% Later: MAE = 624.3 kg/ha MAPE = 12.6%
[110]	2019	YP	Rice grain	Acquire important features associated with rice grain yield	CNN	RGB and multispectral images: $R^2 = 0.464 \sim 0.499$ MAPE = 26.61%
[111]	2019	YP	Strawberry	Strawberry flower detection system for YP	R-CNN	Av. accuracy = 84.1% Av. occlusion = 13.5%
[112]	2019	YP	Wheat	Predict wheat yield across Australia	SVM RF NN	YP at the statistical division level: $R^2 \sim 0.75$
[113]	2020	YP	Wheat	Winter wheat YP based on multi-source data	SVM GPR RF	$R^2 > 0.75$ Yield error < 10%
[114]	2019	YP	Maize	Meta models for the prediction of crop model outputs	XG Boost RF	$R^2 > 0.96$
[115]	2020	CDD	Walnut	Classify leaves to healthy and infected	CNN	Accuracies ranging from 92.4% to 98.7%
[116]	2019	CDD	Vine	Identify crop disease on leaf sample images	One Class SVM model	Total success rate of 95%
[117]	2018	CDD	Vine	Identify infected areas of grapevines	CNN	Accuracy more than 95.8%
[118]	2018	CDD	Maize leaf	Improve identification accuracy of maize leaf diseases	CNN	Accuracy: 98.9% for GoogleLeNet 98.8% for Cifar10
[119]	2018	CDD	Citrus	Detect and classify diseases in citrus plants	M-SVM	Accuracy: 97% on image gallery dataset 90.4% on local dataset
[120]	2018	CDD	Papaya	System that determines the papaya diseases	Clustering SVM	More than 90% classification accuracy

(continued)

Table 1 (continued)

Ref	Year	Cat.	Crop	Purpose	Algor.	Results
[121]	2019	CDD	Citrus	Remote sensing technique to detect citrus canker	KNN	Accuracy: 94% healthy and asymptomatic trees 96% healthy and canker-infected trees
[122]	2019	CDD	Pearl millet	Identify mildew disease in pearl millet	CNN	95% accuracy 90.50% precision 94.5% recall 91.75% f1-score
[123]	2019	CDD	Wheat	Detect many plant diseases in real conditions	RNN	Accuracy: 0.87 under exhaustive testing 0.96 on a pilot test
[125]	2019	WD	Bermuda grass	Detect weed in bermudagrass	CNN	F1 score value >0.99
[126]	2018	WD	Sugar beet	Create pattern based on shape features for different weeds	ANN SVM	Correctly classified weeds: ANN:92.50% SVM:93.33% Plants: ANN:93.33% SVM:96.67%
[127]	2018	WD	Bean spinach	Learning method with unsupervised training data collection	CNN	Differences in accuracy: 1.5% in spinach, 6% in beam compared to supervised training data labeling
[128]	2018	WD	Maize	Weed detection in early season maize field	RF	Accuracy = 0.945 Kappa value = 0.912
[129]	2018	WD	Cotton sunflower	Design prescription maps	RF	Accuracy: 84% of weeds in cotton field, 81.1% in sunflower field
[130]	2018	WD	Sugar beet	Crop-weed classification system	FCN	Precision: 98.3%, 99.1% and 85.5% for crop, weed and intra weed
[131]	2018	WD	Rice	Generate a weed cover map	FCN	Overall accuracy: 0.935
[132]	2018	QA	Pepper	Classify seeds to high and low quality	MLP	99.4% stability rate
[133]	2019	QA	Soybean	Classify 10 soybean varieties	GA-BP and T-S fuzzy NN	Av. accuracy: 96% for training set, 84% for test set
[134]	2018	QA	Beer	Evaluate intensity levels of sensory descriptors in beer	ANN	High correlation ($R = 0.91$) to predict the intensity levels of 10 sensory descriptors

(continued)

Table 1 (continued)

Ref	Year	Cat.	Crop	Purpose	Algor.	Results

Algor. algorithms, *ANN* artificial neural network, *Av* average, *Cat.* categories, *CDD* crop disease detection, *CNN* convolutional neural networks, *DNN* deep neural network, *FCN* fully convolutional network, *GPR* Gaussian process regression, *KNN* k-nearest neighbors, *MAE* mean absolute error, *MLP* multilayer perceptron, *NN* neural network, *R-CNN* region-based convolutional network, *RF* random forest, *RNN* recurrent neural network, *RMSE* root mean square error, *STD* standard deviation, *SVM* support vector machine, *WD* weed detection, *QA* quality assessment, *XG-Boost* extreme gradient boosting, *YP* yield prediction

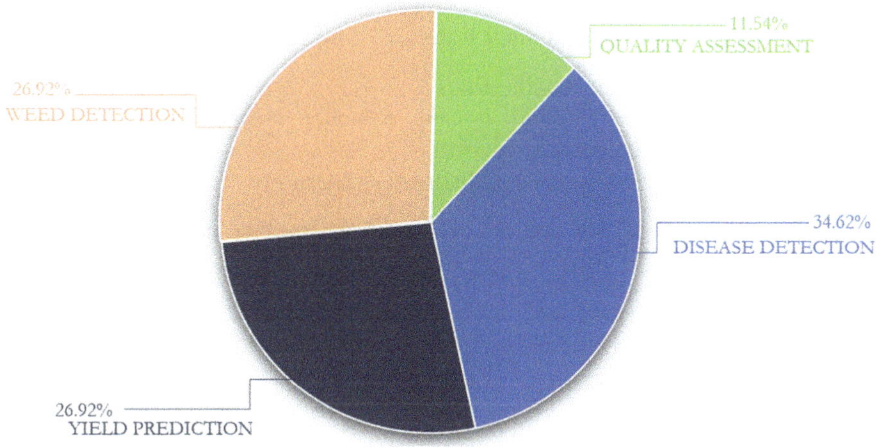

Fig. 14 Distribution of the four categories of crop management

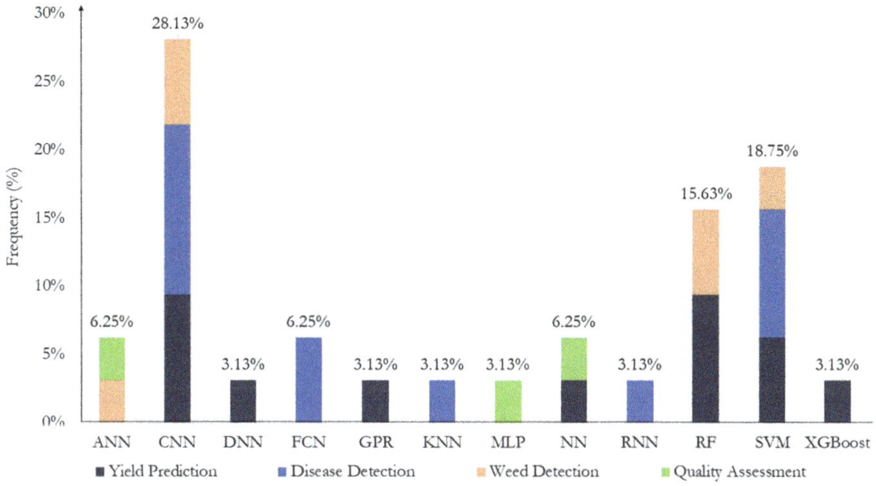

Fig. 15 Frequency (%) of machine learning algorithms appearing in the reviewed studies

8 Conclusions

This chapter focused on the analysis of the fundamental features of machine learning and its application on the domain of agriculture. Machine Learning is a rapidly developing scientific field that pertains to the programming of machines with intention of achieving intelligence. In fact, machine learning enables computers to learn from data, and improve the model's performance based on given inputs and outputs. In turn, the trained model is essentially the outcome of the generated rules, which lead from a particular input to a particular output. Subsequently, it was elaborated how machine learning works and what are the basic machine learning methods. In particular, these methods include:

(a) Supervised (regression, classification)
(b) Unsupervised (clustering, dimensionality reduction, association)
(c) Reinforcement (classification, control)
(d) Active learning (content-based, collaborative filtering)

The above methods are realized by different machine learning algorithms which are:

(a) Regression
(b) Regularization
(c) Bayesian
(d) Instance-based
(e) Decision tree
(f) Ensemble
(g) Clustering
(h) Dimensionality reduction
(i) Association rule
(j) Artificial neural networks
(k) Deep neural networks

To focus on agricultural applications of machine learning, a preliminary review was also performed on the relative literature for the sake of capturing the current trends in this application domain. The up-to-date progress was investigated by considering only recent journal papers concerning crop management. In total, 26 articles were identified which mainly deal with disease detection, yield prediction, weed detection, and quality assessment. It was deduced that disease detection was the most prevalent among all categories, thus, indicating the major importance of this procedure in agriculture. The most commonly used algorithms that were implemented for the purpose of conducting the above agricultural operations were in descending order of incidence CNN, SVM, and RF. As highlighted above, CNN ordinary use is justified due to its capability to reduce the number of parameters without losing considerable information.

In a nutshell, machine learning is an area of high interest in modern times. In the past years, there has been an exponential growth on the scientific breakthroughs, the technological implementations, and the wide adaptation of machine learning science. However, new algorithms that outperform previous benchmarks are developed daily and different types of items receive "smart" capabilities. This is something that only a few people were able to predict 20 years ago. In such a developing sector, no one can be sure what the future holds. The way this new tool will be used in the future depends completely on the prioritization of global needs.

Nowadays, even the most skeptical people have been convinced that artificial intelligence, and more specifically machine learning, has a great potential in the field of agriculture. There are many factors such as climate changes and increase of population, which create uncertainty in farming hence, farmers need a new ally to help them confront these problems. Machine learning solutions are expected to increase production levels and quality of products. However, these are not standalone solutions, but they must intertwine with the existing agricultural operations technology [135].

Currently, machine learning tries to find solutions in particular problems, namely the detection of specific weeds, diseases, or the yield prediction of a specific crop. In the future, it is expected that a fully interconnected system will be created that provides automated data recording, analysis, use of machine learning algorithms, and decision-making. After all, one thing is certain; in the next years machine learning will be a significant part of agricultural decision-making process. Aim of this study is to contribute towards a more systematic investigation on machine learning, especially in agricultural sector.

References

1. Turing A. M. (2012). *Computing machinery and intelligence*. In: Machine intelligence: Perspectives on the computational model. pp 1–28.
2. Solomonoff, R. J. (1985). The time scale of artificial intelligence: Reflections on social effects. *Human Systems Management, 5*, 149–153. https://doi.org/10.3233/HSM-1985-5207
3. Stern, E. (2017). Individual differences in the learning potential of human beings. *NPJ Science of Learning, 2*. https://doi.org/10.1038/s41539-016-0003-0
4. Marinoudi, V., Sørensen, C. G., Pearson, S., & Bochtis, D. (2019). Robotics and labour in agriculture. A context consideration. *Biosystems Engineering, 184*, 111–121. https://doi.org/10.1016/J.BIOSYSTEMSENG.2019.06.013
5. Hassabis, D., Kumaran, D., Summerfield, C., & Botvinick, M. (2017). Neuroscience-inspired artificial intelligence. *Neuron, 95*, 245–258.
6. Kriegeskorte, N., & Douglas, P. K. (2018). Cognitive computational neuroscience. *Nature Neuroscience, 21*, 1148–1160.
7. Makridakis, S. (2017). The forthcoming artificial intelligence (AI) revolution: Its impact on society and firms. *Futures, 90*, 46–60.
8. Agrawal, A., Gans, J., & Goldfarb, A. (2019). The impact of artificial intelligence on innovation. In: *The economics of artificial intelligence*. University of Chicago Press. pp 115–148.
9. Skyttner, L. (2006). Artificial intelligence and life. In: *General systems theory*. World Scientific. pp 319–351.

10. Farkas, I. (2003). Artificial intelligence in agriculture. In: *Computers and electronics in agriculture*. pp 1–3.
11. Hamet, P., & Tremblay, J. (2017). Artificial intelligence in medicine. *Metabolism, 69*, S36–S40. https://doi.org/10.1016/j.metabol.2017.01.011
12. Kalogirou, S. A. (2006). Introduction to artificial intelligence technology. *Artificial Intelligence in Energy & Renewable Energy Systems*, 1–46.
13. Hu, L. B., Cun, H. B., Tao, Y. W., et al. (2017). Applications of artificial intelligence in intelligent manufacturing: A review. *Frontiers of Information Technology & Electronic Engineering, 18*, 86–96.
14. Cook, D. J., Augusto, J. C., & Jakkula, V. R. (2009). Ambient intelligence: Technologies, applications, and opportunities. *Pervasive and Mobile Computing, 5*, 277–298.
15. Jordan, M. I., & Mitchell, T. M. (2015). Machine learning: Trends, perspectives, and prospects. *Science (80-.), 349*, 255–260.
16. Campbell, J. A. (1986). On artificial intelligence. *Artificial Intelligence Review, 1*, 3–9. https://doi.org/10.1007/BF01988524
17. Goldberg, D. E., & Holland, J. H. (1988). Genetic algorithms and machine learning. *Machine Learning, 3*, 95–99.
18. Tiwari, A. K. (2017). Introduction to machine learning. *Ubiquitous Machine Learning and Its Applications*, 1–14.
19. Roth, D. (2006). Learning based programming. *Studies in Fuzziness and Soft Computing, 194*, 73–95. https://doi.org/10.1007/10985687_3
20. Goertzel, T. (2014). The path to more general artificial intelligence. *Journal of Experimental and Theoretical Artificial Intelligence*, *26*, 343–354.
21. Goertzel, B., & Pennachin, C. (2007). Artificial general intelligence. *Cognition, Technology, 8*.
22. Cunningham, P., Cord, M., & Delany, S. J. (2008). *Supervised learning* (pp. 21–49). Cognitive Technologies.
23. Francis, L. (2014). Unsupervised learning. In *Predictive modeling applications in actuarial science: Volume I: Predictive modeling techniques* (pp. 280–312). Cambridge University Press.
24. Aggarwal, C. C. (2014). Educational and software resources for data classification. In *Data classification: Algorithms and applications* (pp. 657–665). Chapman and Hall/CRC.
25. Silver, D., Schrittwieser, J., Simonyan, K., et al. (2017). Mastering the game of go without human knowledge. *Nature, 550*, 354–359. https://doi.org/10.1038/nature24270
26. Haarnoja, T., Ha, S., Zhou, A., et al (2019). Learning to walk via deep reinforcement learning.
27. Liao, Y., Yi, K., & Yang, Z. (2012). *CS229 final report reinforcement learning to play Mario*. StanfordEdu.
28. Jacobson, K., Murali, V., Newett, E., et al (2016) Music personalization at Spotify. Proceedings of the 10th ACM Conference on Recommender Systems. pp 373–373. Association for Computing Machinery, New York, NY, United States.
29. Zhou, Y., Wilkinson, D., Schreiber, R., & Pan, R. (2008). Large-scale parallel collaborative filtering for the Netflix prize. In R. Fleischer & J. Xu (Eds.), *Algorithmic aspects in information and management* (Lecture notes in computer science (including subseries lecture notes in artificial intelligence and lecture notes in bioinformatics)) (pp. 337–348). Springer, Berlin, Heidelberg.
30. Linden, G., Smith, B., & York, J. (2003). Amazon.Com recommendations: Item-to-item collaborative filtering. *IEEE Internet Computing, 7*, 76–80. https://doi.org/10.1109/MIC.2003.1167344
31. Chen, Y., Tsai, F. S., & Chan, K. L. (2008). Machine learning techniques for business blog search and mining. *Expert Systems with Applications, 35*, 581–590. https://doi.org/10.1016/j.eswa.2007.07.015

32. Olive, David J. (2017). *Multiple linear regression*. Linear regression. Springer, Cham, 17–83.
33. Grégoire, G. (2015). *Multiple linear regression* (EAS publications series) (pp. 45–72). European Astronomical Society Publications Series 66.
34. Davis, L. J., & Offord, K. P. (2013). Logistic regression. In *Emerging issues and methods in personality assessment* (pp. 273–283). Routledge.
35. Gooch, J. W. (2011). Stepwise regression. In *Encyclopedic dictionary of polymers* (pp. 998–998). Springer.
36. Moutinho, L., Hutcheson, G., Hutcheson, G., & Hutcheson, G. (2014). Ordinary least-squares regression. In *The SAGE dictionary of quantitative management research* (pp. 225–228). The SAGE Publications Ltd.
37. Friedman, J. H. (1991). Multivariate adaptive regression splines. *The Annals of Statistics, 19*, 1–67. https://doi.org/10.1214/aos/1176347963
38. Cleveland, W. S. (1979). Robust locally weighted regression and smoothing scatterplots. *Journal of the American Statistical Association, 74*, 829–836. https://doi.org/10.1080/01621459.1979.10481038
39. Efron, B., Hastie, T., Johnstone, I., et al. (2004). Least angle regression. *The Annals of Statistics, 32*, 407–499. https://doi.org/10.1214/009053604000000067
40. McDonald, G. C. (2009). Ridge regression. *Wiley Interdisciplinary Reviews: Computational Statistics, 1*, 93–100. https://doi.org/10.1002/wics.14
41. De Mol, C., De Vito, E., & Rosasco, L. (2009). Elastic-net regularization in learning theory. *Journal of Complexity, 25*, 201–230. https://doi.org/10.1016/j.jco.2009.01.002
42. Kukreja, S. L., Löfberg, J., & Brenner, M. J. (2006). A least absolute shrinkage and selection operator (Lasso) for nonlinear system identification. *IFAC Proceedings Volumes, 39*, 814–819. https://doi.org/10.3182/20060329-3-au-2901.00128
43. Hartono, P. (2009). Bayes theorem. *Kyokai Joho Imeji Zasshi/The Journal of The Institute of Image Information and Television Engineers, 63*, 52–54. https://doi.org/10.3169/itej.63.52
44. Stern, H. S. (2015). Bayesian statistics. In *International encyclopedia of the social & behavioral sciences: Second edition* (pp. 373–377). Amsterdam: Elsevier.
45. Ye, N., & Ye, N. (2020). Naïve Bayes classifier. In *Data mining* (pp. 31–36).
46. Zhang, H. (2004). *The optimality of Naive Bayes*. In: Proceedings of the seventeenth international florida artificial intelligence research society conference, FLAIRS 2004.
47. Friedman, N., Geiger, D., & Goldszmidt, M. (1997). Bayesian network classifiers. *Machine Learning, 29*, 131–163. https://doi.org/10.1002/9780470400531.eorms0099
48. Cooper, G. F., & Herskovits, E. (1992). A Bayesian method for the induction of probabilistic networks from data. *Machine Learning, 9*, 309–347. https://doi.org/10.1007/bf00994110
49. Cortes, C., & Vapnik, V. (1995). Support-vector networks. *Machine Learning*. https://doi.org/10.1023/A:1022627411411
50. Altman, N. S. (1992). An introduction to kernel and nearest-neighbor nonparametric regression. *The American Statistician*. https://doi.org/10.1080/00031305.1992.10475879
51. Kohonen, T. (1990). The self-organizing map. *Proceedings of the IEEE, 78*, 1464–1480. https://doi.org/10.1109/5.58325
52. Seo, S., & Obermayer, K. (2003). Soft learning vector quantization. *Neural Computation, 15*, 1589–1604. https://doi.org/10.1162/089976603321891819
53. Atkeson, C. G., Moorey, A. W., Schaalz, S., et al. (1997). Locally weighted learning. *Artificial Intelligence, 11*, 11–73. https://doi.org/10.1023/A:1006559212014
54. Breiman, L., Friedman, J. H., Olshen, R. A., & Stone, C. J. (2017). *Classification and regression trees*. Routledge.
55. Hothorn, T., Hornik, K., & Zeileis, A. (2015). Ctree: Conditional inference trees. *Comprehensive R Archive Network, 8*, 1–34.
56. Quinlan, J. R. (1986). Induction of decision trees. *Machine Learning, 1*, 81–106. https://doi.org/10.1023/A:1022643204877

57. Salzberg, S. L. (1994). C4.5: Programs for machine learning by J. Ross Quinlan. Morgan Kaufmann Publishers, Inc., 1993. *Machine Learning, 16*, 235–240. https://doi.org/10.1007/bf00993309
58. Breiman, L. (2001). Random forrest. *Machine Learning*. https://doi.org/10.1023/A:1010933404324
59. Freund, Y., & Schapire R. E. (1996). Experiments with a new boosting algorithm. In Proceedings of the 37th international conference on machine learning. 10.1.1.51.6252 as retrieved from (https://citeseerx.ist.psu.edu/viewdoc/download?doi=10.1.1.51.6252&rep=rep1&type=pdf).
60. Breiman, L. (1997). Arcing the edge. *Statistics (Berlin)*. 10.1.1.62.8173 as retrieved from (https://citeseerx.ist.psu.edu/viewdoc/download?doi=10.1.1.62.8173&rep=rep1&type=pdf).
61. Breiman, L. (1996). Bagging predictors. *Machine Learning, 24*, 123–140. https://doi.org/10.1007/BF00058655
62. Wolpert, D. H. (1992). Stacked generalization. *Neural Networks, 5*, 241–259. https://doi.org/10.1016/S0893-6080(05)80023-1
63. Hartigan, J. A., & Wong, M. A. (1979). Algorithm AS 136: A K-means clustering algorithm. *Applied Statistics, 28*, 100. https://doi.org/10.2307/2346830
64. Small, C. G. (1990). A survey of multidimensional medians. *International Statistical Review, 58*, 263. https://doi.org/10.2307/1403809
65. Abdi, H., & Williams, L. J. (2010). Principal component analysis. *Wiley Interdisciplinary Reviews: Computational Statistics, 2*, 433–459.
66. De Cheveigné, A. (2012). Quadratic component analysis. *NeuroImage, 59*, 3838–3844. https://doi.org/10.1016/j.neuroimage.2011.10.084
67. Tipping, M. E., & Bishop, C. M. (1999). Mixtures of probabilistic principal component analyzers. *Neural Computation, 11*, 443–482. https://doi.org/10.1162/089976699300016728
68. Hastie, T., Tibshirani, R., & Buja, A. (1994). Flexible discriminant analysis by optimal scoring. *Journal of the American Statistical Association, 89*, 1255–1270. https://doi.org/10.1080/01621459.1994.10476866
69. Geladi, P., & Kowalski, B. R. (1986). Partial least-squares regression: A tutorial. *Analytica Chimica Acta, 185*, 1–17. https://doi.org/10.1016/0003-2670(86)80028-9
70. Kramer, R. (1998). Principal component regression. In *Chemometric techniques for quantitative analysis* (pp. 99–110). CRC Press,
71. Bowen, W. M. (2009). Multidimensional scaling. In *International encyclopedia of human geography* (pp. 216–221).
72. Friedman, J. H., & Stuetzle, W. (1981). Projection pursuit regression. *Journal of the American Statistical Association, 76*, 817. https://doi.org/10.2307/2287576
73. Agrawal, R., & Srikant, R. (2013). Fast algorithms for mining association rules in datamining. *International Journal of Scientific & Technology Research, 1215*, 13–24.
74. Zaki, M. J. (2000). Scalable algorithms for association mining. *IEEE Transactions on Knowledge and Data Engineering, 12*, 372–390. https://doi.org/10.1109/69.846291
75. Chen, Y. Y., Lin, Y. H., Kung, C. C., et al. (2019). Design and implementation of cloud analytics-assisted smart power meters considering advanced artificial intelligence as edge analytics in demand-side management for smart homes. *Sensors (Switzerland), 19*. https://doi.org/10.3390/s19092047
76. Rosenblatt, F. (1958). The perceptron: A probabilistic model for information storage and organization in the brain. *Psychological Review, 65*, 386–408. https://doi.org/10.1037/h0042519
77. Hastie, T., Tibshirani, R., & Friedman, J. (2009). Springer series in statistics. *The Elements of Statistical Learning, 27*, 83–85. https://doi.org/10.1007/b94608
78. Zinkevich, M. A., Weimer, M., Smola, A., & Li, L. (2010). *Parallelized stochastic gradient descent*. In: Advances in neural information processing systems 23: 24th annual conference on neural information processing systems 2010, NIPS 2010.

79. Rumelhart, D. E., Hinton, G. E., & Williams, R. J. (1986). Learning representations by back-propagating errors. *Nature, 323*, 533–536. https://doi.org/10.1038/323533a0
80. Broomhead, D., & Lowe, D. (1988). Multivariable functional interpolation and adaptive networks. *Complex Systems*. https://doi.org/10.1126/science.1179047
81. Hopfield, J. J. (1982). Neural networks and physical systems with emergent collective computational abilities (associative memory/parallel processing/categorization/content-addressable memory/fail-soft devices). *Proceedings of the National Academy of Sciences of the United States of America*. https://doi.org/10.1073/pnas.79.8.2554
82. Bengio, Y., Courville, A., & Vincent, P. (2013). Representation learning: A review and new perspectives. *IEEE Transactions on Pattern Analysis and Machine Intelligence, 35*, 1798–1828. https://doi.org/10.1109/TPAMI.2013.50
83. Hinton, G. (2014). Where do features come from? *Cognitive Science, 38*, 1078–1101. https://doi.org/10.1111/cogs.12049
84. Lecun, Y., Bengio, Y., & Hinton, G. (2015). Deep learning. *Nature*. pp. 436–444.
85. Caterini, A. L., & Chang, D. E. (2018). *Recurrent neural networks* (SpringerBriefs in computer science) (pp. 59–79). Springer, Cham.
86. Hochreiter, S., & Schmidhuber, J. (1997). Long short-term memory. *Neural Computation*. https://doi.org/10.1162/neco.1997.9.8.1735
87. Cho, K., Van Merriënboer, B., Gulcehre, C., et al (2014). *Learning phrase representations using RNN encoder-decoder for statistical machine translation*. In: EMNLP 2014–2014 conference on empirical methods in natural language processing, proceedings of the conference. pp 1724–1734.
88. Günther, J., Pilarski, P. M., Helfrich, G., et al. (2014). First steps towards an intelligent laser welding architecture using deep neural networks and reinforcement learning. *Procedia Technology, 15*, 474–483. https://doi.org/10.1016/j.protcy.2014.09.007
89. Hu, J., Shen, L., & Sun, G. (2018). *Squeeze-and-excitation networks*. In: Proceedings of the IEEE Computer Society conference on computer vision and pattern recognition. pp. 7132–7141.
90. Zhao, Z. Q., Zheng, P., Xu, S. T., & Wu, X. (2019). Object detection with deep learning: A review. *IEEE Transactions on Neural Networks and Learning Systems, 30*, 3212–3232.
91. Li, Y., Qi, H., Dai, J., et al (2017). *Fully convolutional instance-aware semantic segmentation*. In: Proceedings - 30th IEEE conference on computer vision and pattern recognition, CVPR 2017. pp 4438–4446.
92. Kingma, D. P., Welling, M. (2014). *Auto-encoding variational bayes*. In: 2nd international conference on learning representations, ICLR 2014 - conference track proceedings.
93. Goodfellow, I. J., Pouget-Abadie, J., Mirza, M., et al. (2014). *Generative adversarial nets* (Advances in neural information processing systems) (pp. 2672–2680). Red Hook, NY Curran.
94. Hinton, G. (2009). Deep belief networks. *Scholarpedia, 4*, 5947. https://doi.org/10.4249/scholarpedia.5947
95. Salakhutdinov, R., & Hinton, G. (2009). Deep Boltzmann machines. *Aistats, 1*, 448–455. https://doi.org/10.1109/CVPRW.2009.5206577
96. Aznar-Sánchez, J. A., Piquer-Rodríguez, M., Velasco-Muñoz, J. F., & Manzano-Agugliaro, F. (2019). Worldwide research trends on sustainable land use in agriculture. *Land Use Policy*. https://doi.org/10.1016/j.landusepol.2019.104069
97. Lampridi, M., Kateris, D., Sørensen, C. G., & Bochtis, D. (2020). Energy footprint of mechanized agricultural operations. *Energies, 13*, 769. https://doi.org/10.3390/en13030769
98. Gomiero, T., Paoletti, M. G., & Pimentel, D. (2008). Energy and environmental issues in organic and conventional agriculture. *CRC Critical Reviews in Plant Sciences*. pp 239–254.
99. Lampridi, M. G., Sørensen, C. G., & Bochtis, D. (2019). Agricultural sustainability: A review of concepts and methods. *Sustainability, 11*, 5120. https://doi.org/10.3390/su11185120

100. Kamilaris, A., & Prenafeta-Boldú, F. X. (2018). Deep learning in agriculture: A survey. *Computers and Electronics in Agriculture, 147*, 70–90. https://doi.org/10.1016/J.COMPAG.2018.02.016
101. Liakos, K., Moustakidis, S., Tsiotra, G., et al (2017). *Machine learning based computational analysis method for cattle lameness prediction*. In: CEUR Workshop Proceedings.
102. Bochtis, D. D., Sørensen, C. G. C., & Busato, P. (2014). Advances in agricultural machinery management: A review. *Biosystems Engineering, 126*, 69–81. https://doi.org/10.1016/j.biosystemseng.2014.07.012
103. Chlingaryan, A., Sukkarieh, S., & Whelan, B. (2018). Machine learning approaches for crop yield prediction and nitrogen status estimation in precision agriculture: A review. *Computers and Electronics in Agriculture*. pp 61–69
104. Anagnostis, A., Benos, L., Tsaopoulos, D., et al. (2021). Human activity recognition through recurrent neural networks for human–robot interaction in agriculture. *Applied Sciences, 11*, 2188. https://doi.org/10.3390/app11052188
105. Liakos, K., Busato, P., Moshou, D., et al. (2018). Machine learning in agriculture: A review. *Sensors, 18*, 2674. https://doi.org/10.3390/s18082674
106. Anagnostis, A., Tagarakis, A. C., Asiminari, G., et al. (2021). A deep learning approach for anthracnose infected trees classification in walnut orchards. *Computers and Electronics in Agriculture, 182*, 105998. https://doi.org/10.1016/j.compag.2021.105998
107. M.J., McKenzie, J.E., Bossuyt, P.M. et al. (2021). The PRISMA 2020 statement: an updated guideline for reporting systematic reviews. Syst Rev 10, 89. https://doi.org/10.1186/s13643-021-01626-4.
108. Khaki, S., & Wang, L. (2019). Crop yield prediction using deep neural networks. *Frontiers in Plant Science*. https://doi.org/10.3389/fpls.2019.00621
109. Nevavuori, P., Narra, N., & Lipping, T. (2019). Crop yield prediction with deep convolutional neural networks. *Computers and Electronics in Agriculture, 163*. https://doi.org/10.1016/j.compag.2019.104859
110. Yang, Q., Shi, L., Han, J., et al. (2019). Deep convolutional neural networks for rice grain yield estimation at the ripening stage using UAV-based remotely sensed images. *Field Crops Research*. https://doi.org/10.1016/j.fcr.2019.02.022
111. Chen, Y., Lee, W. S., Gan, H., et al. (2019). Strawberry yield prediction based on a deep neural network using high-resolution aerial orthoimages. *Remote Sensing*. https://doi.org/10.3390/rs11131584
112. Cai, Y., Guan, K., Lobell, D., et al. (2019). Integrating satellite and climate data to predict wheat yield in Australia using machine learning approaches. *Agricultural and Forest Meteorology*. https://doi.org/10.1016/j.agrformet.2019.03.010
113. Han, J., Zhang, Z., Cao, J., et al. (2020). Prediction of winter wheat yield based on multi-source data and machine learning in China. *Remote Sensing*. https://doi.org/10.3390/rs12020236
114. Folberth, C., Baklanov, A., Balkovič, J., et al. (2019). Spatio-temporal downscaling of gridded crop model yield estimates based on machine learning. *Agricultural and Forest Meteorology*. https://doi.org/10.1016/j.agrformet.2018.09.021
115. Anagnostis, A., Asiminari, G., Papageorgiou, E., & Bochtis, D. (2020). A convolutional neural networks based method for anthracnose infected walnut tree leaves identification. *Applied Sciences*, 10. https://doi.org/10.3390/app10020469
116. Pantazi, X. E., Moshou, D., & Tamouridou, A. A. (2019). Automated leaf disease detection in different crop species through image features analysis and One Class Classifiers. *Computers and Electronics in Agriculture, 156*, 96–104. https://doi.org/10.1016/j.compag.2018.11.005
117. Kerkech, M., Hafiane, A., & Canals, R. (2018). Deep leaning approach with colorimetric spaces and vegetation indices for vine diseases detection in UAV images. *Computers and Electronics in Agriculture*. https://doi.org/10.1016/j.compag.2018.10.006

118. Zhang, X., Qiao, Y., Meng, F., et al. (2018). Identification of maize leaf diseases using improved deep convolutional neural networks. *IEEE Access*. https://doi.org/10.1109/ACCESS.2018.2844405
119. Sharif, M., Khan, M. A., Iqbal, Z., et al. (2018). Detection and classification of citrus diseases in agriculture based on optimized weighted segmentation and feature selection. *Computers and Electronics in Agriculture*. https://doi.org/10.1016/j.compag.2018.04.023
120. Habib, M. T., Majumder, A., Jakaria, A. Z. M., et al. (2020). Machine vision based papaya disease recognition. *Journal of King Saud University – Computer and Information*. https://doi.org/10.1016/j.jksuci.2018.06.006
121. Abdulridha, J., Batuman, O., & Ampatzidis, Y. (2019). UAV-based remote sensing technique to detect citrus canker disease utilizing hyperspectral imaging and machine learning. *Remote Sensing*. https://doi.org/10.3390/rs11111373
122. Coulibaly, S., Kamsu-Foguem, B., Kamissoko, D., & Traore, D. (2019). Deep neural networks with transfer learning in millet crop images. *Computers in Industry*. https://doi.org/10.1016/j.compind.2019.02.003
123. Picon, A., Alvarez-Gila, A., Seitz, M., et al. (2018). Deep convolutional neural networks for mobile capture device-based crop disease classification in the wild. *Computers and Electronics in Agriculture*. https://doi.org/10.1016/j.compag.2018.04.002
124. Liu, B., & Bruch, R. (2020). Weed detection for selective spraying: A review. *Current Robot Reports*. https://doi.org/10.1007/s43154-020-00001-w
125. Yu, J., Sharpe, S. M., Schumann, A. W., & Boyd, N. S. (2019). Deep learning for image-based weed detection in turfgrass. *European Journal of Agronomy*. https://doi.org/10.1016/j.eja.2019.01.004
126. Bakhshipour, A., & Jafari, A. (2018). Evaluation of support vector machine and artificial neural networks in weed detection using shape features. *Computers and Electronics in Agriculture*. https://doi.org/10.1016/j.compag.2017.12.032
127. Dian Bah, M., Hafiane, A., & Canals, R. (2018). Deep learning with unsupervised data labeling for weed detection in line crops in UAV images. *Remote Sensing*. https://doi.org/10.3390/rs10111690
128. Gao, J., Liao, W., Nuyttens, D., et al. (2018). Fusion of pixel and object-based features for weed mapping using unmanned aerial vehicle imagery. *International Journal of Applied Earth Observation and Geoinformation*. https://doi.org/10.1016/j.jag.2017.12.012
129. de Castro, A. I., Torres-Sánchez, J., Peña, J. M., et al. (2018). An automatic random forest-OBIA algorithm for early weed mapping between and within crop rows using UAV imagery. *Remote Sensing*. https://doi.org/10.3390/rs10020285
130. Lottes, P., Behley, J., Milioto, A., & Stachniss, C. (2018). Fully convolutional networks with sequential information for robust crop and weed detection in precision farming. *IEEE Robotics and Automation Letters*. https://doi.org/10.1109/LRA.2018.2846289
131. Huang, H., Deng, J., Lan, Y., et al. (2018). A fully convolutional network for weed mapping of unmanned aerial vehicle (UAV) imagery. *PLoS One*. https://doi.org/10.1371/journal.pone.0196302
132. Ke-ling, T. U., Lin-juan, L. I., Li-ming, Y., et al. (2018). Selection for high quality pepper seeds by machine vision and classifiers. *Journal of Integrative Agriculture*. https://doi.org/10.1016/S2095-3119(18)62031-3
133. Tan, K., Wang, R., Li, M., & Gong, Z. (2019). Discriminating soybean seed varieties using hyperspectral imaging and machine learning. *Journal of Computational Methods in Science and Engineering*. https://doi.org/10.3233/JCM-193562
134. Gonzalez Viejo, C., Fuentes, S., Torrico, D., et al. (2018). Assessment of beer quality based on foamability and chemical composition using computer vision algorithms, near infrared spectroscopy and machine learning algorithms. *Journal of the Science of Food and Agriculture*. https://doi.org/10.1002/jsfa.8506
135. Bochtis, D., Sørensen, C. A. G., & Kateris, D. (2018). *Operations management in agriculture*. Elsevier.

Part II
Applications

Application Possibilities of IoT-based Management Systems in Agriculture

Mihály Tóth, János Felföldi, László Várallyai, and Róbert Szilágyi

1 Introduction

Nowadays, data has become a crucial resource, the gathering of which is facilitated by new solutions in technology and digitalization. The agricultural sector has implemented technological innovations over the years, providing data regarding the production process, including yield mapping, weed sensing, crop status sensing, and soil sensing, forming the concept of precision agriculture [1]. Such practices could potentially fulfill the requirements of the increasing population demand [2], assist in higher quality standards, improve economic efficiency [3], and reduce the environmental impact of agriculture. The main elements of precision agriculture include sensing, control, the concept of management zones, and calculations, resulting in prescription maps [4]. The site-specific knowledge, derived from these technologies, may enhance sustainability by reducing environmental footprint due to variable-rate fertilizer and pesticide application [5] as well as by improving cost efficiency [6].

Considering the latest innovations, based on internet technologies, digitalization, and other future-oriented technologies, a paradigm shift was established, forming the Industry 4.0 concept, referred often as the fourth industrial revolution [7]. The

M. Tóth (✉)
Faculty of Economics and Business, University of Debrecen, Debrecen, Hungary

Károly Ihrig Doctoral School of Management and Business, University of Debrecen, Debrecen, Hungary
e-mail: toth.mihaly@econ.unideb.hu

J. Felföldi · L. Várallyai · R. Szilágyi
Faculty of Economics and Business, University of Debrecen, Debrecen, Hungary
e-mail: felfoldi.janos@econ.unideb.hu; varallyai.laszlo@econ.unideb.hu; szilagyi.robert@econ.unideb.hu

D. D. Bochtis et al. (eds.), *Information and Communication Technologies for Agriculture—Theme II: Data*, Springer Optimization and Its Applications 183,
https://doi.org/10.1007/978-3-030-84148-5_4

concept was defined as a strategic initiative by the German government, facilitating factors, including horizontal and vertical integration [8]. The Industry 4.0 concept defines nine main pillars, including the IoT (Internet of Things) concept, the Big Data concept, robotics, simulation, horizontal and vertical integration, cybersecurity, cloud computing, additive manufacturing, and augmented reality [9], forming the technological concept of cyber physical systems (CPS) [10]. Some of the mentioned concepts provide new methods for data acquisition, decision support (data analysis and visualization), and process control. Regarding data acquisition, the IoT concept and the sensor networks have gained increasing popularity in the industry as well, of which agriculture is no exception [11].

As the technologies, defined in the Industry 4.0 have been implemented in agriculture, literature presented the concept of Farming 4.0 [12] or Smart Farming, extending the concept of the conventional precision agriculture with new technologies and application areas; the latter being specialized to meet the requirements of the sector, tackling the challenges of agricultural production regarding productivity, environmental impacts, food security, and sustainability [1]. With respect to the technologies, the Internet of Things concept (sensor networks), the Big Data concept (data management), cloud computing, robotics, and artificial intelligence are being emphasized to support decision-making [13]. In the sector, the efficient value creation along the whole supply chain, considering all levels is also of great importance [14], supported by horizontal and vertical integration, resulting from digitalization and standardization.

1.1 Data Acquisition and Management in Agriculture

The natural conditions of agricultural fields and surrounding areas are characterized by meteorological, climatic, and soil attributes [15]. The appearance of various sensors, actuators, and wired or wireless embedded systems in arable crop production, in husbandry, in greenhouses, and in farm management provides new possibilities not just in process control but in decision support as well [16]. The IoT concept describes a new paradigm, combining aspects and technologies from different approaches [17], that defines a set of technology-enabled entities, including objects (sensors, actuators, tags, labels), software services as well as systems, connecting and working together [18]. At the beginning of its appearance, IoT was referred to uniquely identifiable interoperable connected objects, using radio-frequency based identification (RFID) [19], but later on, the concept begun to implement other technologies, including sensors, actuators, Global Navigation Satellite Systems (GNSS), and mobile devices, operated via Wi-Fi, Bluetooth, cellular networks, or Near Field Communication (NFC) [20]. RFID and Wireless sensor networks are considered as the most important aspect of the IoT concept.

A sensor network consists of various nodes, equipped with a microcontroller, communication device, and a number of dedicated sensors [21], capable of measuring environmental factors. We can distinguish between heterogeneous and homogeneous networks, describing the protocol of the networks. Heterogeneous networks support diverse protocols and software patterns for communication, while homogenous networks are relying on custom solutions, which are important when considering integration. Particularly, easy integration is an important aspect of Industry 4.0 due to the fact that heterogeneous networks complicate network design, as they need bi-directional translation to support different formats [22]. Their practical use in agricultural production encompasses environmental data acquisition based on spatial parameters, including air, soil, water, plant, animal monitoring [23], forest monitoring, and climate monitoring [24]. In addition to production, the concept also influences agri-food supply chain management, focusing on food safety and quality, considering (cold chain) logistics and condition monitoring as well as early warning systems [25].

A popular topic is greenhouse monitoring and controlling [26]. Parameters such as temperature, humidity, CO_2 concentration [27], luminous intensity, and soil moisture [28]can be measured in spatial or volumetric form [29]. After production, there are options for inventory management [30] and for vertically spanning processes as well, including food supply chain tracking [11] due to RFID technologies. Based on the characteristics of the system, computer vision can also be considered as an IoT component as the basic principle of having an interconnected sensor and controller, which can be applied for weed identification [31], disease identification [32], quality evaluation [33], grading [34], and selective harvesting [35]. Of course, computer vision is only a tool and the relevant information is only obtained using Big Data and artificial intelligence. The images are mostly processed server-side as simple data, with workflow, including pre-processing (uniformization, based on white balance, gamma levels, contrast, etc.), post-processing (segmentation, feature extraction) and finally, classification (CNN – Convolutional neural networks or SVM – Support vector machines).

In order to utilize the new data sources and analytical methods, Big Data, as a data management concept supports the optimization of agricultural processes [13]. In this field, we can distinguish between machine data, including fuel rate, speed, direction, hydraulics, and diagnostics, as well as agronomic data, including planting and fertilizing, spacing, total acres, moisture, and grain temperature [36]. To handle the increasing volume of data, due to the measurements and the new management systems as well as due to their variety, regarding the structure, it is necessary to implement a method to structure, process, store, and analyze the data. The data is being stored in structured, semi-structured, and unstructured form [37] and requires conversion in order to use it as an integrated dataset. The issues are addressed in the Big Data concept, describing how to capture, curate, store, search, share, transfer, analyze, and visualize such amounts of data [38].

In 2010, Apache Hadoop, as one of the dominant platforms, defined Big Data as datasets that cannot be captured, managed, and processed by general computers within an acceptable scope [39]. The typical characteristics of big data are defined using a 3-V concept, which includes volume, velocity, and variety [40]. As new aspects have appeared in researches, veracity has been included, in 2012, as an attribute, referring to the biases, noise, and abnormalities [41]. The actual 7-V model includes furthermore the value, variability, and visualization [42]. The concept relies on knowledge fusion as well, integrating the vast volume of data in a single database [43], leveraged by combining external sources, including weather data, market data, or benchmarks with other farms [13]. Big Data defines descriptive, predictive, and prescriptive analytics, where artificial intelligence is a crucial and nowadays frequent topic.

In the following, machine learning (ML) will be discussed as a subtopic of artificial intelligence. ML has emerged together with Big Data technologies and computing to create new opportunities to unravel, quantify, and understand data-intensive processes [44]. With the combination of the mentioned methods and considering only sensory data, there are advantages that can be achieved in agricultural production, including field monitoring and automation, air temperature forecasting [45], disease forecasting [46], and yield prediction [47]. Utilizing other variables, including yield, weather, soil parameters, etc., enables crop planning [48] as well, facilitates strategical aspects of production.

2 Methodology

As the scope of a long-term research, several multifunctional, modular data acquisition and management systems have been used based on a production system to perform small-scale experiments. The system comprised of sensor networks, databases, application programming interfaces (API), and desktop or web-based management applications [49]. However, in order to utilize the full potential of the Industry 4.0 concept, not only data acquisition, but the whole workflow should be addressed from data management, up to analytics, as it was mentioned as a conclusion in previous papers. Taking this into account, the following chapter discusses the summary and current progression of the data acquisition (IoT – Internet of Things) and data management methods (Big Data). The goal of this chapter is to determine some possible development branches of the subsystems in order to progress the design of the new iteration, which will be used in various experiments to get comparable results for research in decision support methods. In the following paragraphs, production and experimental systems are mentioned, described in the literature, highlighting environmental data acquisition in agricultural production for managerial decision support.

The research consists of two main parts (Fig. 1) of which the former one utilizes various methods for quantitative bibliometric analysis, while the latter part mainly focuses on the development of the current iteration of a data acquisition system, based on a production system. The iteration is influenced by the current literature as it was analyzed in the first step. The current iteration was evaluated as well, by presenting a practical experiment in data acquisition and data management, considering the characteristics of the sensory data, typical of agricultural production. Moreover, the result was compared with relevant research in the literature, performed with similar aim, but using different approaches.

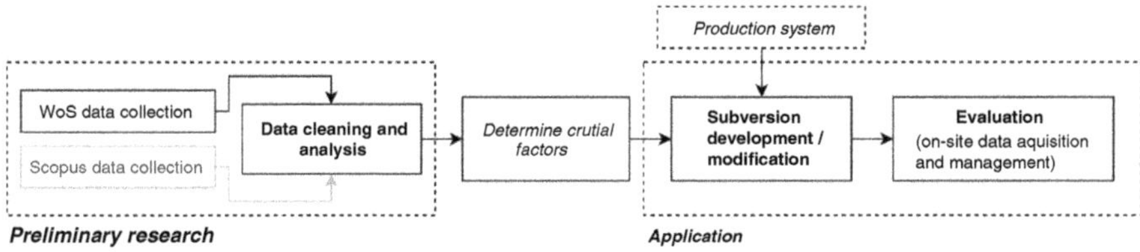

Fig. 1 Main steps of the research

It has become a common practice to analyze the literature in a quantitative manner since it may facilitate the discovery of unknown relationships, factors, and topics that should be considered later in the study. Bibliographic analysis was utilized in order to identify the development possibilities and current trends using network analysis, as well as conceptual structure map, based on multiple correspondence analysis (MCA) in multiple passes, continuously expanding the keywords to set the right directions. As a third approach of this section, thematic evolution was performed, increasing the available information about the field based on historical attributes, expressing the changes in the trends over the years.

In order to examine the field, a dataset was downloaded from the Web of Science database, containing metadata of scientific articles, based on only one keyword combination, describing the two topics, as the scope of this research: *((“IoT” OR “sensor network*” OR “Internet of Things”) AND (“Agriculture” OR “Farm*” OR “Agro*”))*. The bibliographic dataset contains metadata of scientific articles, the most important attributes being the keywords and abstracts. As we can see, the keyword combination focuses on the occurrence of the IoT concept (as well as sensor networks, as a tool for data acquisition) and agriculture. This keyword combination resulted in 3386 records (defined by 71 variables); however, despite the simple definition of the keywords, the occurrence of irrelevant records needs to be considered. In order to solve this issue, the data needed to be cleansed by handling the possible synonyms, as well as by removing irrelevant keywords or entire records. The data cleansing was performed using a custom script, relying on text mining, which examines the title, topics, and abstract of the articles and performs the necessary operation, based on a predefined dictionary.

After this process, the data volume was reduced in this iteration of this analysis, resulting in 2842 records. Thus the dataset was pre-processed using a custom script in Python programming environment, which was followed by three analyses, performed in R using a modified, custom package, based on the open-source Bibliometrix package [50]. The modification was required in order for the library to accept the pre-processed data frame, as well as to enhance the functionality. The network analysis is based on the co-occurrence matrix of every possible keyword in the downloaded dataset, forming a matrix. The visualization was generated based on a force-directed graph drawing, graph layout algorithm [51] using WOSviewer. The MCA analysis was also applied for further analysis, which result was analyzed further, using K-means clustering to determine articles that express similar concepts. The thematic evolution analysis is based on co-word network analysis and clustering

[52], visualized using a Sankey diagram. Its aim is to show the changes in the relationship between different fields by projecting previous results onto a time axis. Every column describes a specific time interval, presenting the dominant keywords for that period.

The system to be presented in the second part of this chapter was improved using an incremental build model, which combines the elements of the waterfall model and prototyping [53], where the product is developed, implemented, and evaluated incrementally, carrying out minor changes with each version through iterative process, allowing the development to benefit from the experience of previous implementations [54]. This implementation was based on a production system, but due to the fact that the development belongs to an author, full access was granted to the source, allowing profound modifications. The possible modifications and improvements were determined by the aforementioned bibliometric research which provided a highlight of highly researched factors of the field, allowing the implementation of current solutions. This final design of this iteration was determined by joint consideration of the mentioned bibliometric information and a subjective opinion, based on previous evaluation sessions. Due to the nature of the incremental build model, some part of the data acquisition system and the management systems were improved separately to meet new requirements, without changing the system architecture. In order to obtain relevant experience, a limited number of factors were selected from the literature to be used for this iteration. However, instead of development-related details, the focus in this chapter was on the determination of these factors and on the adaptation of current technologies through an example of a long-term experiment.

3 Progression and Evaluation of the System

The result, described in this work mainly focuses on data acquisition and management systems, used in agricultural production. As it was mentioned previously, the iterative development of the system over the years divided it to various generations. For every generation of production system, there is at least one experimental version that is based on that system. This current iteration was based on own experience of practical application and on the literature in order to present the progression as well as to define a recommendation with the result. To present the use cases of data acquisition in agriculture, the common steps were examined and explicated, beginning with the measurement of environmental factors up to their utilization in decision support through data management, relying on current solutions, defined in the Big Data concept.

Based on previous practical experience, the data in decision support can be classified into three main groups, formed by environmental (sensory), economic (business-related), and agricultural data groups, classified further by the place of acquisition as internal and external data. In the following, the focus will strictly be on environmental (sensory) data, considering the three main data groups that can be obtained using sensor networks, based on the IoT concept. Other factors, including business (input costs, commodity market prices, etc.) and agronomic (plant-specific data), will be mentioned but only perfunctorily, to explain the concept of horizontal and vertical integration that the research aspires to achieve in the future.

3.1 The Main Characteristics Based on the Literature

The first analysis seeks to discover key relationships based on keywords and abstracts defined in the literature. The analysis was based on the co-occurrence of the available keywords, considering the 150 most frequent elements from the dataset. The keywords were defined in three columns, as three variables, consisting of author keywords, automatically added keywords (during the indexing) as abstract keywords that were added during pre-processing, text-based on mining, using the custom solutions, mentioned in the methodology section.

Based on the result (Fig. 2) the centrality of the IoT concept and sensor networks can be seen, which is inevitable as they serve as a key element of this study. The clustering in this case was influenced by significant noise due to the low threshold of co-occurrence applied, in order to achieve the sample size. However, using the result we can examine the direct and indirect connections, typical of the field. Among these, we can see a dominant connection with sensor networks as the tools to put the

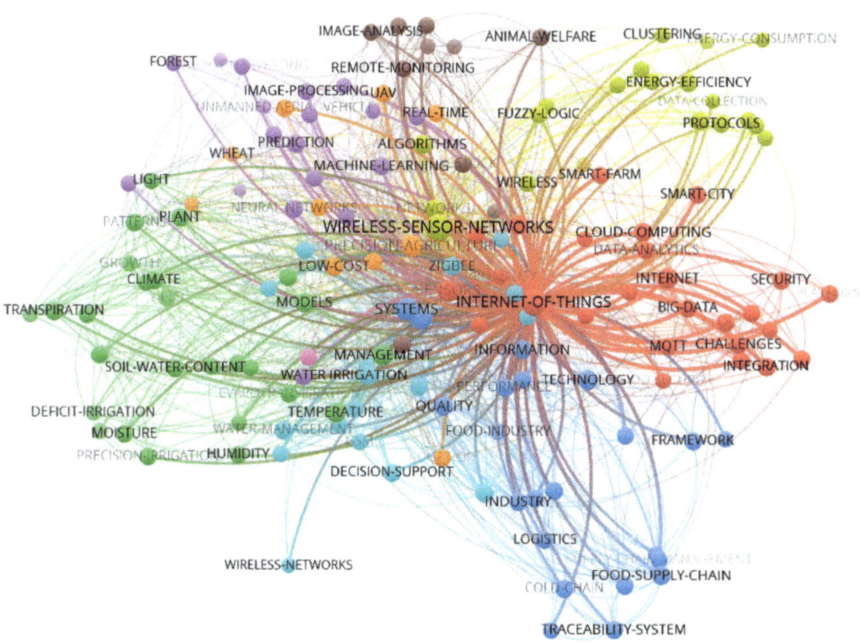

Fig. 2 Structure of the field

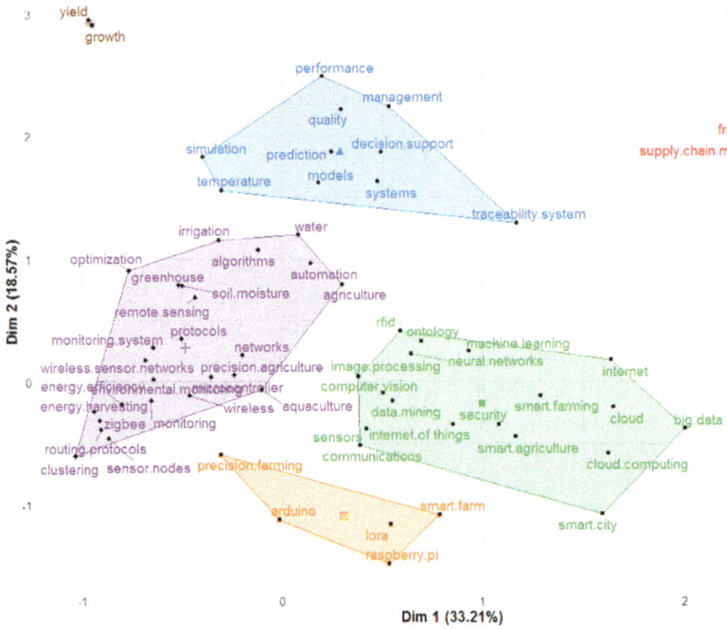

Fig. 3 Clustering of the main topics

concept into practice. Big Data (in connection with data analytics and cloud computing) also serves an important role as a data management concept. The latter relates to the Hadoop framework as a practical implementation of the concept. Machine learning on the other hand (in connection with neural networks, fuzzy logic, prediction, regression, and classification) serves as an analytical method. Precision agriculture appears as a dominant topic as well, along with Industry 4.0, indirectly connected with the IoT concept through sensor networks. Focusing on the two main components of the IoT concept and sensor networks, we can see solutions, trending on this field, including Arduino and Raspberry Pi (prototyping platforms), ZigBee, LoRa (wireless protocols), MQTT (Message Queuing Telemetry Transport protocol) and blockchain as well as concerns, like reliability, cost, and energy efficiency. From the perspective of agriculture, we can see correspondent topics of water management, smart irrigation, yield, growth, crops, greenhouse (in connection with temperature, humidity), and quality, which probably relate to the product.

In order to examine the field from another perspective, six main clusters were defined using K-means clustering, based on MCA, considering the keywords (Fig. 3). The visualized result presents a progress, where higher threshold was used regarding the minimum connections between the keywords, resulting in lower sample size, which reduced the available information, but increased readability. The green cluster mainly represents solutions in the field; however, moving in direction of the purple cluster, data acquisition-related keywords are beginning to

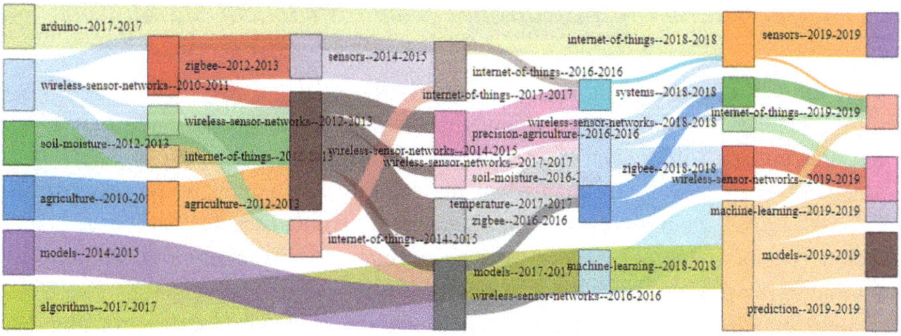

Fig. 4 Thematic evolution

appear. The blue cluster mainly represents data analytics and use cases that should be considered. The red cluster represents verticality, with supply chain and standardized frameworks. The yellow cluster represents data acquisition from a development standpoint (with development boards); however, the cluster contains precision agriculture that tends to lean toward data acquisition, while smart farming leans toward the solutions. Finally, the brown cluster contains agriculture-specific factors, but due to the lower sample size, the cluster is not properly expressed.

The result of the following thematic evolution represents the experienced changes in the presented connections over the years, determining the current and constantly dominant topics. For this analysis, seven breakpoints were determined, representing the current states in 2009, 2011, 2013, 2015, 2016, 2017, 2018, and 2019. The required minimum count of the co-occurrence was set to eight records to achieve sufficient sample size for each time periods at the breakpoints.

Based on the result (Fig. 4), we can see the appearance of wireless sensor networks in 2010, connected to ZigBee and Internet of Things in 2012. In 2014, ZigBee, soil moisture, and agriculture become connected as well, resulting in precision agriculture in 2016. In 2016, we can see the connection of precision agriculture with the IoT concept through the systems, but directly connected to the sensor networks. Models and machine learning become connected as well in 2016. Based on the current articles, machine learning (along with model and predictions) become connected to Internet of Things in 2019, representing the use of machine learning for sensory data. Arduino, as a user-friendly development board follows the entire timeline, connecting to the IoT concept in 2018, as well as indirectly connecting to precision agriculture and sensors. If we follow the diagram, we can examine the evolution of the sensor networks, becoming part of the IoT concept as well as precision agriculture. Moreover, we can see the connection of machine learning, models, and prediction along with the topic of sensor networks.

3.2 Determining the Possibilities from a Practical Standpoint

Based on the overall result of the bibliometric analysis, the focus was on the communication and data management due to the increasing presence of the Big Data concept and other related concepts in sensory data acquisition. The related frameworks not only assist data management but also connectivity between the physical devices, enabling the standardization of the data transfer that was achieved previously using custom protocol. Moreover, open-source nature was also considered due to the popularity of the Arduino platform.

There are various long-term, own experiments in the topic of environmental data acquisition, based on the IoT concept, developed specifically for agricultural production [29, 49, 55]. An advantage is that practical experience (from agricultural production) can be used while developing the system due to the fact that it is always gained and evaluated in a farm. All the components, including the data acquisition device, data models, databases, and management applications, are being developed based on a production system to achieve the highest integration possible, as an important aspect of the Industry 4.0 concept. These iterations are necessary in order to optimize the attributes for academic use (short-term experiments, enabling rapid deployment and programming with reduced reliability) over the production system, which is focused more on reliability. These two systems are compatible on the surface, but the underlying structure is entirely different. In the following, the former system (for academic use) is used to present the characteristics and possibilities of data acquisition, and more importantly, data management in agricultural production, based on the IoT concept.

As a result, the current development iteration of data acquisition and management system is presented, based on the current trends of the literature to aid further development. The area to be presented includes the steps of *data acquisition*, *data management* (structure and storage) as well as *data utilization* to encompass the tasks that usually need to be performed to facilitate data-driven decision support. Before these tasks are explained in more detail, the basic hierarchy should be presented for clear understanding. The production version of the data acquisition system is directly connected with a server-side API (application programming interface) via TCP/IP protocol which also serves as an entry point to access the databases in order to exchange data (measurements, settings, validation data). The system also includes a web and desktop-based management application for basic visualization and analytics that uses the same API to communicate with the databases. In the following, these three components are the base of the comparison and discussion (Fig. 5), including data acquisition, data management, and data utilization. The section highlighted with blue defines the new iteration, while the section highlighted with red defines the previous methods, used by the previous commercial version.

Regarding agricultural processes, the management functions can be classified into three categories, including sensing and monitoring, analysis and decision making as well as intervention [56]. In the following, sensing and monitoring, as well as

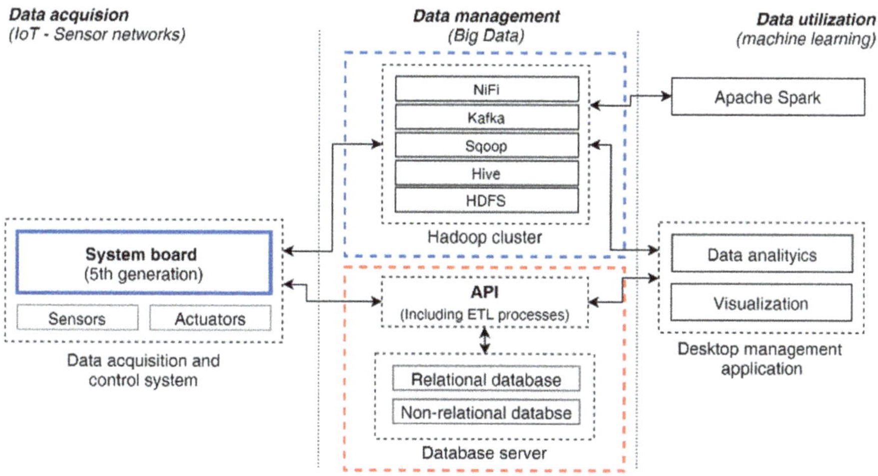

Fig. 5 The basic architecture of the system

decision support are highlighted, not focusing on process control, as it highly diverges from the topic of managerial decision support.

Data Acquisition Systems

In the following sections, the current iteration of the data acquisition system will be presented briefly. The second experimental iteration based on the 5th generation of the commercial system has been changed regarding its architecture based on the previous experiences and the literature. Instead of using task-specific modules (main controller, user interface, sensor boards, external device controller, wireless controller) in a hierarchic structure, constructed by separate hardware, this iteration uses a single board, capable of adapting to various tasks, thereby simplifying the architecture and reducing costs, which is now estimated to be 30 Euro per node, one-third of the previous one. However, the new controller board uses the same unified connector (based on the RJ45 physical connector), thereby facilitating backward compatibility.

This version of the device is not microcontroller unit (MCU) independent, unlike previous generations or the production version, meaning that it is only compatible with Arduino boards based on the Cortex M0 architecture (Atmel D21 series MCU). For this academic version, the use of Arduino nodes proved to be useful due to the nature of the experiments that are usually performed using the system, due to the faster development, compared to native solutions (embedded programming without libraries) when robust firmware is not a necessity. Based on the literature, several experiments were performed using Arduino developer boards; however, in this case, a custom solution was implemented, ensuring the characteristics of a task-specific system.

Based on the experience, open-source libraries that are available for the platform, tend to be developed based on different approaches, resulting in unstable operation (unoptimized methods, unexpected downtime, unhandled errors) in some cases, thus a custom library was used based on top of the Arduino hardware abstraction layer to solve this issue by implementing the compatible sensor drivers according to their datasheet, considering the requirements. The system is capable of offline and online measurements as well, the former performed using Wi-Fi connection while the latter option, using the integrated SD (Secure digital) card reader. The wireless connection between the controller and the server is achieved using TCP/IP protocol, using an integrated Wi-Fi controller, utilizing custom data protocol but for this research, instead of using the existing framework, an open-source solution will be presented in the next section as one of the countless alternatives available.

In case of the sensor connectivity, this version of the system lacks wireless options, whereas it utilizes point-to-point, wired connection due to the nature of the experiments it usually being used. The sensors are connected using a converter board which provides a direct connection with the MCU, enabling efficient setup in every location. The commercial version uses active translations in case of sensor connectors with pre-defined sensor boards, while the presented academic version uses sensor converters, capable of converting sensor connections physically, reducing the logical complexity, as a favorable trait in ad-hoc experiments. The implemented sensors for the test system are capable of measuring temperature (ambient and soil), relative humidity, lighting intensity, UV radiation, soil moisture, rain intensity, wind speed and direction. In addition to the standardized connections, the printed circuit board (PCB) contains integrated components for basic operation, including a real-time clock (RTC), a LED driver, power supply, and a set of indicators, representing the current state of the system. In contrast with the production system, graphical user interface was not implemented in this iteration due to the lack of demand considering the possible use-cases. The enclosure was designed based on the PCB for appropriate integration, with 3D printing in mind to support prototyping (Fig. 6) as well as to comply with additive manufacturing, defined in the Industry 4.0 concept.

Every new generation is being tested in a greenhouse or on the field at the Abádi Major family farm (Abádszalók, Hungary), where agricultural production takes place. Depending on the version (or iteration), other locations were also considered including food processing facilities and warehouses. To validate the proper operation, a short session of a long-term experiment will be presented, performed in a greenhouse, where 37 sensors were used with the new iteration, capable of measuring environmental factors in multiple spatial points. Typical parameters include ambient temperature (21 internal and one external point), humidity (20 internal and one external point), barometric pressure (one external point), luminosity (four internal points), UV radiation (two internal points), soil moisture (one internal points), soil temperature (six internal points), CO_2 concentration (one external point), and rain intensity (one external point). Multiple sensors were used to discover local differences, which may influence the result of the production on the long term. The experiment is performed constantly to expand the dataset and test analytical

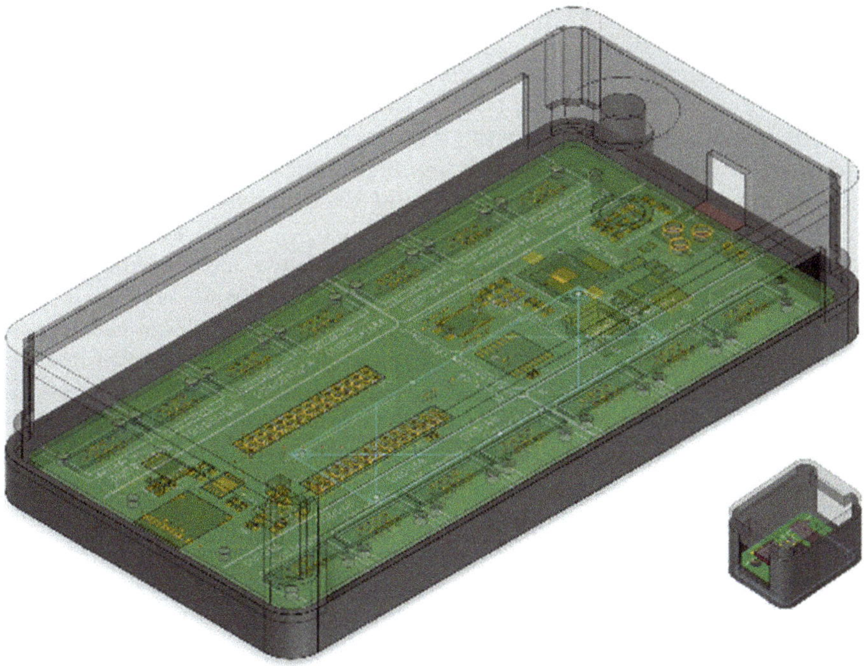

Fig. 6 Design, based on the 5th generation data acquisition device

capabilities in further research. To present the operation two main datasets were chosen from the database, of which one was performed on the 21st of June 2019, while the latter was performed on the 12th of January 2020 to create a diverse dataset for the next phase of the chapter, regarding data management. The measurements were performed in the same greenhouse, however, using a different set of sensors.

In recent measurement sessions, higher spatial resolution was used, compared to the previous sessions regarding the placements of the sensors, where diagonal structure was implemented to reduce costs [29]. The reason for the higher resolution is to research the practicality of volumetric data acquisition and to experimentalize with scalability in the future once sufficient data volume becomes available. Scalability is a solved issue with the new production system due to the wireless option, but due to the fact that this iteration uses wired connections, limited options were available.

Based on the literature, the attributes of controllers and sensors, the communication protocol should be considered from a technical point of view, while sensor placement, aggregation should be considered from an application point of view as these attributes appear to be a common key factor in research. Regarding the *controllers*, some research is performed using development boards including Arduino [57–60], Raspberry Pi [61], STM32 Nucleo [62, 63], or MicaZ nodes [64, 65], resulting in a less complex system, but some of the research presents custom made, multifunctional [66], or a task-specific solutions [21, 67], based on hierarchical [68] structure.

During this iteration of the development, multifunctionality has become a priority, but experience suggests so far that the development of compatible task-specific modules (collection of sensors in a single board) is needed for cost-effective use in agriculture. Considering the previously mentioned research, the use application of these devices includes greenhouse management. Regarding *communication technologies*, the current trend focuses on wireless technologies, which are also a factor to be improved, for them to facilitate the development of task-specific modules. Current trends are showing the increased diffusion of LoRaWAN network protocol [59, 62, 69] along with the usual ZigBee protocol [70–72], which can be integrated without modification by the expansion connector of the board for further research. Based on the characteristics, ZigBee network can be used to create a local network in the greenhouse using mesh topology, while LoRaWAN network can be used to connect these subsystems in larger distances. Sensor placement can also be a subject for debate, as some studies use homogenous grid topology [72]. Advanced strategies involve geostatistical analysis, Monte Carlo theory, or Gaussian Processes (GPs) to develop optimal placement [73]. In addition to the communication between the devices, communication between the devices and the server is also a crucial factor, which can be improved based on the literature as it can be seen in the following section of this chapter.

Data Management Methods and Applications

The production system uses a custom data-model, commissioned in a separate database server, containing five main databases to store sensory, business, and agronomic data. The database server utilized relational and non-relational databases to adapt to the requirements of the data, but measurements were still stored in a relational database. From the perspective of this chapter, we concentrate on the database, related to sensory data, specialized for measurements of sensor networks, without involving databases for other categories (considering the mentioned three data groups) due to the highly differing structure. The database is capable of handling metadata as well, including device (controller and sensor) hierarchy description and validation data, among others. The measurements were originally stored in a transaction-oriented structure, meaning that every instance of a variable was stored in a single record with its corresponding ID, variable UUID (Universally unique identifier) and timestamp. Of course, other parameters of the system components, including location (three-dimensional local and global coordinates), type, etc., were also stored in the database. To ensure compatibility, existing data storage was not modified (it was only extended with new solutions), thus only the data pipeline was modified, where measurements and the corresponding metadata are merged. Since the measurements were stored in a transaction-oriented structure, transformation was needed in order to match the timestamps, thus create a columnal structure that can be used in analytics. In the case of the production system, this process was performed using a custom data processing algorithm, integrated into the API, by matching the timestamps based on a threshold as well as considering local coordinates.

An important aspect of the study was to enable horizontal and vertical integration to facilitate data exchange inside, but also outside the farm for collaborations to provide a shared data pool, that can be utilized for machine learning as training data. As previously mentioned, the initial data model of the production system was designed with a set of compatible, *custom ETL (extract, transform, load) software*, capable of structuring internal measurement data and external (mainly business-related) data to provide a single interface serving various data for decision support, based on the concept of data lakes. This includes the unification of the spatial and temporal dimensions, as well as data labels. In order to facilitate the integration, *standardization is needed* which was achieved in a horizontal dimension, but in case of vertical integration, widely available tools were needed without relying on custom solutions. To solve this issue, the Hadoop ecosystem was implemented previously for business-related data and now, for sensory data as well in order to accept, transform, and store the data as well as provide an interface, to structure them according to the requirements. In order to assess the capabilities of the ecosystem for sensory data, especially for the IoT concept, a new cluster was created based on three virtual servers (on one physical server) to imitate a real business scenario.

Hadoop ecosystem is a parallel, distributed platform, utilizing multiples servers for big data processing [40], founded in 2008 at Yahoo and the University of Michigan as an Apache project [74]. Hadoop uses a three-layered model, handling storage, processing, and management separately [75] for distributed processing. It is a collection of services and development frameworks, providing a toolkit for data management from importing, up to analytics for structured, semi-structured, or unstructured data. The components are based on the HDFS filesystem, providing scalable, fault-tolerant, thus reliable, and cost-effective storage. On top of the filesystem, YARN is responsible for resource management, which now includes MapReduce framework as one of the core features, providing parallel data processing, using mappers and reducers.

The ecosystem also provides integrated non-relational database services, including HBase, designed to store columnal data, which is optimal for sensory data. Sqoop can be used for importing structured data, while enabling export as well. However, based on the scope of this paper, other integrated solutions, capable of handling stream data, like MQTT (Message Queuing Telemetry Transport protocol), Apache NiFi (ingestion layer), and Apache Kafka, should also be addressed as a typical pipeline. The processing of stored data can be performed using Hive, providing an SQL-like interface to define MapReduce jobs. As an in-memory solution, Apache Spark can also be used, which is able to read, pre-process, and analyze stored and streamed data as well [76].

In order to facilitate the possibilities, multiple approaches were defined (Fig. 7) in addition to the conventional method as an initial experiment, the first being a real-time connection, handling the data as a stream using Apache NiFi and Apache Kafka, while the latter being an offline solution, importing the content of the

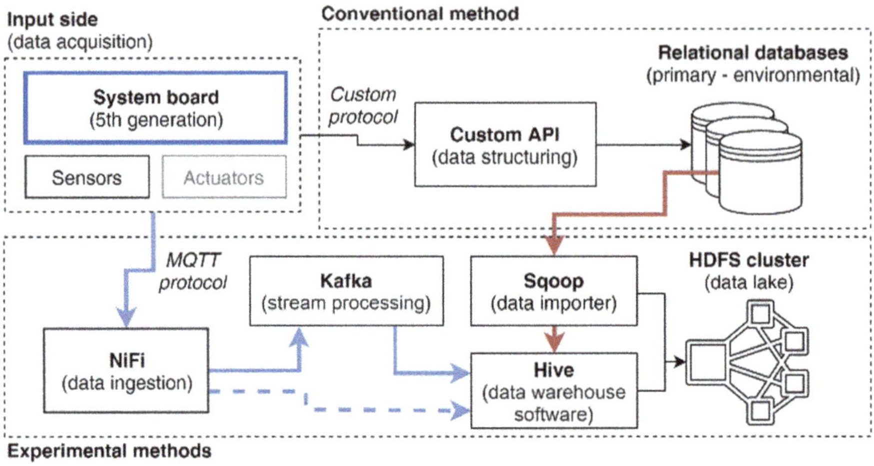

Fig. 7 Importing sensory data to the HDFS cluster

relational database onto the HDFS (Hadoop distributed file system) cluster using Apache Squoop. In a similar way it was performed in case of business-related data in former research, in order to store and analyze the data. The first method requires the modification of the firmware of the data acquisition device as it now uses a proprietary protocol to communicate with the custom API, while the latter one required the definition of the non-relational data model that will represent the measurements. Previously, it was mentioned that two datasets are used in this research, measured at a different period. The first dataset was used to assess the former method (importing from a relational database), while the new measurements were performed in order to experiment with the MQTT protocol and NiFi, as a built-in service in Hadoop for IoT devices.

In the case of *offline data import* (highlighted with blue color), Sqoop was used in order to mirror the full database, containing the measurements to the HDFS cluster, enabling to use the data inside of the Hadoop ecosystem. To import the data, the full load was used as an option, meaning that the database was mirrored without any restrictions. After importing the data, multiple blocks were generated, corresponding with the number of mappers and reducers, used to import the data. Due to the fact the data was stored as text files in the cluster, the definition of a Hive table was required.

In order to utilize *streaming data* (highlighted with red color) in Hadoop, we need to define an open-source protocol instead of the proprietary protocol that was used in the previous iteration in the case of the production system. As it was mentioned, the system was using custom data protocol, forwarded through TCP/IP connection. The server-side API received the message, containing the measurement of one cycle, from one controller; then processed it to upload to the relational database, handling every received variable separately. This protocol is not recognizable by the components of the Hadoop ecosystem, in order to integrate the existing data acquisition system, thus a new layer was implemented for the MCU in order to support MQTT

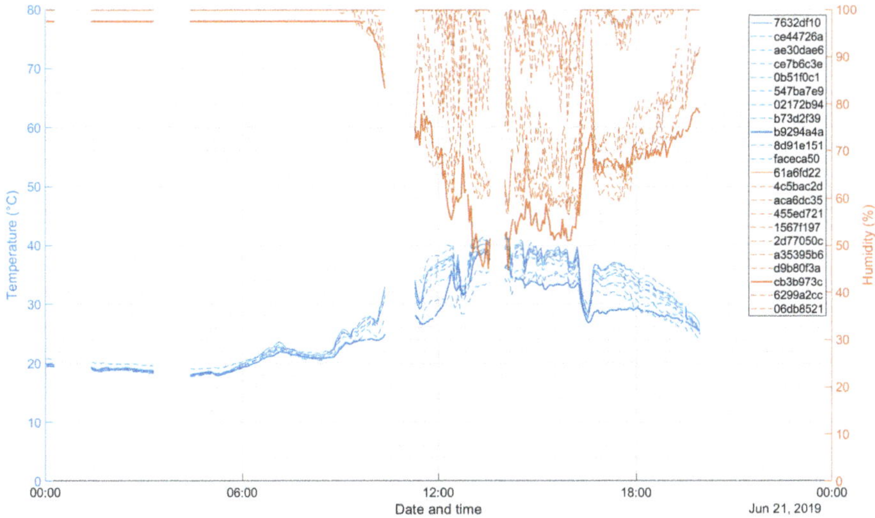

Fig. 8 Time-series measurements of temperature and humidity in a greenhouse

messaging protocol. Due to the fact the first application serves only as an assessment, the protocol was implemented using an open-source library. In case of positive experiences, custom implementation is required to match the structure of the existing firmware.

Considering the Hadoop ecosystem, the devices serve as "producer", while the server serves as a broker, serving data to the "consumers", which can be an application or storage. On the server-side, Apache NiFi was commissioned on one virtual server. The architecture utilized topics, called "Environmental Parameters", which consists of subtopics, including "Temperature", "Humidity", "Luminosity", "Pressure", "Moisture", "Wind", and "Rain". The structure of topics is convenient, since the consumer can select the data based on hierarchical properties. This function was also implemented in the original API but based on other principles.

After the two datasets were imported, Apache Spark was used, instead of the management application of the production system in order to finalize the data frame, containing the measurements for the defined measurement sessions. Instead of using proprietary algorithms, the data was structured internally, to create columnal structure from the transaction-oriented offline (one variable per transaction) and batched online (one controller per transaction) data. The most important is that Spark is capable of merging all measurements in one single data frame, in an integrated aspect, meaning that it can be utilized in machine learning task. The data transformation was performed using efficient, built-in functions for merging and projecting the data into a predefined time axis, handling different temporal resolution. The following figure represents a former measurement session, now integrated using Squoop, resulting in a structure, similar to previous sessions (Fig. 8).

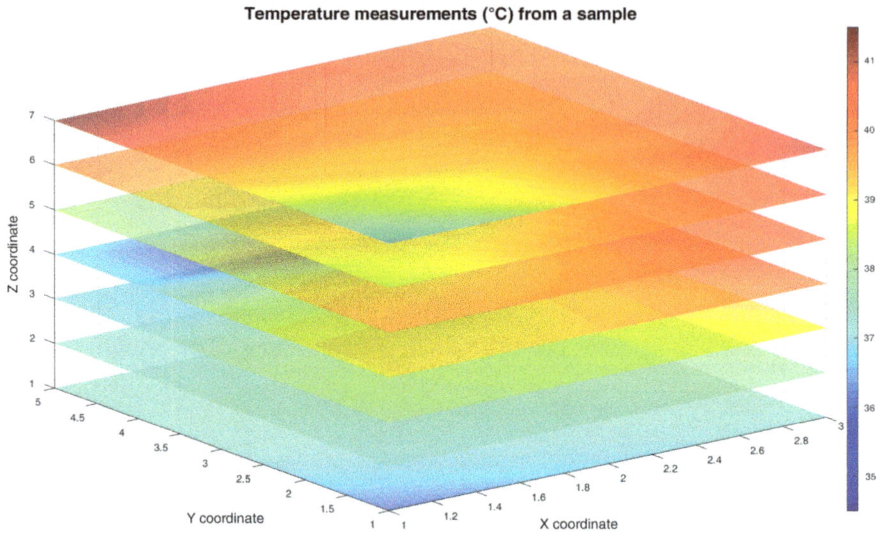

Fig. 9 Volumetric representation of temperature data

Integrating the data of the two-measurement session is not as complex as in the case of business-related data due to the already standardized structure. The data is required to be transformed based on location and time. They were given variables (time, coordinates, UUID), making them effective to use as keys when merging datasets. In this case, the mentioned three variables were used to merge the datasets. For machine learning, constant scales are required, thus resampling was used to achieve one-minute resolution. In case of three consecutive missing values, linear interpolation is being applied automatically in case of a query. As we can see on the figure, longer sections are not handled automatically based on the algorithm of the new API, to avoid distortion, but it can be handled manually, using Apache Spark, based on more advanced methods.

The dataset for volumetric visualization (Fig. 9) can be generated using Apache Spark as well, without relying on the custom solution, considering the coordinates, stored as metadata in the database. The visualization represents the spatial differences in the greenhouse, facilitating decision support and process control as well. The data is interpolated along all axes, to represent the gradient characteristics of the measurements. Due to the increased number of sensors, the data could be transformed to more layers than before.

Based on the literature, we can find a wide range of data management strategies, handling sensory data in agriculture regarding weather and climate data [77], but also for remote sensing [78]. Data integration is also considered in research, utilizing the Hadoop ecosystem to integrate weather stations, satellite images, sensors, and existing relational databases in a similar manner [79]. Wider integration also considers supply chain integration, utilizing a sensor network in the Hadoop ecosystem to provide additional information about the production [80]. In the case of communication protocols, MQTT is a trusted method by researchers as well [81, 82]. For sensor connection, utilizing the Hadoop ecosystem, some research uses Kafka for data ingestion [83] as opposed to the presented experiment, where NiFi was used to forward messages to Kafka or NiFi was directly writing into the database. The available databases encompass task-specific [84] and multifunctional [85] structures, the latter typical for the presented system as well.

Data Utilization

It is difficult to describe the business value in IoT (sensory data), without involving other variables (business or agronomic related data) since it provides indirect advantages over conventional methods. Based on the experiences and the literature, data utilization varies greatly between simple descriptive statistics up to machine learning, both resulting in decision support (in form of report or visualization) or process control. In the following, we will focus mainly on decision support. In order to comply with the standards of the Industry 4.0 concept, data visualization and analytics is considered, the latter based on artificial intelligence. The database used by the system stores time-series measurement data and farm-related data among others. This data can be transformed into a volumetric data set using ETL processes, while maintaining the time-series quality. Before using the data for training purposes, some research integrates data cleansing and averaging [86] to match time periods, which should be considered before developing the analytics module. The measurements are stored in a structured database, but in case of integrating other, in many cases unstructured data, some researchers recommend using Hadoop-based data management during transformation [40] as it now was achieved in this research as well as the part of this iteration.

Up until this point, only business-related data was utilized in machine learning application considering the system, due to the inconsistency of the sensory data. However, the aim was to implement the devices and the data into the Hadoop ecosystem, enabling easy access to the available data. Using the standardized data frame, containing the two mentioned datasets (measured during the two sessions), a model was created for time-series forecasts; however, the volume of the data did not provide sufficient information for generalization, resulting in overfitting. However, clustering does not require time dimension on case of volumetric data, unlike the mentioned time-series analytics, thus an initial clustering was already performed, based on the presented pipeline, demonstrating the feasibility of the data acquisition and management.

In the literature the current trend focuses on optimization, prediction, and anomaly detection, using measurements, supported by artificial intelligence, which is important in our case to complement the existing modules. Based on the literature, methods such as recurrent neural networks, deep neural networks, and feedforward neural networks can be used in greenhouses for environmental data due to the time-series nature of the data. Moreover, deep neural networks are also suitable for classification in machine vision [87], which is also a dominant topic in greenhouses.

In the case of prediction, a recurrent neural network was capable of forecasting internal temperature for a year based on just four variables, after two weeks of training [88]. A research was performed using a feedforward neural network to predict cuticle cracking up to four weeks in advance using nine measured environmental variables [89]. The result of the production is also predictable, as it was presented in research, determining the yields of tomato using a feedforward neural network and four environmental variables and previous yield, showing a fundamental relationship between environmental and economic data. The literature also describes soil moisture prediction [90]. In the case of process control, like irrigation, researchers tend to prefer fuzzy logic [91]. As can be seen, none of the mentioned experiments used volumetric data, which raises the question of how the accuracy can be improved by a greater spatial resolution. In the case of visualization, there was no suitable research, considering the field.

4 Discussion

In the following, the possibilities for the next iteration are presented. An IoT-based system consists of subsystems, performing tasks of sensing and actuation, system management and storage, modeling and optimal planning as well as monitoring and visualization [92]. In the present study, sensing (data acquisition) and planning (decision support using analytics and visualization) were of paramount importance in defining further development possibilities. As a discussion, alternatives were defined based on the literature for each task of data acquisition, in order to facilitate further progression.

Regarding *data acquisition*, the current system provides a robust architecture to perform experiments in the agricultural field. However, based on the trends in the literature, the integration of wireless options should be considered in the future for the academic version as well, to broaden the possibilities. The main concerns of the implementation were cost efficiency (based on this system, only one controller board is needed for more than 25 sensors) as well as power management. Due to the fact that the measurements usually take place in a greenhouse, the build of a wired network is convenient, because the sensors have permanent placing, enabling proper cable management to be fixed on the frame of the greenhouse.

Regarding *data management*, the incremental adaptation of the Hadoop ecosystem provides current and diverse solutions for issues in data management. In that section, the MQTT protocol was implemented for IoT-based devices, along with core features of the Hadoop ecosystem, including NiFi and Apache Kafka. The experiences show that Apache NiFi provides a user-friendly option for ingesting data, created by embedded systems (including sensor networks). However, it proves to be useful if the use of Hadoop ecosystem is necessary or a highly scalable interface is required. A possible use case would be the integration of computer vision (using on ARM-based embedded system) on the same network. It would make the permanent implementation of this data integration method more sensible, as the original API has no function in this regard, especially considering direct connection with machine learning applications through Apache Spark, which can be achieved almost effortlessly considering the implemented structure. Considering a permanent implementation, the development of an alternative database on the Hadoop cluster, based on HBase is required for better consistency.

Regarding *data utilization*, there are positive experiences in 3D visualization, shown above, thus further development is needed, including the integration of 3D models, using layers, representing the objects (the greenhouse, plants, sensors, and other relevant objects) on the location of the measurements, providing marked points during visualization. In addition to visualization, there is a high demand for artificial intelligence applications, for data analytics. In previous research, machine learning, including deep learning models, based on LSTM (long short-term memory) layers were implemented for time-series forecasts using the same data model, however, only using business-related data. In case of a consistent data source, describing long-term measurements, an assessment would be required, examining the usability of these methods on spatiotemporal data.

5 Conclusions

In the age of digitization, the application of the pillars defined by Industry 4.0 is becoming increasingly important to optimize production and business processes in agriculture. Based on the bibliometric analysis and previous experience, Big Data and the corresponding data management as well as open-source solutions are considered to be crucial factors in the field of agricultural data acquisition considering the IoT concept. This chapter describes the current iteration of a development based on a production system. This includes the modification of the data acquisition device to support Arduino platform instead of using native solutions, as it is one of the generally known open-source solutions in the literature. The methods of data management were also modified, as it is now facilitated by open-source frameworks in addition to the previously developed custom solutions, considering the importance of vertical and horizontal integration.

Based on the experiments, performed in a greenhouse, the implementation of the Hadoop framework proves to be an efficient alternative to custom solutions to achieve standardized communication between the data acquisition devices and the servers. Moreover, it also provides standardized interface for data management, integrating the environmental data from various sources, creating a unified time-series dataset which can be used directly in decision support. The further development possibilities are also described, based on the current literature, in order to support the design of the next iteration, highlighting data acquisition, data management, and data utilization separately. Crucial factors to be addressed include the implementation of wireless connectivity, by following the progression of the production system. The incremental build model allows finding the development directions at regular intervals, to adapt to needs and trends, which is also supported by the presented overview.

Acknowledgments This paper was supported by EFOP3.6.3-VEKOP-16-2017-00007—"Young researchers for talent"—Supporting careers in research activities in higher education program.

References

1. Gebbers, R., & Adamchuk, V. I. (2010). Precision agriculture and food security. *Science (80), 327*, 828–831. https://doi.org/10.1126/science.1183899
2. FAO (2009). Global agriculture towards 2050.
3. Lee, M., Hwang, J., & Yoe, H. (2013). Agricultural production system based on IoT. *Proceedings of the 2013 IEEE 16th International Conference on Computational Science and Engineering CSE*, 833–837. https://doi.org/10.1109/CSE.2013.126
4. Schuster E, Kumar S, Sarma S, et al. (2011). Infrastructure for data-driven agriculture: Identifying management zones for cotton using statistical modeling and machine learning techniques. In: 2011 8th International Conference & Expo on emerging Technologies for a Smarter World. IEEE, pp. 1–6.
5. Bongiovanni, R., & Lowenberg-Deboer, J. (2004). Precision agriculture and sustainability. *Precision Agriculture, 5*, 359–387. https://doi.org/10.1023/B:PRAG.0000040806.39604.aa
6. Schimmelpfennig, D., & Ebel, R. (2016). Sequential adoption and cost savings from precision agriculture. *Journal of Agricultural and Resource Economics, 41*, 97–115.
7. Lasi, H., Fettke, P., Kemper, H.-G., et al. (2014). Industry 4.0. *Business and Information Systems Engineering, 6*, 239–242. https://doi.org/10.1007/s12599-014-0334-4
8. Bartodziej, C. J. (2017). *The concept industry 4.0*. Springer Fachmedien Wiesbaden.
9. Vaidya, S., Ambad, P., & Bhosle, S. (2018). Industry 4.0 - a glimpse. *Procedia Manufacturing, 20*, 233–238. https://doi.org/10.1016/j.promfg.2018.02.034
10. Babiceanu, R. F., & Seker, R. (2016). Big data and virtualization for manufacturing cyber-physical systems: A survey of the current status and future outlook. *Computers in Industry, 81*, 128–137. https://doi.org/10.1016/j.compind.2016.02.004
11. Tzounis, A., Katsoulas, N., Bartzanas, T., & Kittas, C. (2017). Internet of things in agriculture, recent advances and future challenges. *Biosystems Engineering, 164*, 31–48. https://doi.org/10.1016/j.biosystemseng.2017.09.007
12. Clasen, M. (2016). Farming 4.0 und andere Anwendungen des Internet der Dinge. *Inform der Land-, Forst- und Ernährungswirtschaft, 2016*, 33–36.
13. Wolfert, S., Ge, L., Verdouw, C., & Bogaardt, M. J. (2017). Big data in smart farming – A review. *Agricultural Systems, 153*, 69–80. https://doi.org/10.1016/j.agsy.2017.01.023
14. Braun, A. T., Colangelo, E., & Steckel, T. (2018). Farming in the era of Industrie 4.0. *Procedia CIRP, 72*, 979–984. https://doi.org/10.1016/j.procir.2018.03.176
15. Voutos, Y., Mylonas, P., Katheniotis, J., & Sofou, A. (2019). A survey on intelligent agricultural information handling methodologies. *Sustainability, 11*, 3278. https://doi.org/10.3390/su11123278
16. Elijah, O., Rahman, T. A., Orikumhi, I., et al. (2018). An overview of internet of things (IoT) and data analytics in agriculture: Benefits and challenges. *IEEE Internet of Things Journal, 5*, 3758–3773. https://doi.org/10.1109/JIOT.2018.2844296
17. Borgia, E. (2014). The internet of things vision: Key features, applications and open issues. *Computer Communications, 54*, 1–31. https://doi.org/10.1016/j.comcom.2014.09.008
18. Fahmideh, M., & Zowghi, D. (2020). An exploration of IoT platform development. *Information Systems, 87*, 101409. https://doi.org/10.1016/j.is.2019.06.005
19. Ashton, K. (2009). That "internet of things" thing. *RFiD Journal, 22*, 97–114.

20. Da Xu, L., Xu, E. L., & Li, L. (2018). Industry 4.0: State of the art and future trends. *International Journal of Production Research, 56*, 2941–2962. https://doi.org/10.1080/00207543.2018.1444806
21. Stamenkovi, Z., Randji, S., Santamaria, I., et al. (2016). Advanced wireless sensor nodes and networks for agricultural applications. https://doi.org/10.1109/TELFOR.2016.7818709
22. Gilchrist, A. (2016). *Industry 4.0*. Apress.
23. Talavera, J. M., Tobón, L. E., Gómez, J. A., et al. (2017). Review of IoT applications in agro-industrial and environmental fields. *Computers and Electronics in Agriculture, 142*, 283–297. https://doi.org/10.1016/j.compag.2017.09.015
24. Othman, M. F., & Shazali, K. (2012). Wireless sensor network applications: A study in environment monitoring system. *Procedia Engineering, 41*, 1204–1210. https://doi.org/10.1016/j.proeng.2012.07.302
25. Verdouw, C. (2016). Internet of things in agriculture. *CAB Reviews: Perspectives in Agriculture, Veterinary Science Nutrition and Natural Resources, 11*, 1–12. https://doi.org/10.1079/PAVSNNR201611035
26. Park, D. H., & Park, J. W. (2011). Wireless sensor network-based greenhouse environment monitoring and automatic control system for dew condensation prevention. *Sensors, 11*, 3640–3651. https://doi.org/10.3390/s110403640
27. Su, Y., & Xu, L. (2017). Towards discrete time model for greenhouse climate control. *Engineering in Agriculture, Environment and Food, 10*, 157–170. https://doi.org/10.1016/j.eaef.2017.01.001
28. Baviskar J, Mulla A, Baviskar A, et al (2014) Real time monitoring and control system for green house based on 802.15.4 wireless sensor network. *2014 Fourth Int Conf Commun Syst Netw Technol*. 98–103. https://doi.org/10.1109/CSNT.2014.28.
29. Tóth, M., Felföldi, J., & Szilágyi, R. (2019). Possibilities of IoT based management system in greenhouses. *Georg Agricultural, 23*, 43–62.
30. Barreto, L., Amaral, A., & Pereira, T. (2017). Industry 4.0 implications in logistics: An overview. *Procedia Manufacturing, 13*, 1245–1252. https://doi.org/10.1016/j.promfg.2017.09.045
31. Wu, X., Xu, W., Song, Y., & Cai, M. (2011). A detection method of weed in wheat field on machine vision. *Procedia Engineering, 15*, 1998–2003. https://doi.org/10.1016/j.proeng.2011.08.373
32. Habib, M. T., Majumder, A., Jakaria, A. Z. M., et al. (2018). Machine vision based papaya disease recognition. *Journal of King Saud University Computer and Information Sciences, 32*, 300–309. https://doi.org/10.1016/j.jksuci.2018.06.006
33. Momin, M. A., Yamamoto, K., Miyamoto, M., et al. (2017). Machine vision based soybean quality evaluation. *Computers and Electronics in Agriculture, 140*, 452–460. https://doi.org/10.1016/j.compag.2017.06.023
34. Unay, D., Gosselin, B., Kleynen, O., et al. (2011). Automatic grading of bi-colored apples by multispectral machine vision. *Computers and Electronics in Agriculture, 75*, 204–212. https://doi.org/10.1016/j.compag.2010.11.006
35. Blok, P. M., Barth, R., & van den Berg, W. (2016). Machine vision for a selective broccoli harvesting robot. *IFAC-PapersOnLine, 49*, 66–71. https://doi.org/10.1016/j.ifacol.2016.10.013
36. Pham, X., & Stack, M. (2018). How data analytics is transforming agriculture. *Business Horizons, 61*, 125–133. https://doi.org/10.1016/j.bushor.2017.09.011
37. Ghiwari, S., Sambrekar, K., & Rajpurohit, V. S. (2018). Hierarchical storage for agro informatics system using NoSQL technology. *2017 International Conference on Computing, Communication, Control and Automation ICCUBEA, 2017*, 1–5. https://doi.org/10.1109/ICCUBEA.2017.8463693
38. Hammer, B., He, H., & Martinetz, T. (2014). Learning and modeling big data. *22th European Symposium on Artificial Neural Networks*, 23–25.
39. Chen, M., Mao, S., & Liu, Y. (2014). Big data: A survey. *Mobile Networks and Applications, 19*, 171–209. https://doi.org/10.1007/s11036-013-0489-0

40. Bendre, M. R., Thool, R. C., & Thool, V. R. (2016, 2015). Big data in precision agriculture: Weather forecasting for future farming. *Proceedings of 2015 1st International Conference on Next Generation Computing Technologies NGCT*, 744–750. https://doi.org/10.1109/NGCT.2015.7375220
41. Saggi, M. K., & Jain, S. (2018). A survey towards an integration of big data analytics to big insights for value-creation. *Information Processing and Management, 54*, 758–790. https://doi.org/10.1016/j.ipm.2018.01.010
42. Seddon, J. J. J. M., & Currie, W. L. (2017). A model for unpacking big data analytics in high-frequency trading. *Journal of Business Research, 70*, 300–307. https://doi.org/10.1016/j.jbusres.2016.08.003
43. Xie, N., Wang, W., Ma, B., et al. (2015). Research on an agricultural knowledge fusion method for big data. *Data Science Journal, 14*, 7. https://doi.org/10.5334/dsj-2015-007
44. Liakos, K. G., Busato, P., Moshou, D., et al. (2018). Machine learning in agriculture: A review. *Sensors (Switzerland), 18*, 1–29. https://doi.org/10.3390/s18082674
45. Shin, J.-Y., Kim, K. R., & Ha, J.-C. (2020). Seasonal forecasting of daily mean air temperatures using a coupled global climate model and machine learning algorithm for field-scale agricultural management. *Agricultural and Forest Meteorology, 281*, 107858. https://doi.org/10.1016/j.agrformet.2019.107858
46. Sannakki, S., Rajpurohit, V. S., Sumira, F., & Venkatesh, H. (2013). A neural network approach for disease forecasting in grapes using weather parameters. *2013 Fourth International Conference on Computing, Communications and Networking TechnologiesICCCNT, 2013*, 2–6. https://doi.org/10.1109/ICCCNT.2013.6726613
47. Veenadhari, S., Misra, B., & Singh, C. D. (2014). Machine learning approach for forecasting crop yield based on climatic parameters. *2014 International Conference on Computer Communication and Informatics Usher Technologies Tomorrow, Today, ICCCI, 2014*, 1–5. https://doi.org/10.1109/ICCCI.2014.6921718
48. Himesh, S., Prakasa Rao, E. V. S., Gouda, K. C., et al. (2018). Digital revolution and big data: A new revolution in agriculture. *CAB Reviews: Perspectives in Agriculture, Veterinary Science Nutrition and Natural Resource, 13*. https://doi.org/10.1079/PAVSNNR201813021
49. Tóth, M., & Szilágyi, R. (2017). Development and testing experiences of a management supporting data acquisition system. *Journal of Agricultural Informatics, 8*, 55–70. https://doi.org/10.17700/jai.2017.8.2.382
50. Aria, M., & Cuccurullo, C. (2017). Bibliometrix: An R-tool for comprehensive science mapping analysis. *Journal of Informetrics, 11*, 959–975. https://doi.org/10.1016/j.joi.2017.08.007
51. Kamada, T., & Kawai, S. (1989). An algorithm for drawing general undirected graphs. *Information Processing Letters, 31*, 7–15. https://doi.org/10.1016/0020-0190(89)90102-6
52. Cobo, M. J., López-Herrera, A. G., Herrera-Viedma, E., & Herrera, F. (2011). An approach for detecting, quantifying, and visualizing the evolution of a research field: A practical application to the fuzzy sets theory field. *Journal of Informetrics, 5*, 146–166. https://doi.org/10.1016/j.joi.2010.10.002
53. Nabil Mohammed, M. A., & Govardhan, A. (2010). A comparison between five models of software engineering. *International Journal of Computer Science, 7*, 94–101.
54. Axinte S-D, Petrica G, Barbu I-D (2017) E-learning platform development model. In: 2017 10th international symposium on advanced topics in electrical engineering (ATEE). IEEE, pp. 687–692.
55. Tóth, M., & Szilágyi, R. (2017). Gazdálkodást támogató adatgyűjtő rendszer fejlesztése és tesztelése. In *Informatika a felsőoktatásban 2017 konferencia* (pp. 453–461). Debrecen.
56. Verdouw, C. N., Beulens, A. J. M., Reijers, H. A., & Van Der Vorst, J. G. A. J. (2015). A control model for object virtualization in supply chain management. *Computers in Industry, 68*, 116–131. https://doi.org/10.1016/j.compind.2014.12.011
57. Bin, J. I., Raihana, K., Bhowmik, S., & Shakil, S. R. (2014). Wireless monitoring system and controlling software for smart greenhouse management. 2014 Int Conf informatics. *Electron Vision, ICIEV, 2014*, 1–5. https://doi.org/10.1109/ICIEV.2014.6850748

58. Bajer, L., & Krejcar, O. (2015). Design and realization of low cost control for greenhouse environment with remote control. *IFAC-PapersOnLine, 28*, 368–373. https://doi.org/10.1016/j.ifacol.2015.07.062
59. dos Santos, U. J. L., Pessin, G., da Costa, C. A., & da Rosa, R. R. (2019). AgriPrediction: A proactive internet of things model to anticipate problems and improve production in agricultural crops. *Computers and Electronics in Agriculture, 161*, 202–213. https://doi.org/10.1016/j.compag.2018.10.010
60. Yoon, C., Huh, M., Kang, S. G., et al. (2018). Implement smart farm with IoT technology. *International Conference on Advanced Communication Technology ICACT, 2018*, 749–752. https://doi.org/10.23919/ICACT.2018.8323908
61. Gorrepotu, R., Korivi, N. S., Chandu, K., & Deb, S. (2018). Sub-1GHz miniature wireless sensor node for IoT applications. *Internet of Things, 1–2*, 27–39. https://doi.org/10.1016/j.iot.2018.08.002
62. Reka, S. S., Chezian, B. K., & Chandra, S. S. (2019). *A novel approach of IoT-based smart greenhouse farming system* (pp. 227–235). Springer.
63. Azaza, M., Tanougast, C., Fabrizio, E., & Mami, A. (2016). Smart greenhouse fuzzy logic based control system enhanced with wireless data monitoring. *ISA Transactions, 61*, 297–307. https://doi.org/10.1016/j.isatra.2015.12.006
64. Akkaş, M. A., & Sokullu, R. (2017). An IoT-based greenhouse monitoring system with Micaz motes. *Procedia Computer Science, 113*, 603–608. https://doi.org/10.1016/j.procs.2017.08.300
65. Abd El-Kader, S. M., & Mohammad El-Basioni, B. M. (2013). Precision farming solution in Egypt using the wireless sensor network technology. *Egyptian Informatics Journal, 14*, 221–233. https://doi.org/10.1016/j.eij.2013.06.004
66. Goumopoulos, C., O'Flynn, B., & Kameas, A. (2014). Automated zone-specific irrigation with wireless sensor/actuator network and adaptable decision support. *Computers and Electronics in Agriculture, 105*, 20–33. https://doi.org/10.1016/j.compag.2014.03.012
67. Suresh VM, Sidhu R, Karkare P, et al (2018) Powering the IoT through embedded machine learning and LoRa. IEEE World Forum Internet Things, WF-IoT 2018 - Proc 2018-Janua: 349–354. https://doi.org/10.1109/WF-IoT.2018.8355177.
68. Yiming, Z., Xianglong, Y., Xishan, G., et al. (2007). A design of greenhouse monitoring & control system based on ZigBee wireless sensor network. *Wireless Communications, Networking and Mobile Computing 2007 WiCom 2007 International Conference*, 2563–2567. https://doi.org/10.1109/WICOM.2007.638
69. Stočes, M., Vaněk, J., Masner, J., & Pavlík, J. (2016). Internet of things (IoT) in agriculture - selected aspects. *Agris On-line Papers in Economics and Informatics, 8*, 83–88. https://doi.org/10.7160/aol.2016.080108
70. Keshtgari, M., & Deljoo, A. (2012). A wireless sensor network solution for precision agriculture based on Zigbee technology. *Wireless Sensor Network, 04*, 25–30. https://doi.org/10.4236/wsn.2012.41004
71. Tafa, Z., Ramadani, F., & Cakolli, B. (2018). The design of a ZigBee-based greenhouse monitoring system. In *2018 7th Mediterranean conference on embedded computing (MECO)* (pp. 1–4). IEEE.
72. Dan, L., Jianmei, S., Yang, Y., & Jianqiu, X. (2017). Precise agricultural greenhouses based on the IoT and fuzzy control. In *Proceedings - 2016 international conference on intelligent transportation, big data and smart city*. ICITBS 2016.
73. Castello, C. C., Fan, J., Davari, A., & Chen, R. X. (2010). Optimal sensor placement strategy for environmental monitoring using wireless sensor networks. *Proceedings of the Annual Southeastern Symposium on System Theory*, 275–279. https://doi.org/10.1109/SSST.2010.5442825
74. Polato, I., Ré, R., Goldman, A., & Kon, F. (2014). A comprehensive view of Hadoop research - a systematic literature review. *Journal of Network and Computer Applications, 46*, 1–25. https://doi.org/10.1016/j.jnca.2014.07.022

75. Hadi, M. S., Lawey, A. Q., El-Gorashi, T. E. H., & Elmirghani, J. M. H. (2018). Big data analytics for wireless and wired network design: A survey. *Computer Networks, 132*, 180–199. https://doi.org/10.1016/j.comnet.2018.01.016
76. Buyya, R., Calheiros, R. N., & Vahid Dastjerdi, A. (2016). *Big data: Principles and paradigms*. Morgan Kaufmann.
77. Pandey, A. K., Agrawal, C. P., & Agrawal, M. (2017). A hadoop based weather prediction model for classification of weather data. In *2017 second international conference on electrical* (pp. 1–5). Computer and Communication Technologies (ICECCT).
78. Chi, M., Plaza, A., Benediktsson, J. A., et al. (2016). Big data for remote sensing: Challenges and opportunities. *Proceedings of the IEEE, 104*, 2207–2219. https://doi.org/10.1109/JPROC.2016.2598228
79. Lamrhari, S., Elghazi, H., Sadiki, T., & El Faker, A. (2016). A profile-based big data architecture for agricultural context. *Proc 2016 Int Conf Electr Inf Technol ICEIT 2016*. 22–27. https://doi.org/10.1109/EITech.2016.7519585
80. Yan, J., Xin, S., Liu, Q., et al. (2014). Intelligent supply chain integration and management based on cloud of things. *International Journal of Distributed Sensor Networks*, 2014. https://doi.org/10.1155/2014/624839
81. Minh QT, Phan TN, Takahashi A, et al (2017) A cost-effective smart farming system with knowledge base. In: *ACM International Conference Proceeding Series*. pp. 309–316.
82. Kamilaris, A., Kartakoullis, A., & Prenafeta-Boldú, F. X. (2017). A review on the practice of big data analysis in agriculture. *Computers and Electronics in Agriculture, 143*, 23–37. https://doi.org/10.1016/j.compag.2017.09.037
83. Wiska R, Habibie N, Wibisono A, et al (2017) Big sensor-generated data streaming using Kafka and Impala for data storage in Wireless Sensor Network for CO2 monitoring. *2016 Int Work Big Data Inf Secur IWBIS*. 97–101. https://doi.org/10.1109/IWBIS.2016.7872896.
84. Gallinucci, E., Golfarelli, M., & Rizzi, S. (2019). *A hybrid architecture for tactical and strategic precision agriculture* (pp. 13–23). Springer International Publishing.
85. Jardak, C., Riihijärvi, J., Oldewurtel, F., & Mähönen, P. (2010). Parallel processing of data from very large-scale wireless sensor networks. *HPDC 2010 - Proceedings of the 19th ACM International Symposium on High Performance Distributed Computing*, 787–794. https://doi.org/10.1145/1851476.1851590
86. Goldstein, A., Fink, L., Meitin, A., et al. (2018). Applying machine learning on sensor data for irrigation recommendations: Revealing the agronomist's tacit knowledge. *Precision Agriculture, 19*, 421–444. https://doi.org/10.1007/s11119-017-9527-4
87. Habaragamuwa, H., Ogawa, Y., Suzuki, T., et al. (2018). Detecting greenhouse strawberries (mature and immature), using deep convolutional neural network. *Engineering in Agriculture, Environment and Food, 11*, 127–138. https://doi.org/10.1016/j.eaef.2018.03.001
88. Uchida Frausto, H., & Pieters, J. G. (2004). Modelling greenhouse temperature using system identification by means of neural networks. *Neurocomputing, 56*, 423–428. https://doi.org/10.1016/j.neucom.2003.08.001
89. Ehret, D. L., Hill, B. D., Raworth, D. A., & Estergaard, B. (2008). Artificial neural network modelling to predict cuticle cracking in greenhouse peppers and tomatoes. *Computers and Electronics in Agriculture, 61*, 108–116. https://doi.org/10.1016/j.compag.2007.09.011
90. Song, X., Zhang, G., Liu, F., et al. (2016). Modeling spatio-temporal distribution of soil moisture by deep learning-based cellular automata model. *Journal of Arid Land, 8*, 734–748. https://doi.org/10.1007/s40333-016-0049-0
91. Gao, L., Zhang, M., & Chen, G. (2013). An intelligent irrigation system based on wireless sensor network and fuzzy control. *Journal of Networks, 8*, 1080–1087. https://doi.org/10.4304/jnw.8.5.1080-1087
92. Somov, A., Seledets, I., Matveev, S., et al. (2019). Pervasive agriculture: IoT-enabled greenhouse for plant growth control. *IEEE Pervasive Computing, 17*, 65–75. https://doi.org/10.1109/mprv.2018.2873849

Plant Species Detection Using Image Processing and Deep Learning: A Mobile-Based Application

Eleni Mangina, Elizabeth Burke, Ronan Matson, Rossa O'Briain, Joe M. Caffrey, and Mohammad Saffari

1 Introduction

Plant identification is a task that extends far beyond the traditional role of botanists and ecologists [1]. It is relevant to a much larger portion of society, from livelihoods such as horticulturalists, farmers, and biologists to nature lovers and eco-tourists alike. However, current methods of plant identification can be both time-consuming for those with limited expert knowledge, and completely inaccessible to those without it, due to the complex nomenclature common to botanical keys [2]. Recent advances in the fields of computer vision [3] and highly accurate pattern recognition algorithms have considerable potential in the field of automated plant species recognition [4] using image processing and deep learning. Accurate automatic plant identification will support ecological monitoring and as a result biodiversity conservation.

E. Mangina (✉)
School of Computer Science, University College Dublin, Dublin, Ireland

School of Mechanical and Materials Engineering, UCD Energy Institute, University College Dublin, Dublin, Ireland
e-mail: eleni.mangina@ucd.ie

E. Burke
School of Computer Science, University College Dublin, Dublin, Ireland

R. Matson · R. O'Briain
Inland Fisheries Ireland, Dublin, Ireland

J. M. Caffrey
INVAS Biosecurity Ltd., Dublin, Ireland

M. Saffari
School of Mechanical and Materials Engineering, UCD Energy Institute, University College Dublin, Dublin, Ireland

D. D. Bochtis et al. (eds.), *Information and Communication Technologies for Agriculture—Theme II: Data*, Springer Optimization and Its Applications 183,
https://doi.org/10.1007/978-3-030-84148-5_5

The ability to automate categorization and identification of these plant species via image processing would be highly valuable. Currently, manual classification by botanists is both time-consuming and costly. Plant species identification requires advanced botanical identification skills, an understanding of specialist terminology, and an ability to use complex identification keys reliably. Experts and amateurs alike could recognize the merit in an app that can classify these images accurately and in real time, making both the work of botanists more precise, and plant classification more accessible to non-botanists and the general public. Plant biodiversity and ecological function is directly linked to the health of the environment around it [5]. Adverse changes to an environment, such as pollution, chemical damage, or drought, may all be reflected in a general deterioration of plant health, reduced biodiversity, and changes in abundance. By tracking both the quality and quantity of flora, in a given ecosystem or habitat, valuable insights can be harnessed, and actions taken to investigate/treat any issues within the environment. The ability to monitor both native and introduced species within a system would be beneficial to ecologists.

Therefore, creating a system with which plant species can be accurately recognized by Mobile Imagery utilizing Deep Learning models [6] would be valuable for three main reasons: plant species identification, plant health and condition, and general ecological monitoring. The research described in this chapter focuses on the process which involves selecting and testing a number of deep learning models, training them with the image dataset, and lastly documenting the speed and robustness of species identification using the selected deep learning models. Regarding the more advanced tasks of this study, the initial focus is on the development of a classification algorithm based on the given image dataset and the calculation of the accuracy. The subsequent task involves creating an online system to allow for documentation to be uploaded and classified.

2 Background Research

There are many different approaches to automatic plant identification and due to the different industries involved, a diverse range of strategies have been adopted. In a recent research carried out by Lee et al. [7], the methods used for early detection of plant diseases using sophisticated automatic image recognition systems based on deep learning were studied. These methods are very important for controlling plant diseases and ensuring a sustainable and secured food and agriculture sector. However, despite the diverse strategies applied, within the overall picture for the most accurate identification scores, certain trends emerge which will be discussed in the following sections.

2.1 Deep Learning

Deep learning is considered as a particular type of machine learning. The term machine learning itself refers to the automated detection of meaningful patterns in data [8]. The main principle of machine learning (ML) is the use of learning algorithms. An algorithm in ML can learn from data [9].

There are many types of machine learning models that can perform classification tasks. In recent years, deep learning Convolutional Neural Networks (CNNs) have seen a significant breakthrough in computer vision due to the availability of efficient and massively parallel computing on graphics processing units (GPUs) and the availability of large-scale image data necessary for training deep CNNs with millions of parameters [10].

One of the main trends that have been found is that the vast majority of studies were undertaken on plant identification, all focused specifically on leaf-based analysis. For instance, the study of Barre et al. [11] aimed to develop a deep learning system (LeafNet) to learn discriminative features from leaf images along with a classifier for species identification of plants. By comparing results with customized systems like LeafSnap [12], it was intended to show that learning the features by a CNN can provide better feature representation for leaf images compared to hand-crafted features. Their results reveal a better performance of LeafNet compared to hand-crafted customized systems, when evaluating the recognition accuracy of LeafNet (a CNN-based plant identification system), on the LeafSnap [9], Flavia, and Foliage datasets.

In another study authored by Larese et al. [13], an interesting re-prioritization of features for segmenting and classifying scanned legume leaves based only on the analysis of their veins was proposed (leaf shape, size, texture, and color are discarded). This study achieved 87% accuracy with the PDA classifier for scanned leaves of soybean, red, and white beans. Promising results were presented, revealing the proposed approach to be an effective and economical alternative, which outperform the manual recognition by an expert.

Mohanty et al. [14], however, focused more on the condition of the leaf, rather than the actual leaf itself. This study used a public dataset of 54,306 images of diseased and healthy plant leaves collected under controlled conditions, before training a deep CNN to identify 14 crop species and 26 diseases (or absence thereof). The trained model achieved an accuracy of 99.35% on a held-out test set, demonstrating the feasibility of this approach.

Moving on from leaves, another very popular focus feature for plant identification is perhaps the one most similar to how humans identify plants – the flower. Seeland et al. [15] examine methods spanning from detection, extraction, fusion, pooling, to encoding of local features for quantifying shape and color information of flower images. Findings show large differences among the various studied techniques and that their wisely chosen orchestration allows for high accuracy in species classification. Color was also found to be an indispensable feature for high classification results, especially while preserving spatial correspondence to grey-level features.

Building on artificial neural networks has allowed deep approaches to have many more hidden layers in the network, and hence have greater discriminative and predictive power, as stated by Pound et al. [16]. This study demonstrates the use of such approaches as part of a plant phenotyping pipeline. It shows the success offered by such techniques when applied to the challenging problem of image-based plant phenotyping and demonstrates state-of-the-art results (>97% accuracy) for root and shoot feature identification and localization. It can fully automate trait identification using deep learning to identify quantitative trait loci in root architecture datasets. The study of Pound et al. [16] was particularly interesting as it singles out parts of the plant to look for specific features, i.e. root and shoot, which may be applicable for some of the classes in the dataset of the current study, where the initial analysis indicates there are multiple angles of specific plants, often revealing different subsets of plant anatomy/features.

In a recent study, a fusion method has been proposed by Zhang et al. [17] for recognition of large-scale plant species. A fixed plant taxonomy has been developed to automatically define the inter-species relations in the plant world. In this study, an attention-based deep hierarchical multi-task learning algorithm is proposed to recognize fine-grained plant species belonging to the same task group by learning more discriminative deep features and classifiers mutually. An outstanding performance has been reported on recognizing 1001 fine-grained plant species. In addition, the experimental results on Orchid 2608 dataset and Plantae subset of iNaturalist2018 [18] also justified the capability and strength of their proposed methodology [17].

In addition to plant species classification research, there have been many research studies published on data mining and machine learning in recent years [19–21]. One of the most technologically advanced papers was researched by Sandino et al. [22]. This paper presents a pipeline process to detect and generate a pixel-wise segmentation of invasive grasses, using buffel grass (*Cenchrus ciliaris*) and spinifex (*Triodia* sp.) as examples. The process integrates unmanned aerial vehicles (UAVs) also commonly known as drones, high-resolution red, green, blue color model (RGB) cameras, and a data processing approach based on machine learning algorithms. In total, 342,626 samples were extracted from the obtained dataset and labeled into six classes. Segmentation results provided an individual detection rate of 97% for buffel grass and 96% for spinifex, with a global multiclass pixel-wise detection rate of 97%. Obtained results were robust against illumination changes, object rotation, occlusion, background cluttering, and floral density variation.

The research performed by Sun et al. [23] focuses on the work of plant species' identification in their natural environment using deep learning models, which is very similar to the current study. The model used in this research was the Resnet model as shown in Fig. 1.

Based on the conclusions from the background research study, CNNs were found to overwhelmingly be the best performing model for plant species identification [7]. Given the challenging nature of plant species image classification, which is the main scope of this study, CNN is the most suitable model to use. CNNs are a specialized kind of neural network for processing data that has a known grid-like topology [24]. For example, in this study, the topology data is the available plant

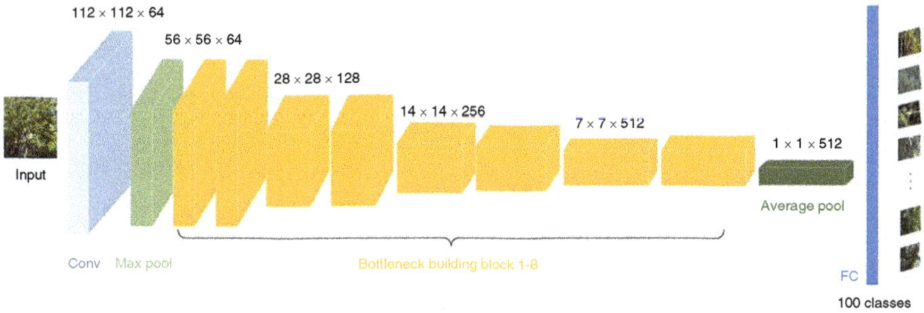

Fig. 1 Architecture of Resnet model [23]

image data, which can be thought of as a 2-D grid of pixels. The name "convolutional neural network" indicates that the network employs a linear mathematical operation called convolution. Convolutional networks are simply neural networks that use convolution in place of general matrix multiplication in at least one of their layers [24]. The key characteristic of CNNs is their multi-layer architecture. This involves the inclusion of several convolutional and subsampling layers, potentially including an additional fully connected layer.

3 Methodology

3.1 Dataset and Data Preparation

The floral image dataset used in this analysis had approximately 28,000 RGB images (over 112 GB) and was categorized into 380 species by scientific name. The data presents flora captured from an average of four different viewpoints, as well as varying backgrounds, proximities, and quantities. Each plant species also has several different parts of the plant photographed (flower, leaf, stem, seed). Among the 380 species photographed, there are sometimes more than one species within the same genus. Many of our identified species in the dataset have been broken down into genus or species. A unique characteristic of this very rich dataset is the provision of over 4000 images of unidentified species, ranging across 150 different categories. The categorized data was used to train a deep learning identification model that was developed. The model was used to classify images fed into it and return the species in either real time or near real time.

Originally, it was expected that the dataset would also contain GPS (Global Positioning System) metadata about where each plant image was captured. A significant portion of images were collected before GPS metadata was standard on regular digital cameras. The equipment used to capture the data images at that time did not have the capabilities of recording GPS data. This reduces the scope of the

Fig. 2 Example images from the dataset: (1) Mentha aquatica, (2) Hypericum pulchrum, (3) Achillea millefolium, (4) Aponogeton distachyos, (5) Dactylorhiza fuchsii, (6) Angelica sylvestris, (7) Bellis perennis, (8) Lythrum salicaria, (9) Mimulus moschatus, (10) Crocosmia x crocosmiflora, (11) Butomus umbellatus, (12) Barbarea vulgaris, (13) Allium triquetrum, (14) Caltha palustris, (15) Scrophularia auriculata, (16) Menyanthes trifoliata, (17) Arum maculatum, (18) Hypericum androsaemum, (19) Eupatorium cannibinum, (20) Gunnera tinctoria, (21) Digitalis purpurea, (22) Lysimachia vulgaris, (23) Ranunculus peltatus-penicillatus, (24) Epilobium hirsutum, (25) Iris pseudacorus, (26) Impatiens glandulifera, (27) Geranium robertianum, (28) Ficaria verna, (29) Oenanthe crocata, (30) Prunella vulgaris, (31) Mimulus guttatus, (32) Myosotis scorpioides, (33) Persicaria maculosa, (34) Nymphaea alba, (35) Rumex hydrolapathum, (36) Senecio jacobaea x aquaticus, (37) Scrophularia nodosa, (38) Heracleum mantegazzianum, (39) Lotus corniculatus, (40) Heracleum sphondylium, (41) Hypericum tetrapterum, (42) Solanum dulcamara

study, eliminating the proposed feature of species location mapping. Examples of dataset images are shown in Fig. 2.

Figure 2 shows the reference to the species and their names associated with a number. Any data classes that had either the species or genus missing from the class label were removed manually. To limit the imbalanced data classes and maintain a minimum level of image samples for each class when using the images for our model training data, bash Scripts were written to automatically classify every plant species class. This reduced the dataset to 312 plant species classes.

The data within each plant species class was unstructured (i.e., pictures of individual parts of the plant mixed in together). To create a combined classifier, the contents of each plant species were sorted into subcategories: flower, leaf, stem, and seed. This was a manual task, sorting through all pictures in each class and distributing the image of each plant component into the relevant sub-category. Then, the bash script was re-run to remove sub-folders with less than 15 images. This further reduced our dataset to 220 plant species.

Background Removal

The next task was to perform some image pre-processing on the data. Initially, the goal had been to create a combined classifier, with a specific classifier each for the flowers, leaves, stems, and seeds subclasses. A major feature of all previous research conducted in this field is the removal of plant image backgrounds. According to the work of Sun et al. [23], it is essential to remove the background of the training data to improve learning process and relevant features.

PlantCV [25] is an open-source image analysis software package designed for plant phenotyping and was used to remove plant image background. It is a collection of modular Python functions, which are reusable units of Python code with defined inputs and outputs. PlantCV [25] functions can be assembled into simple sequential or branching/merging pipelines. A pipeline can be as long or as short as it needs to be, allowing for maximum flexibility for users using different imaging systems and analyzing features of seed, shoot, root, or other plant systems [26].

PlantCV [25] contains a VIS Image Pipeline (Visible Infrared) for background removal of RGB images. The VIS Image Pipeline can process images regardless of what type of VIS camera was used (digital camera, mobile phone camera). A key constraint of this pipeline requires plant material to be a distinctly different color from the image background. This meant the PlantCV software [25] was effective for the flower data, but not for any other of the plant components. Background removal tests and configurations were tested for numerous data removal pipelines and were trialed with the stem, leaf, and seed image data subclasses, but performance was poor, so the focus was given mainly on flower data.

Upon further manual quality control checks on the flower data available, this was further filtered down from 91 to 31 classes. The issues with the unusable data ranged across blurred images, extreme lighting, and poorly framed images. Choosing to focus exclusively on these 31 high-quality species classes of flower data, there were several steps in the pipeline of the background removal (Fig. 3).

Step 1 was to crop the original image as shown in Fig. 4. This enabled retention of only the area of interest (the flower itself), being centralized in the picture and reducing the dimensions of the unnecessary data in the wider background as can be seen in Fig. 4.

Step 2 in this specific pipeline was to perform pre-masking of the background. The goal is to remove as much background as possible without losing any plant information. In order to perform a binary threshold on an image you need to select one of the color channels H, S, V, L, A, B, R, or G. Here we converted the RGB image to HSV color space then extracted the '*S*' or saturation channel, but any channel can be selected based on user need (as shown in Fig. 5). If some part of the plant is missed or not visible, then thresholded channels may be combined [27].

Fig. 3 Filter steps in VIS PlantCV pipeline

Fig. 4 Original image (left) – cropped image (right)

Fig. 5 HSV filter image

Step 3 was thresholding the saturation channel. Thresholding involves the creation of a binary image from a grey image, based on the specified threshold values. The threshold can be selected to target either the light or dark objects in the image. In this study, the threshold selected was *'light'*, as the dim green backgrounds are always darker than the actual object target (flower) color in this dataset. An intermediary step between Step 3 and Step 4 was to apply a median blur filter to reduce noise in the image. It operates by applying a median value to the central pixel within a kernel size. It is important to minimize the use of median blur-type steps, as they can cause plant material loss if the blur is too intense. The effect is so subtle that it is difficult for the human eye to distinguish, as seen in Fig. 6.

Step 4 returns to the original unfiltered cropped image, creating a new branch within the pipeline where a new filter was applied to convert the image from the RGB to the LAB color space, in which the *'B'* or blue-yellow channel is chosen for extraction, as seen in Fig. 7.

Fig. 6 Median blur filter

Fig. 7 LAB filter image

Step 5 involved a repeat of Step 3, thresholding the saturation channel, this time on the LAB filtered image. Again, the threshold option selected was '*light*'.

Step 6 was a combination step, in which the two separate filtered branches in the pipeline, Step 3 (HSV) and Step 5 (LAB), are merged by combining the resulting saturation thresholded images for both. To achieve this, the two images are joined using the bitwise '*OR*' operator. The only constraint to performing this operation is that the two image sizes must be identical.

Step 7 was to apply the joined threshold binary image (Step 6) as a mask over the original image (Step 1) (see Fig. 8). This mask functions as a separator, excluding the maximum amount of background without omitting any of the target object plant material.

This pipeline process was repeated for each of our 31 plant classes. Minor variable parameter adjustments were needed for each individual class, in the form or selecting different color channels to extract (H, S, or V for the HSV filter and L, A, or B for the LAB filter) depending on the actual color of the flower.

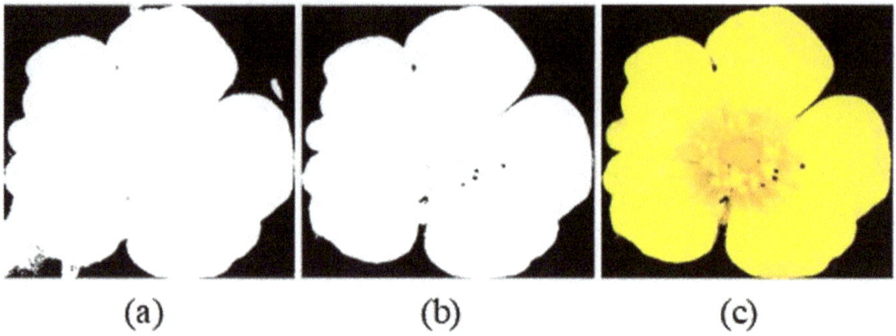

Fig. 8 (a) Saturation thresholded image, (b) joined threshold image, (c) final image with joined threshold mask applied

Data Augmentation

The next issue that had to be addressed in the pre-processing phase of this study was the number of pictures in each class. Each of the 31 species of flower contained between 15 and 20 pictures per class.

To improve overall model accuracy, by increasing the number of pictures in each class, data augmentation was performed. Data augmentation, in machine learning terms, involves predominantly data warping, which is an approach which seeks to directly augment the input data to the model in data space. A very generic and accepted current practice for augmenting image data is to perform geometric and color augmentations, such as reflecting the image, cropping, and translating the image, and changing the color palette of the image [28].

The augmentation approach used in this research was achieved using an Open-Source Computer Vision library called OpenCV [23]. The OpenCV library [29] contains over 500 functions, one of which is the opencv-createsamples library. This library was used to create two different variations of augmented datasets, both outlined below.

These two versions of data augmentation were undertaken in order to test the speed and accuracy of training the model with random image backgrounds (Version one of augmentation), which would be consistent with our desired project use-case of photographing a plant in its natural environment and instantly being provided a classification prediction for that plant class. This is compared with the speed and accuracy of training a model with all backgrounds removed/black backgrounds (Version two of augmentation), which would involve our use-case now including the pre-processing of the user-input image to remove the background before classifying and returning them a result. The performance results of training using both these augmentation techniques are outlined and compared in later sections, during the model evaluations.

The opencv-createsamples library [29] operates on the premise of '*positive*' and '*negative*' image datasets. We consider a positive image (Fig. 8c) to be a picture containing the target object, which in this case is a flower. For this study, the positive

Fig. 9 Examples of 'negative' images

Fig. 10 Examples of version one augmented

image dataset is the data we currently already have (i.e., the 31 classes of plants with backgrounds removed).

A negative image (Fig. 9), in the context of this library, is an image that does not contain the target object (i.e., an image not containing the flower). For the first version data augmentation in this study, the first step was creating a supplementary dataset of 600 simple abstract pictures (none of which contained any of our existing classes of flower). The size of 600 was selected in order to minimize repetition of backgrounds used and allow a large enough pool for random selection from the dataset to be viable.

The opencv-createsamples library [29] operates by individually taking one of the 15–20 positive pictures in a particular plant class, and performing transformations on the actual flower object, before superimposing it onto a negative or '*random*' image

Fig. 11 Examples of version two augmented

(Fig. 10). This process of transforming and superimposing onto a negative background image is then repeated for a user-specified number of times for that particular positive image. This is then repeated for every image, in every plant class. For the purpose of minimizing bias, the classes were balanced by tailoring the number of augmented images needed for each particular class, based on creating the necessary amount to supplement the data in each class to a total of 400 positive images per class.

The second version of data augmentation tested for this study involved the same process as above, but with key difference of the '*negative*' image dataset consisting exclusively of plain black images instead of random backgrounds (Fig. 11). Both versions of the augmented dataset were tested and evaluated when training predictive models in a later section of this chapter.

4 Software Development and Analysis

The software development for this study can be broken down into two individual software components. The first is the software to build and train a deep learning CNN classifier using the training data pre-processed earlier discussed. The second component is an online classification system that will enable the user to interact with the actual classifier.

The first component developed, and the core priority task for the study overall, was the machine learning algorithm. Software was written to handle the entire pipeline process: loading the training data images and applying machine learning techniques, before creating a final machine learning algorithm that is highly accurate. As mentioned previously, a CNN classifier was chosen as the model for development. To briefly focus more on that architecture, and actual software deliverable extracted from that process, see the diagram in Fig. 12.

The actual software that was developed using libraries from a deep learning framework was simply a means of training and testing the model. The data was split at the start of the process: 70% training data and 30% testing data. This will enable accurate evaluations of '*unseen*' data (i.e., our use-case). Code was written to load in all the training data and labels, split it, automatically extract feature vectors, before being passed to the CNN model layers, and trained to create a machine learning algorithm that can classify plant images. Evaluation code was written to

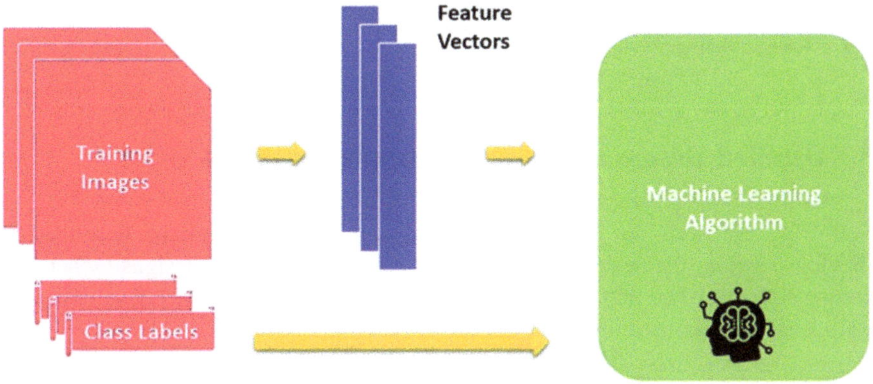

Fig. 12 CNN process software design

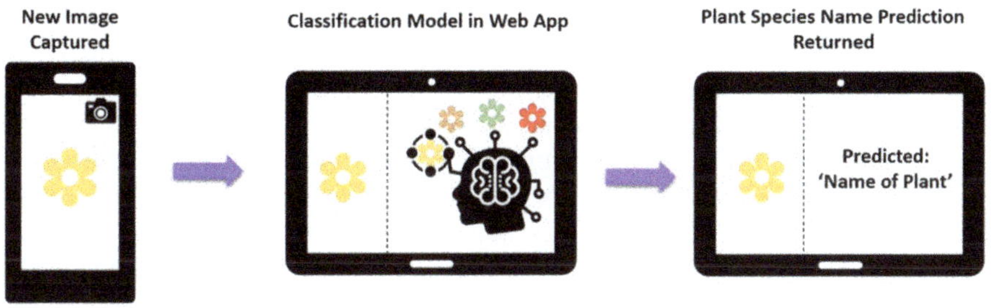

Fig. 13 Software system core use case

test the classification accuracy of the algorithms. Once a high accuracy of classification performance was achieved, the trained machine learning algorithm is exported as a plant classification model with stored feature weights.

The training engine refers to the software developed using the Tensorflow [30] and Keras [31] libraries in order to create CNN models. Each variation of model generated was evaluated based on classification accuracy scores. The most accurately trained model was then exported in a *.h* file format (standard format for multidimensional arrays of scientific data) and comprised the classified model, which formed the back end to the second software component, the "Online Identification System" for the work described in this chapter.

Figure 13 shows a high-level overview of the software system use case in this study. A new image is captured, it is uploaded to the classification model, and the classification model then returns its prediction for that new image. That is the core functionality of classification this study had to achieve and in the simplest terms possible, that was our target deliverable.

The current state-of-the-art approach to supervised machine learning for image plant species identification, based on the concluding findings of the research, is summarized in Wäldchen et al. [2], and the supervised CNN model architecture is outlined above. That model and the online classification web app are both key deliverables in this overall software system.

5 Detailed Design and Software Implementation

Following on from the high-level software architecture diagrams described in previous section, the next task was to start implementing these designs by actually coding software. The first phase of implementation undertaken was developing the deep learning classification model.

5.1 Developing Convolutional Neural Network

There are many different frameworks widely available for deep learning. The chosen framework used for this study is TensorFlow [30], which is a machine learning system that operates at large scale and in a heterogeneous environment. This tool uses dataflow graphs to represent computation, shared state, and the operations that mutate that state. It maps the nodes of a dataflow graph across many machines in a cluster, and within a machine across multiple computational devices, including multicore CPUs, general-purpose GPUs, and custom-designed ASICs known as Tensor Processing Units (TPUs). It enables experimentation with novel optimizations and training algorithms. TensorFlow supports a variety of applications, with a focus on training and inference on deep neural networks.

The first step in this implementation is to create a simple CNN model and compare how the model performs on both versions of data, before choosing to move forward with the most suitable version of data for this study. The model begins by importing the TensorFlow and related libraries. The next step is the load data function; this function reads all the image data and their corresponding species-name as images and label arrays. These arrays are then divided into training-data and test-data sub-arrays, which are loaded into the model. The model itself is a very simple iteration, consisting of just one fully connected layer, an accuracy and loss metrics calculator, and a training optimizer. The model will train using the training-data (images, labels) arrays and validate using the test-data arrays. Both versions of augmented data (random background and black background) were used to train models with this architecture. A visual summary of their performance results is illustrated in Figs. 20 and 21 in the following section.

While the model trained with version two of augmented data (black background) has a higher accuracy score at training time, and it is more accurate at identifying new images as long as the background is also removed, performance greatly deteriorates when this model is used to identify new plant images in their natural background (the normal study use-case of a plant image captured in its natural habitat with green foliage in the background). In contrast, the model trained with

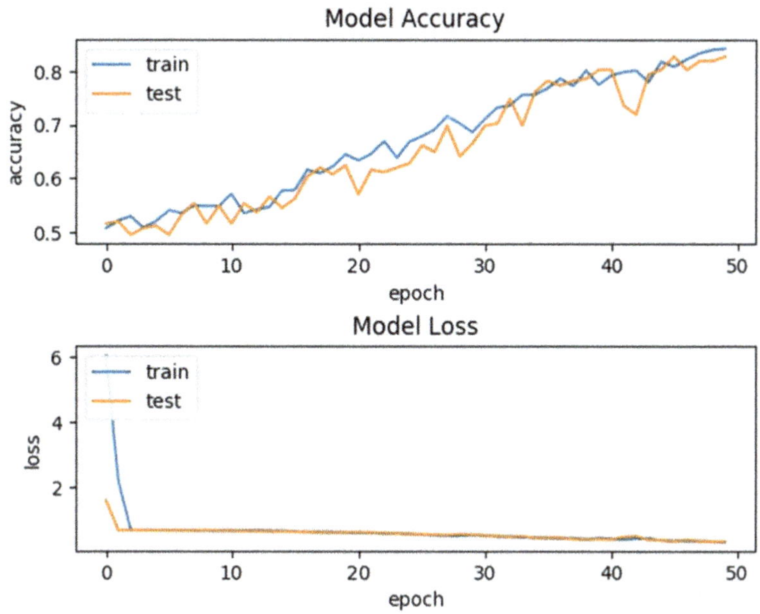

Fig. 20 Accuracy of data augmentation: version one – random background

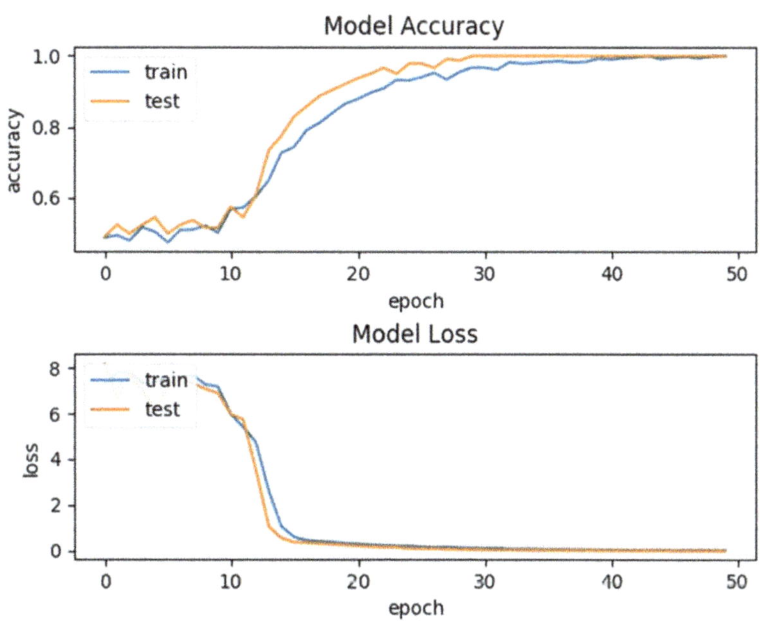

Fig. 21 Accuracy of data augmentation: version two – black background

version one of augmented data (random picture backgrounds), while having lower accuracy at training time, performed significantly better on plant images in a natural background. Its identification accuracy of new unseen images averages approximately the same as its accuracy scores on the random background test data.

Therefore, an important design decision had to be made. Option one was prioritizing the higher classification accuracy by adding image processing pipeline features to the online identification system, where each new user image would have the background removal process applied prior to classification by the model trained with black backgrounds (version two of data augmentation). Option two was prioritizing user experience, specifically the speed of instant classification by the model trained with version one of the data augmented with random backgrounds (version one of data augmentation) without needing to spend time pre-processing and removing the background before every classification prediction is returned. Both design options were considered equally. After much strategic analysis, Option two was decided as the most feasible implementation.

As outlined in the Data Pre-processing section (Sect. 3.1), particularly when describing the use of the PlantCV [27] VIS background removal pipeline, the big obstacle here was the need to customize the pipeline variables to various configurations depending on the color and lighting of that particular plant. This need, to manually adjust the pipeline variables depending on the flora colors, meant building a highly complex background removal pipeline, with several different deployment configurations specific to every color of the plant. However, that type of substantial pre-processing pipeline was beyond the scope of this study. Therefore, the more feasible Option two was pursued, using the data augmented with random background images to train all remaining models evaluated in this study.

The next phase was to increase both the number of plant classes and the overall classification accuracy. This was achieved by adding more layers and creating a deeper model. To move incrementally and preserve a scientific process, layers were added individually. After adding 16 plant classes and another layer to the TensorFlow framework [30], it became stagnant before dropping accuracy. With 16 classes, it was no longer performing above 10% accuracy in classification. Keras [31] was then introduced, which is a high-level neural networks API (Application Programming Interface), written in Python and capable of running on top of TensorFlow [30]. See Table 1 for a comprehensive general structure outline of all models built and tested for this study.

Iterating through deeper Keras models in the format structure outlined in the table above, eventually the final and most successful model is a 5-layer Keras CNN (Model-11) with detailed architecture shown in Fig. 14. The input layer takes a 64×64-pixel flora image from the pre-processed plant species dataset. The first layer has a hidden convolutional and MaxPooling operation. This layer has a filter size of 40. Each filter runs across the input image and creates a 2-D feature activation map. This activation map takes the form of a binary vector. The MaxPooling operation is then performed to reduce the maximum output within a rectangular neighborhood. The common parameters are pooled in order to eliminate noise in the image and focus on key feature identification, which helps to combat overfitting.

Table 1 Summary of deep learning model structures implemented

Model	Framework	Structure	Data augmentation Version	Number of classes
1	TensorFlow	1 FC layer	Version 1	2
2	TensorFlow	1 FC layer	Version 2	2
3	TensorFlow	1 FC layer	Version 1	10
4	TensorFlow	1 FC layer	Version 1	16
5	Keras	2 conv layers, 2 FC layers	Version 1	2
6	Keras	2 conv layers, 2 FC layers	Version 1	5
7	Keras	2 conv layers, 2 FC layers	Version 1	15
8	Keras	3 conv layers, 2 FC layers	Version 1	2
9	Keras	3 conv layers, 2 FC layers	Version 1	5
10	Keras	3 conv layers, 2 FC layers	Version 1	13
11	Keras	3 conv layers, 2 FC layers	Version 1	31

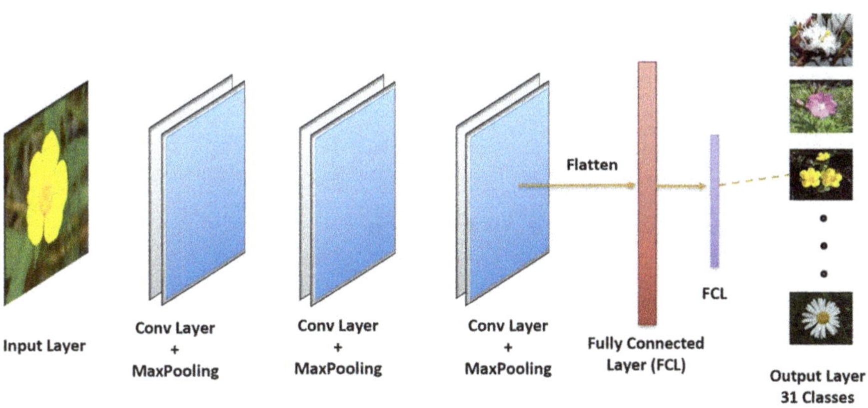

Fig. 14 Keras 5-layer CNN architecture for final model

This output is then fed forward sequentially to the next layer. The second layer is also a convolutional and MaxPooling operations layer with increased filter size of 80 in order to improve accuracy. After MaxPooling application, the output is forwarded to the next layer and so on with the third layer with filter size increased to 164. The output of that layer is then fed to an intermediary layer, called a flatten layer. This flattening operation takes the output dimensions from the convolutional layers and reshapes to match the size of the original input layer dimensions, 64 × 64. This is done in order to connect it to the next layer – a Fully Connected Layer (FCL).

FCL connects every neuron in one layer, to every layer in another. While the layer four FCL is size 64 × 64 in correspondence with the input layer, it is passed onto our fifth and final layer. The size of our second FCL is 31, each node corresponding to one of our 31 plant species classes in the dataset. This final FCL is then connected to our concluding output layer, which returns the predicted classification label for the input image.

A more in-depth and exhaustive outline of the model structure and specific layer configurations is outlined below in Fig. 15. As mentioned already, we start with our image input size 64 × 64 pixels. The first layer is Convolutional layer one that takes the 64 × 64 input and passes it through a filter size of 40. This output is then passed onto MaxPooling that reduces the size to 13 × 13 pixels. Convolutional layer two takes an input size of 13 × 13 and applies a filter of size 80. The next MaxPooling layer takes that input of 13 × 13 and reduces the output size to 3 × 3. Convolutional layer three takes that input and applies a filter of 164. The third MaxPooling operation takes the input of 3 × 3 and reduces it to an output size 1 × 1. This is passed to a dropout layer, in which randomly selected neurons are ignored in training in order to prevent overfitting. This operation does not affect the output size. The flatten layer takes the output dimensions 1 × 1 from the previous layer and reshapes it to match the size of the original input layer dimensions 64 × 64. Next, a fully connected layer (dense layer) takes input 64 × 64 and connects to the next layer. An additional dropout layer is added next for further protection against overfitting. Finally, every neuron from FCL one is connected to every neuron in FCL two. In line with having 31 plant classes in our model, fully connected layer two has 31 neurons corresponding to a flora species, which form the predicted classes in the output layer.

5.2 *Online Classification System App*

The next step, once the model has been trained on the above architecture, is to save it and export it in a *.h* file format. This leads to the concluding step in this overall implementation – creating an online system whereby plant images can be uploaded and classified by a user. To limit the scope of this feature and prioritize fast prototyping and delivering an effective and functional final project, a web application was developed in favor of a mobile application. The architecture of the App can be found in Fig. 16.

The web app was developed using Flask [32], which is a micro web framework written in Python. It is highly compatible with other frameworks and allows for the easy importing of the Keras model [31]. For the design styling of the app, a Bootstrap template [33] was used to give a professional look. Bootstrap is an open-source toolkit for developing with HTML, CSS, and JS.

The online classification system demonstrated in Fig. 17 works as follows: at start time the script (*App.py*) initiates a CNN object to which the Keras model is loaded into. When the user submits the image to be classified (Step 1) it passes that image

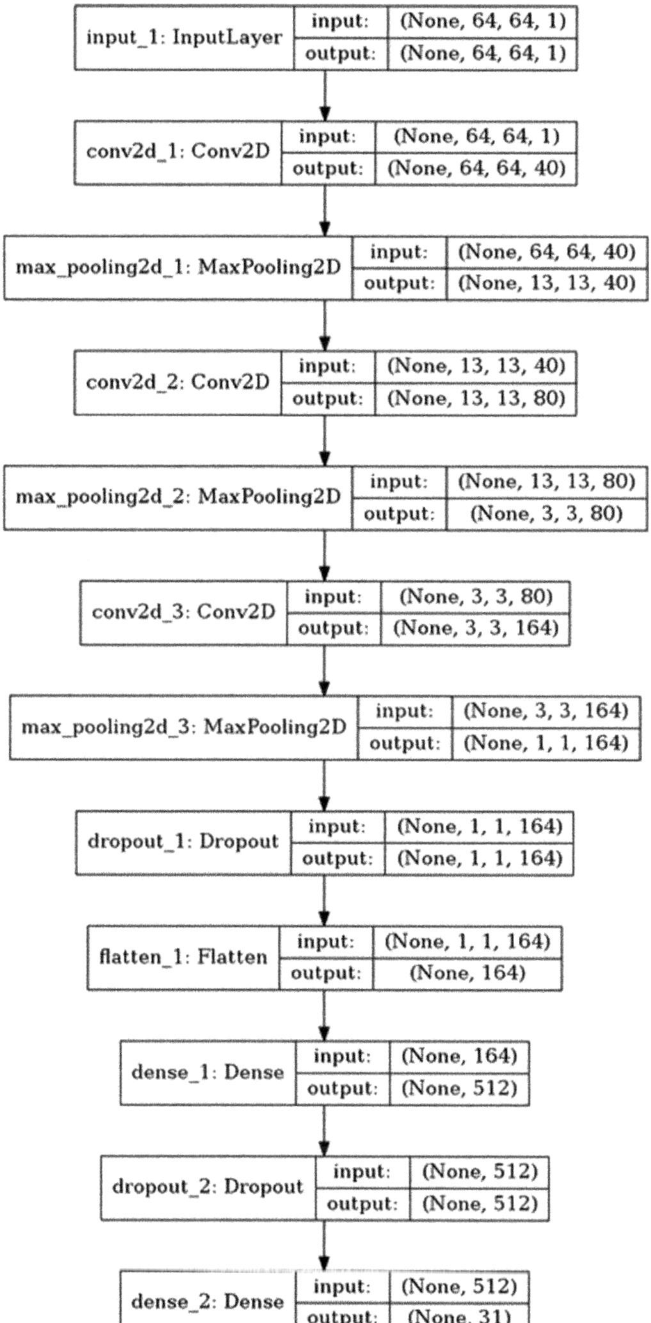

Fig. 15 Final Keras 5-Layer CNN technical model structure specification

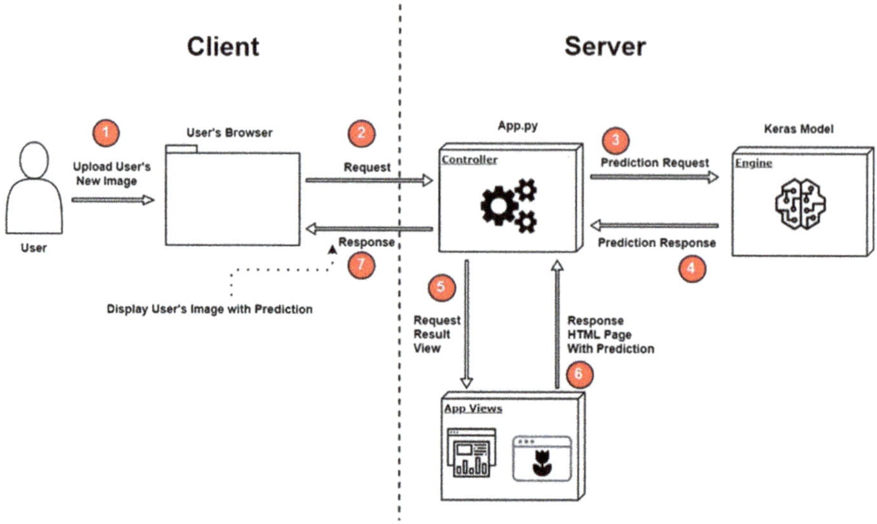

Fig. 16 Online classification system implementation architecture

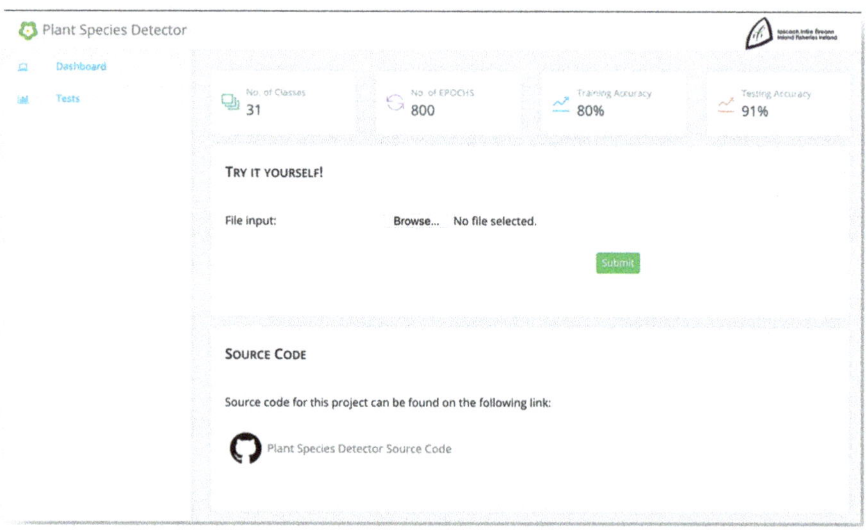

Fig. 17 Screenshot of web application homepage dashboard

through the app controller (Step 2) to the model's prediction function (Step 3). The Keras model returns the prediction response (Step 4) to the app controller. The app then requests the '*Results View*' from the app views (Step 5), which returns a response of an HTML page with the plant species name prediction (Step 6). This is then returned through the app in the browser, which displays the user image and

Fig. 18 Examples of images classified by the online classification system Web App

the predicted plant class (Step 7). Finally, it clears the prediction function of the model, ready for reuse.

The App user experience was designed to be both quick and easy to use. To upload a new image to classify, the user clicks the '*Browse...*' button in the center of the home screen. That opens the user's device file system, allowing them to select an image. Once the selected image is loaded to the web page, the user can click '*Submit*' and the image will be instantly classified. The app architecture itself consists of two pages: a home page and a test page. The home page is shown in Fig. 17, outlining the actual dashboard for interacting with the model. The test page contains the model performance accuracy graphs, as shown later in this section.

The plant species which has the highest probability score is returned as the predicted class. The user's selected image is displayed on the screen, as well as the predicted plant class label and model's percentage confidence in the prediction (Fig. 18). However, should the predicted accuracy be below the reasonable threshold of 60% at which we deem prediction confidence is not high enough, the webpage is programmed to inform you that the species captured in the uploaded image does not yet exist in the database. To classify a new image, one should simply re-click '*Browse*' and repeat the steps.

6 Testing and Evaluation

The main method of evaluation in this study is the classification accuracy of the models implemented. An exhaustive overview of the entire collection of models evaluated during the implementation of this study is summarized in Table 1 in the previous section.

The first element of implementation to evaluate is the type of augmented data used. As was detailed earlier in Sect. 3.1, there were two versions of augmented data used. The first had random backgrounds and the other had plain black backgrounds (Fig. 19).

Both versions of augmented data were tested separately on a single-layer TensorFlow [24] CNN. A visual display of the results is demonstrated in Figs. 20 and 21.

Fig. 19 Data augmentation: version 1-random background (left), version 2-black background (right)

Table 2 Summary of all model accuracy evaluations

Model	Framework	Structure	Data augmentation version	Number of classes	Epochs	Testing accuracy at training
1	TensorFlow	1 FC layer	Version 1	2	50	93%
2	TensorFlow	1 FC layer	Version 2	2	50	100%
3	TensorFlow	1 FC layer	Version 1	10	120	65%
4	TensorFlow	1 FC layer	Version 1	16	300	<10%
5	Keras	2 conv layers, 2 FC layers	Version 1	2	70	86%
6	Keras	2 conv layers, 2 FC layers	Version 1	5	90	80%
7	Keras	2 conv layers, 2 FC layers	Version 1	15	180	77%
8	Keras	3 conv layers, 2 FC layers	Version 1	2	175	99%
9	Keras	3 conv layers, 2 FC layers	Version 1	5	175	96%
10	Keras	3 conv layers, 2 FC layers	Version 1	13	220	93%
11	Keras	3 conv layers, 2 FC layers	Version 1	31	1500	91%

Within the research described in this chapter, the model trained with the black background has 100% accuracy score at training time and higher accuracy identifying new images with background removed, although the accuracy drops to <65% when the background is included.

In contrast, the model (Table 2, *Model* – 1) trained with version one of augmented data (random picture backgrounds), while having lower accuracy (93%) at training time, performed significantly better on plant images in a natural background. Its identification accuracy of new unseen images averages approximately the same as its accuracy scores on the random background test data (>85%).

Based on the justification provided earlier in the section on implementation (Sect. 5), version one (data augmented with random backgrounds) was then used to train and test all subsequent models in this study. The strategy adopted in building these classification models was an iterative process.

Starting with just a single fully connected layer model, it was trained on 10 classes of plant species (Table 2, *Model* − 3), with an increase in epochs to help improve accuracy (65%). The single FC TensorFlow [30] CNN was exhausted with a maximum of 16 plant classes (Table 2, *Model* − 4) before accuracy plummeted (to less than 10%).

Next, again as outlined in the previous section, Keras [31] was chosen as the framework for building a deeper CNN. Starting with just two Convolutional layers and two Fully Connected layers, a model (Table 2, *Model* − 5) was trained on two plant classes, yielding 86% accuracy. While that was lower than our previous initial accuracy scores for *Model* − 1, as classes were added, performance reduced to 80% accuracy for five classes (Table 2, *Model* − 6) and 77% accuracy for 15 classes (Table 2, *Model* − 7). Increasing epochs no longer improved performance for these models, leading to the next step, adding another layer. A third convolution layer was added, meaning the model now comprised of five layers in total (three Conv layers, two FC layers). This new layer model architecture was tested again, initially on two plant classes (Table 2, *Model* − 8). Accuracy was found to be fantastic for this model architecture (99% accuracy), so the iterative process of adding more plant classes began. It started with a small increase, up to five plant species classes (Table 2, *Model* − 9), which yielded an accuracy of 96%. Continuing to add more classes, a model (Table 2, *Model* − 10) was trained on nine classes, and yielded an accuracy of 93%.

Finally, the most accurate model created in this study (Table 2, *Model* − 11) is a five-layer Keras CNN [31]. It contains three Conv layers and two FC layers. This model was trained on 12,400 images (400 images per flora class) to classify all 31 types of plant species in our dataset (Fig. 22), achieving an accuracy score of 91%. Experiments were run, adding additional layers, re-ordering the existing layer configurations, but accuracy did not improve. Various hyper-parameter tuning techniques such as varying the number of epochs ran, as well as testing a range of filter sizes for the Conv layers, ultimately led to no significant increase in accuracy.

Therefore, *Model* − 11, with five layers and 91% testing accuracy is the most accurate model for this study, and thus was the model used to create the online web app classification system. This implementation is subsequently delivering on the core objective of this study, as outlined in the specification 'Create a system whereby flora can be identified by mobile imagery with a high degree of accuracy using deep learning models'.

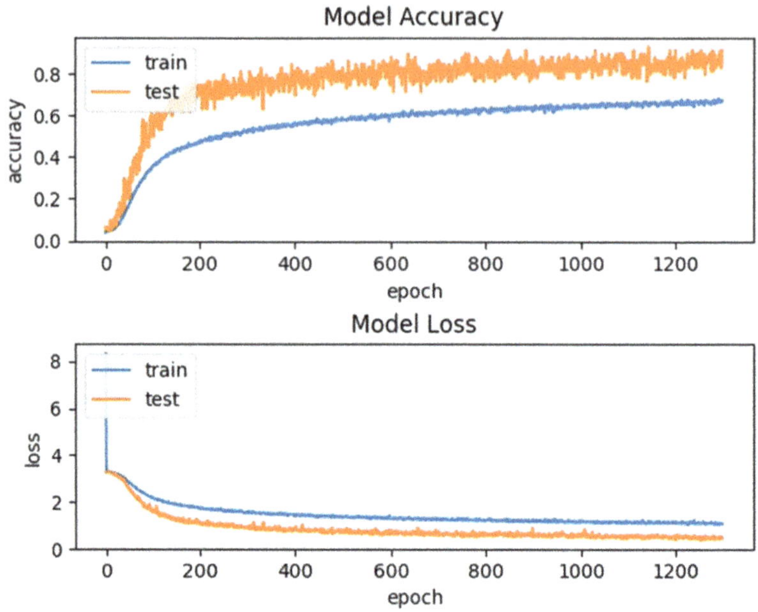

Fig. 22 Evaluation accuracy of five-layer Keras CNN with 31 classes

7 Discussion and Future Work

There are some immediate feature extensions that would advance the quality of this study. First and foremost, the most intuitive next step would be to add more flowering plant species for classification. While the model is capable of identifying the 31 classes of usable data, it would be beneficial to source additional quality training data and continue adding more species. Furthermore, while this study was focused solely on flowering plant identification due to the available dataset. It would be interesting and beneficial to further develop a classifier, that could identify other plant components (e.g., flower, leaf, stem) and calculate the predicted class based on overall plant anatomy together with a combination of other plant features. This would help make the classifier model more robust, especially when flowering structures are not present, most species only flowering for a couple of weeks to months at a time. Another immediate feature which could further enhance the spatial distribution of Irelands' plant communities and their habitats by mapping species occurrence, is by using GPS metadata. Building on the current study, delivery of a species GPS documentation and mapping function would be highly useful for monitoring plant populations throughout Ireland. Invaluable insights could be gained through GPS data associated with plant images for global map visualizations, for example, citizen science facilitated distribution mapping and monitoring the effectiveness of remediation efforts to combat non-native invasive species that can be both ecologically and economically damaging.

On a broader note, there are many exciting areas emerging in field technology adapted for plants surveying. While many current studies operate on relatively small and non-representative datasets such as the one used in this study, only a select few studies train CNN plant species classifiers on much larger floral image datasets. The most recent research trends are now gravitating toward applications of transfer learning. This would mean pre-training a classifier on a large dataset, such as ResNet, before actual training begins. The classifier will then only be fine-tuned to the specific plant species classification problem by the training of a small number of high-level network layers proportional to the amount of available problem-specific training data [26]. Considering the limited size of this dataset, investigating the application of transfer learning with the aim to improve accuracy would be an interesting development in the study.

It is noteworthy to mention that over a decade ago, Wäldchen et al. [2] proposed that successful plant species identification development would rely on interdisciplinary research collaborations between both biologists and computer scientists. The majority of automated plant species identification today is undertaken in academic research surrounding computer vision and machine learning. Research should move toward more interdisciplinary endeavors. Biologists can apply machine learning methods more effectively with the help of computer scientists, and the latter are able to gain the required exhaustive understanding of the problem they are tackling, by working with the former [2].

In terms of taking a closer look at the topic of plant phenotyping, all approaches for improving crops require measurement of morphological and life-history traits (phenotyping) [34]. Our methodology could be adopted for application to plant image classification, in identifying features for stress phenotyping that affects specific plants with a high economic impact for Ireland [35]. Machine learning approaches facilitate the ability to learn patterns from data, while automating decisions in terms of plant's stress without programming explicit rules. As an ultimate future vision of this study, drone-collected data could provide a valuable source of information for flora on earth with further prospects of exploitation for ecological monitoring and biodiversity conservation as well as monitoring the effects of climate change.

8 Conclusions

The core aim of this study was the creation of a system whereby plant species can be identified by imagery with a high degree of accuracy using deep learning models and artificial intelligence. The highest performing model was the Convolutional Neural Network (CNN) with five layers. This 91% accuracy satisfies the core objective of the study, to classify plant species images precisely using deep learning models. Eleven different Deep Learning models were trained and evaluated to document the speed and robustness of the floral identification using the selected deep learning models. The biggest unforeseen challenge in this study was the sheer amount of pre-processing and data cleaning needed to create a usable high-quality production dataset. For the advanced tasks, the initial focus is on the development of a classification algorithm based on the given image dataset and the calculation of the accuracy proved to be the biggest challenge in maintaining high accuracy scores when increasing the number of species to be identified. The more species added, the less accurate the model became. This issue was overcome through significant data pre-processing and adding additional layers to the network depth. An online web page classification system was also successfully implemented for easy use by botanists.

Acknowledgments This publication has been supported by Inland Fisheries Ireland (IFI) and INVAS Biosecurity Ltd. for providing most of the photography dataset.

References

1. Seeland, M., Rzanny, M., Boho, D., Wäldchen, J., & Mäder, P. (2019). Image-based classification of plant genus and family for trained and untrained plant species. *BMC Bioinformatics, 20*, 4. https://doi.org/10.1186/s12859-018-2474-x
2. Wäldchen, J., Rzanny, M., Seeland, M., & Mäder, P. (2018). Automated plant species identification-trends and future directions. *PLoS Computational Biology, 14*(4), e1005993. https://doi.org/10.1371/journal.pcbi.1005993
3. Weiss, Y., Ferrari, V., Sminchisescu, C., & Hebert, M. (2020). Special issue: Advances in architectures and theories for computer vision. *International Journal of Computer Vision*. https://doi.org/10.1007/s11263-019-01289-8
4. Wang, B., Gao, Y., Yuan, X., Xiong, S., & Feng, X. (2020). From species to cultivar: Soybean cultivar recognition using joint leaf image patterns by multiscale sliding chord matching. *Biosystems Engineering, 194*, 99–111. https://doi.org/10.1016/j.biosystemseng.2020.03.019
5. Balvanera, P., Pfisterer, A. B., Buchmann, N., et al. (2006). Quantifying the evidence for biodiversity effects on ecosystem functioning and services. *Ecology Letters, 9*(10), 1146–1156. https://doi.org/10.1111/j.1461-0248.2006.00963.x
6. Kamilaris, A., & Prenafeta-Boldú, F. X. (2018). Deep learning in agriculture: A survey. *Computers and Electronics in Agriculture, 147*, 70–90. https://doi.org/10.1016/j.compag.2018.02.016
7. Lee, S. H., Goëau, H., Bonnet, P., & Joly, A. (2020). New perspectives on plant disease characterization based on deep learning. *Computers and Electronics in Agriculture, 170*, 105220. https://doi.org/10.1016/j.compag.2020.105220
8. Watt J., Borhani R., Katsaggelos A. K. (2016), Machine learning refined: Foundations, algorithms, and applications, Cambridge University, ISBN 978–1–107-12352-6.
9. Witten, I. H., Frank, E. & Hall, M. A. . (2011) doi:https://doi.org/10.1016/B978-0-12-374856-0.00001-8.
10. Zhang, K., Zuo, W., Chen, Y., et al. (2017). Beyond a Gaussian denoiser: Residual learning of deep CNN for image denoising. *IEEE Transactions on Image Processing, 26*, 3142–3155. https://doi.org/10.1109/TIP.2017.2662206
11. Barré, P., Stöver, B. C., Müller, K. F., & Steinhage, V. (2017). LeafNet: A computer vision system for automatic plant species identification. *Ecological Informatics, 40*, 50–56. https://doi.org/10.1016/j.ecoinf.2017.05.005
12. lLeafSnap, (2020). Leafsnap: An Electronic Field Guide, viewed 22nd April 2020. http://leafsnap.com/.

13. Larese, M. G., Namías, R., Craviotto, R. M., et al. (2014). Automatic classification of legumes using leaf vein image features. *Pattern Recognition, 47*(2014), 158–168. https://doi.org/10.1016/j.patcog.2013.06.012
14. Mohanty, S. P., Hughes, D. P., & Salathé, M. (2016). Using deep learning for image-based plant disease detection. *Frontiers in Plant Science, 7*, 1419. https://doi.org/10.3389/fpls.2016.01419
15. Seeland, M., Rzanny, M., Alaqraa, N., et al. (2017). Plant species classification using flower images—A comparative study of local feature representations. *PLoS One, 12*, e0170629. https://doi.org/10.1371/journal.pone.0170629
16. Pound, M. P., Atkinson, J. A., Townsend, A. J., et al. (2017). Deep machine learning provides state-of-the-art performance in image-based plant phenotyping. *Gigascience, 6*, gix083. https://doi.org/10.1093/gigascience/gix083
17. Zhang, H., Kuang, Z., Peng, X., et al. (2020). Aggregating diverse deep attention networks for large-scale plant species identification. *Neurocomputing, 378*, 283–294. https://doi.org/10.1016/j.neucom.2019.10.077
18. Horn G Van, Aodha O Mac, Song Y, et al (2018) The iNaturalist species classification and detection dataset. In: Proceedings of the IEEE Computer Society Conference on Computer Vision and Pattern Recognition.
19. Li, R., Sun, Y., & Sun, Q. (2018). Automated flowering time prediction using data mining and machine learning. In *Lecture notes of the institute for computer sciences, social-informatics and telecommunications engineering, LNICST* (pp. 518–527). Springer Verlag.
20. Rahaman, M. M., Ahsan, M. A., & Chen, M. (2019). Data-mining techniques for image-based plant phenotypic traits identification and classification. *Scientific Reports, 9*, 19526. https://doi.org/10.1038/s41598-019-55609-6
21. Soltis, P. S., Nelson, G., Zare, A., & Meineke, E. K. (2020). Plants meet machines: Prospects in machine learning for plant biology. *Applications in Plant Sciences, 8*(6), e11371. https://doi.org/10.1002/aps3.11371
22. Sandino, J., Gonzalez, F., Mengersen, K., & Gaston, K. J. (2018). UAVs and machine learning revolutionising invasive grass and vegetation surveys in remote arid lands. *Sensors (Switzerland), 18*, 605. https://doi.org/10.3390/s18020605
23. Sun, Y., Liu, Y., Wang, G., & Zhang, H. (2017). Deep learning for plant identification in natural environment. *Computational Intelligence and Neuroscience*. https://doi.org/10.1155/2017/7361042
24. Goodfellow I, Bengio Y, Courville A (2016) Adaptive computation and machine learning, The MIT Press, ISBN-10 : 0262035618.
25. Danforth D. Plant Science Center (2020) PlantCV: Plant phenotyping using computer vision, viewed 24th of January 2020. https://plantcv.danforthcenter.org/.
26. Gehan, M. A., Fahlgren, N., Abbasi, A., et al. (2017). PlantCV v2: Image analysis software for high-throughput plant phenotyping. *PeerJ, 2017*. https://doi.org/10.7717/peerj.4088
27. Danforth D. Plant Science Center (2020), PlantCV Tutorial: VIS Image Workflow, viewed 24th of January 2020, https://plantcv.readthedocs.io/en/latest/vis_tutorial/.
28. Aggarwal st LP. (2019). Data augmentation in dermatology image recognition using machine learning. *Skin Research and Technology, 25*, 815–820. https://doi.org/10.1111/srt.12726
29. OpenCV Team (2020), OpenCV Open Source Computer Vision Library, viewed 21st of April 2020. https://opencv.org/about/.
30. Abadi, M., Agarwal, A., Paul Barham, E. B., et al. (2016). TensorFlow: Large-scale machine learning on heterogeneous systems. *Methods in Enzymology*. https://doi.org/10.1016/0076-6879(83)01039-3
31. Chollet F. (2020), Keras: The Python Deep Learning Library, viewed 21st of April 2020. https://keras.io/.

32. Matt M. (2020), Flask Python Web Framework, viewed 22nd April 2020, https://www.fullstackpython.com/flask.html.
33. Bootstrap Team MIT, 2020, Bootstrap, viewed 24th January 2020, https://getbootstrap.com/.
34. Fahlgren, N., Gehan, M. A., & Baxter, I. (2015). Lights, camera, action: High-throughput plant phenotyping is ready for a close-up. *Current Opinion in Plant Biology*. https://doi.org/10.1016/j.pbi.2015.02.006
35. Thuiller, W., Richardson, D. M., & Midgley, G. F. (2007). Will climate change promote alien plant invasions? In *Biological invasions*. Springer. https://doi.org/10.1007/978-3-540-36920-2_12

Computer Vision-based Detection and Tracking in the Olive Sorting Pipeline

George Georgiou, Petros Karvelis, and Christos Gogos

1 Introduction

Automation in agriculture is mainly associated with higher production volumes, less human involvement, and increased speed. Through the years several tasks that in the past required manual labor are shifted to machines. Recent advances in certain subfields of Artificial Intelligence, like computer vision, robotics, and machine learning, are gradually assimilated in the daily practice of agriculture.

The specific problem that is concerned is the automation of the olive sorting procedure. Olive sorting is a yearly repeated task that has to do with examining large amounts of olives and separating them based on skin color which is a strong indication of each olive's ripeness level.

This chapter examines the possibility of using computer vision-based techniques for the detection and tracking of olive fruit moving on a conveyor belt, a crucial component in the olive sorting pipe-line, which should be executed accurately, fast and on low-end hardware, in the aim of constructing a low budget sorting machine. Industrial sorters do exist, and some of them are briefly presented in the following subsection. Nevertheless, they represent a considerable financial investment that might be difficult or impossible to be undertaken by individual producers. So, a business opportunity might exist for small- or medium-sized companies, given the

G. Georgiou
University of Ioannina, Department of Informatics and Telecommunications, Arta, Greece

P. Karvelis · C. Gogos (✉)
University of Ioannina, Department of Informatics and Telecommunications, Arta, Greece

Laboratory of Knowledge and Intelligent Computing, Department of Informatics and Telecommunications, Arta, Greece
e-mail: cgogos@uoi.gr

D. D. Bochtis et al. (eds.), *Information and Communication Technologies for Agriculture—Theme II: Data*, Springer Optimization and Its Applications 183,
https://doi.org/10.1007/978-3-030-84148-5_6

recent advancements in computer vision and software in accordance with cost decrease of mechanical and electronic parts to produce a budget machine.

1.1 Industrial Sorters

A non-exhaustive list of industrial sorters that currently exist in the market is presented below in alphabetical order:

- Bühler Sorters[1]
- Multiscan Technologies S.L.[2]
- Protec s.r.l. ExtraSorter 3W[3]
- TOMRA Genius[4]

Bühler features a wide range of machines that can sort grain, food, and plastic. Some models of the company that can process olives and in general oilseeds are:

- DA MultiVision: It is labeled as the most cost-effective machine of the company. It can detect up to five color defects.
- Sortex A: It detects color variations alongside with shape, growth, and size defects. It uses high-resolution cameras, broadband LED lighting, and a proprietary air ejection system for sorting (SmartEject).
- Sortex B: This machine is an upgrade to Sortex A. It uses high-resolution monochromatic or color cameras or even InGaAs cameras.

Multiscan Technologies is a company that introduced their first automatic selection machine targeting the olive industry back in 1998. In later years they developed machines for more products including tomatoes, cherries, nuts, and various others. The company currently provides four series of machines with various capabilities:

- V series: They are the most affordable machines provided by Multiscan Technologies. Their transport system is a conveyor belt, and their computer vision system can identify shape and color.
- S series: The transport system uses parallel rollers that allow the full surface of each individual fruit to be examined. A computer vision system is used to identify defects, color, shape, size, and texture.
- I series: Upgrade from S series. The transport system consists of individual rollers that allow more sophisticated selection strategies.
- X series: Machines of this type use X-rays to inspect defects not visible to the naked eye.

---

[1] https://www.buhlergroup.com

[2] http://www.multiscan.eu

[3] https://www.protec-italy.com/en/equipment/extrasorter-3w-optical-sorter

[4] https://www.tomra.com/en/sorting/food/your-produce/fruit/olives

ExtraSorter 3W is a sorting machine manufactured by the PROTEC company. It employs two cameras that can capture the whole surface of the fruit while falling and two consecutive rejection systems. It can sort multiple products including olives and can also detect defects on the fruit and foreign bodies. The 3W in the name of the machine indicates the 3-way sorting that it can achieve in three different classes by color, defects, and foreign bodies. Several models exist based on the width dimension of the machine.

Genius is an "optical belt" sorting machine by the TOMRA company. It processes olives, and other fruit, by removing discolored or pitted fruit, stems, and various types of foreign materials. A set of cameras and lasers monitor each fruit while moving on the belt and during free fall, and sorting is done using precise air guns. TOMRA company manufactures another sorting machine, the IXUS BULK that uses X-rays and can sort fruit based on characteristics invisible to the naked eye.

2 Problem Description

Every fall, after the olive fruit has been collected, olive tree owners employ workers that are assigned the task of olive separation by ripeness. The procedure is costly and time-consuming since each olive must be picked manually and subsequently put in the right basket. Olive skin color can be in several shades of green, greenish-yellow, red, purple, or black indicating the maturity of each individual fruit. Olive sorting aims to generate homogeneous groups of olives that will be needed for the next steps of the olive fruit value chain. Price per kilogram differs among various colors and therefore the process must have high accuracy. Human eyes can easily discriminate different olive shade colors. Nevertheless, in practice misclassifications do occur due to the volume of olives that must be processed and the accumulated fatigue of doing a repetitive task.

The usual practice that is exercised by human workers is using selection tables with raised sides that can hold several olives at each batch, and which are sorted by a team of workers. When a batch is sorted, a new one fills the table, and this is repeated until all olives are examined. Naturally, this procedure is tiresome and time consuming.

As an improvement to the previous setup, another setup involves a conveyor belt. In this setup the conveyor is filled from one side at a steady pace with olives that are passing in front of workers who pick the ones with certain skin colors and let the rest of them pass through. Of course, the speed of the conveyor must be suitable for human workers to be able to sort the olives efficiently and relatively accurately.

2.1 Related Work

A similar problem is that of identifying defects on olives. In [1] olives are classified and evaluated according to quality, using a computer vision system capable of identifying external and internal defects. Three classifications are identified. The first one is according to the ripeness degree of olives collected either directly from the olive trees or from the ground. The second classification occurs after fermentation completes and categorizes olives by quality. The last one occurs after the stone of each olive is eliminated and perhaps after various stuffing occurs. This classification is done according to shade while at the same time failed attempts of stone removal are identified. The work presented in this chapter facilitates the first classification as identified above. In [2] the olive maturity index is estimated using image analysis to classify olives at the reception of olive mills. For the same problem, [3] proposes an image processing system accompanied with classification techniques like discriminant analysis and neural networks.

In [4], green olives were classified in three output classes: solid, bruised, and fly defected. Two near infrared (NIR) spectrometer readings, reflectance, and transmittance were used as input data for an artificial neural network (ANN) classifier and a statistical classifier. The ANN performed better, resulting in 92% success rate in classifying olives to the aforementioned classes.

In [5], size and mass estimation of olives is achieved using computer vision and six different Convolutional Neural Network architectures [6, 7]. In particular, the Inception-ResnetV2 [6] architecture manages to reach accuracy levels of over 95%. In [8], a significant number of olive fruit from various varieties were photographed using a specially designed imaging chamber. Then, image analysis techniques, based on mathematical morphology, and binarization by statistical thresholding techniques were combined in a methodology aiming at automated counting and mass estimation. Error rates close to 1% were reported in segmenting individual olive fruit.

3 The Proposed Olive Separation Approach

The proposed methodology consists of a number of steps, as depicted in Fig. 1, organized in two main phases. The first phase consists of a set of image processing techniques, responsible for detecting the centroids of the olive fruit present on a frame. The second phase aims at providing the best estimate of the positions of the centroids of the olive fruit on the present frame and tracking them in consecutive video frames. The main algorithm for the first phase is based on the Watershed Transform [9], which is a widely used method for decomposing touching objects [10–13]. For accomplishing the tracking task of the second phase, Kalman filtering and the Kuhn-Munkres assignment algorithms are employed for estimation and association, respectively. We provide here below the key steps of the methodology adopted.

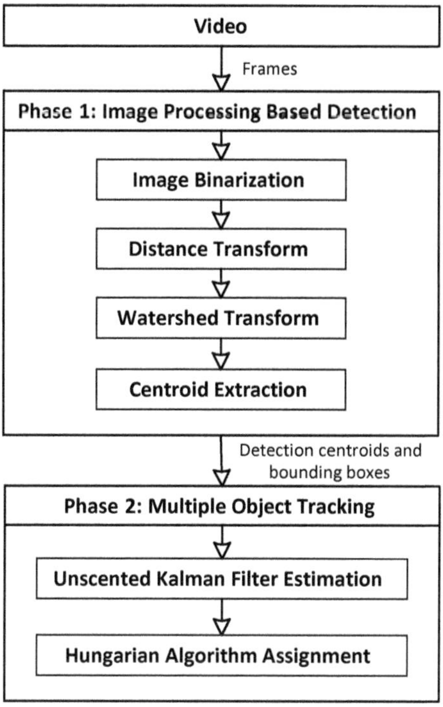

Fig. 1 Flowchart of the methodology employed

During the first, image processing, phase, the video is initially decomposed into a number of frames. Then each frame is converted to a binary image, and in the sequel the computation of the distance transform takes place. Subsequently, the watershed transform is computed, and each touching binary object is disentangled. Finally, the centroid of each binary object is computed, which is then available to the second, tracking phase of the process. During the tracking phase, an Unscented Kalman Filter is responsible for providing the best estimate of each olive fruit's position and velocity, in the presence of noise, and this estimate feeds an assignment algorithm that associates existing olive fruit tracks to detections, in order to track the centroid of the same olive fruit, as it moves on the conveyor belt and appears in consecutive video frames.

3.1 Image Binarization

In the first step of the image processing phase, the frame is converted to a binary one, by using a well-known automated threshold selection process [14, 15]. The threshold operation at grey level l partitions the pixel values of an image into two classes, C_0 and C_1 (representing background and object respectively), i.e., $C_0 = \{1, 2, \ldots, l\}$ and

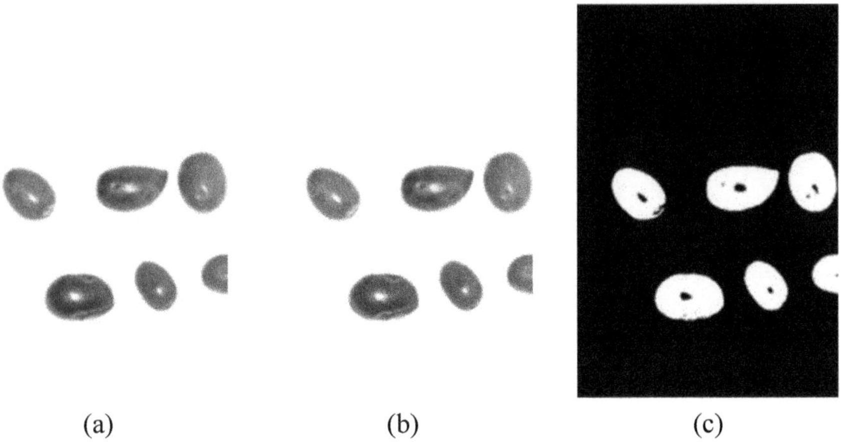

Fig. 2 The thresholding operation. (a) The initial color image, (b) the transformation of the color image to grayscale, and (c) the final binary image

$C_1 = \{l+1, l+2, \ldots, L\}$, where L is the total number of grey levels in the image. Let $\sigma_B^2(l)$ be the between-class variance for the threshold value l:

$$\sigma_B^2(l) = w_0(m_0 - m_T)^2 + w_1(m_1 - m_T)^2 \tag{1}$$

$$w_0 = \sum_{i=0}^{l} P_i, w_1 = 1 - w_0, m_1 = \frac{m_T - m_l}{1 - w_0}, m_0 = \frac{m_l}{w_0} \tag{2}$$

$$m_l = \sum_{i=0}^{l} iP_i, m_T = \sum_{i=0}^{L} iP_i, \quad P_i = \frac{n_i}{n} \tag{3}$$

where n_i is the number of pixels with grey level i, n is the total number of pixels of the image and P_i is the probability of the occurrence level i. Thus, an optimal threshold can be calculated by minimizing the following criterion (4):

$$l^* = \text{argmin}\, \sigma_B^2(l) \tag{4}$$

After the computation of the threshold l^*, a binary image is computed [16]. The binary image, denoted here as B, is an image where each pixel is either "on" (i.e., 1), if the pixel has intensity level above or equal to the computed threshold level l^*, or "off" (i.e., 0) otherwise.

In our case, a color frame from the video is initially transformed into a grey image. This can be achieved by using the average color brightness from each channel of the color image. After the transformation to grayscale, the binary image is computed by using the process described above. An example of the application of the binarization procedure is displayed in Fig. 2.

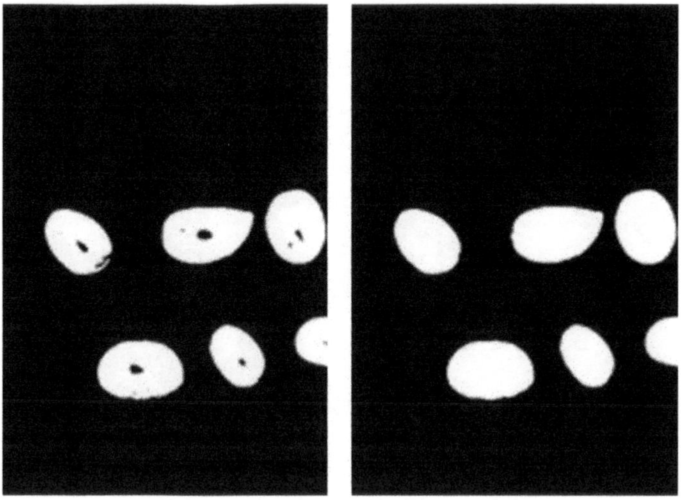
(a) (b)

Fig. 3 The morphological open transformation. (**a**) The initial binary image, and (**b**) the final binary image

The purpose of the binarization step is to detect the objects (olive fruit) in the image. However, and due to uneven illumination problems, as one can see in Fig. 2 (b), holes exist in the center of the positions of the olive fruit in the binary image (Fig. 2(c)). This is a common problem for image binarization algorithms and usually it is solved with morphological transformation methods. Thus, an open morphological operation takes place with a structuring element, e.g., a disk-shaped object with a radius of 10 pixels. The interested reader can refer to [17] for more details on the morphological transformations. The application of the open transformation leads to Fig. 3.

3.2 Distance Transform

It is very common, when applying thresholding methods for the binarization of the initial image, that some objects to either touch or overlap each other [16, 18]. This was also observed in our case, since two or more olives could appear touching each other and are thus presented in the final binary image as one object. An example of two olives touching each other and the resulting binary image is displayed in Fig. 4. Thus, a step was necessary to disentangle these objects [19].

A suitable method is the Watershed Transform (WT) [9], which is a popular segmentation method originated in the field of mathematical morphology. The image is considered as a topographical relief, where the height of each point is related to its grey level. Imaginary rain falls on the terrain and water begins to rise filling the

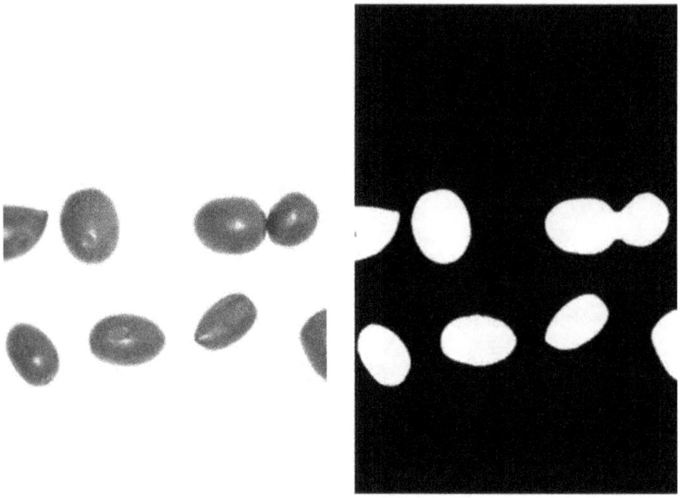

(a) (b)

Fig. 4 An example of two olives touching each other and the resulting binary image. (**a**) The initial image and (**b**) the thresholded binary image displaying two touching olives as one object

different catchment basins. The watersheds are the lines separating the catchment basins that form.

A very common way to apply the WT to an image is to first compute a measure of separation between the binary objects. Such a measure is the Distance Transform (DT). Provided a binary image B of size mxn its distance transform is a map which assigns to every "on" pixel (with coordinates (x_1, y_1)) the distance to its nearest "off" pixel (with coordinates (x_2, y_2)). The distance metric that we used is the Euclidean distance:

$$D(p_1, p_2) = \sqrt{(x_1 - x_2)^2 + (y_1 - y_2)^2} \tag{5}$$

The distance transform plays an important role for the computation of the WT, since every regional minimum [9, 17] will provide an area to the final segmentation map. However, a common problem is that the distance transform contains a large number of such minima, leading the WT to over-segment the initial image. A solution to this problem is to reduce the number of minima, by smoothing the DT image. This can be achieved by applying again a morphological greyscale open transform [18] to the image of the distance transform. Figure 5 displays the computation of the distance transform. Notice, in Fig. 5(b), the minima of the Distance Transform located near the centroid of the olives.

Fig. 5 Computation of the Distance Transform. (**a**) The Distance Transform of the image of Fig. 4a, and (**b**) a contour plot of the Distance Transform (notice that the minima appear correctly at the center of the olive area)

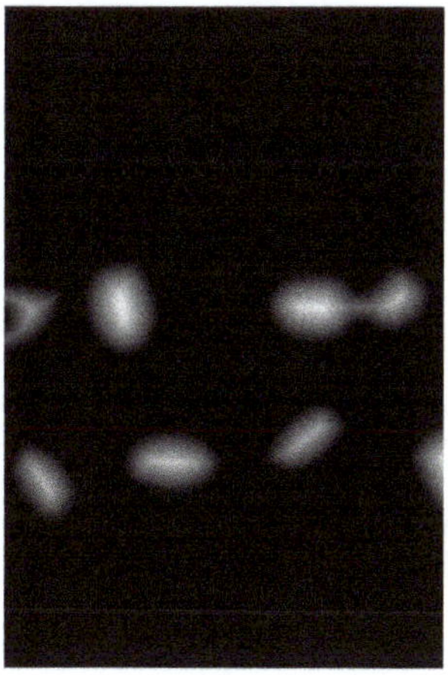

(a)

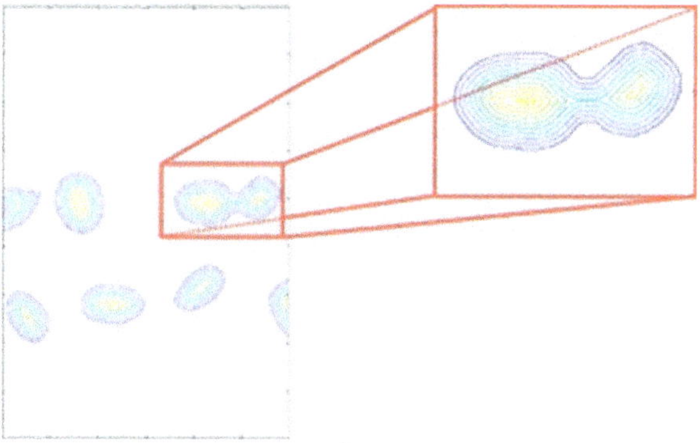

(b)

3.3 *Watershed Transform*

After the computation of the Distance Transform, the Watershed Transform is applied. The Watershed algorithm produces a tessellation of the image into regions that are called watershed regions, and which are depicted in Fig. 6(a). The concept of watersheds in image processing is based on considering an image in the three-dimensional space, with two spatial coordinates versus intensity. The value of the intensity is assumed to be the elevation information. Pixels having the highest intensities correspond to watershed lines, which represent the region boundaries. Water "placed" on any pixel enclosed by a common watershed line flows downhill to a common local intensity minimum. Pixels draining to a common minimum form a catch basin, which represents a region. The result of the watershed transform is a tessellation of the image into regions.

Actually, the watershed transformation assigns to each pixel a number which is the label of the region which it belongs to and the watershed lines are assigned a

(a) (b)

Fig. 6 An example of the application of the Watershed Transform. (**a**) The watershed lines (red lines) delineate the watershed regions and (**b**) the binary image after using the watershed lines to disentangle the touching objects

special number, e.g., zero (0). Thus, the watershed lines can then be used to disentangle the connected/touching regions as displayed in Fig. 6(b). An example of the application of the disentaglement of the touching objects using the watershed transform is shown in Fig. 6.

3.4 Centroid Extraction

After the watershed lines have been superimposed on the binary image and all regions have been disentangled, the computation of the centroids on each binary image follows. The coordinates $(\overline{x}, \overline{y})$ of the centroid C of each binary region R are computed as the center of mass for that binary object, e.g.,

$$\overline{x} = \frac{1}{N}\sum_{i=1}^{N} x_i, \quad \overline{y} = \frac{1}{N}\sum_{i=1}^{N} y_i \tag{6}$$

where N is the number of pixels and (x_i, y_i) are the coordinates of each pixel p_i of the binary region R. An example of the computation of the centroid of each binary region is displayed in Fig. 7. Notice that the centroid of the binary object is not computed if the binary object touches the borderline of the image. Bounding boxes are also defined for each segmented region R, that surround that region, by using the

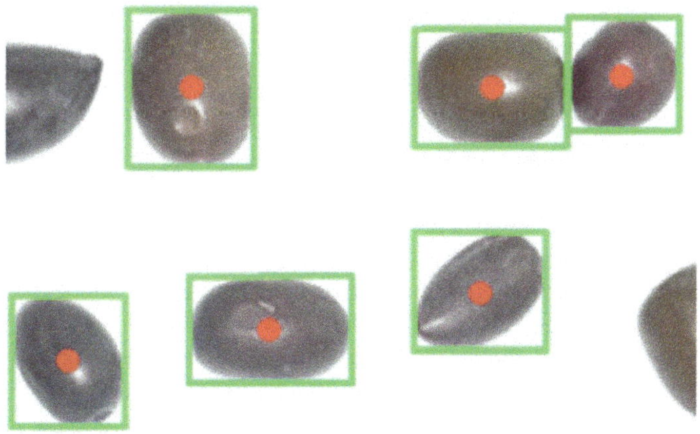

Fig. 7 An example of the computation of the centroid (red dot) and the bounding box (green rectangle) of each binary object, respectively

minimum and maximum coordinate values, in the *x* and *y* axes, of the set of pixels belonging to the region.

3.5 Multiple Object Tracking

Apart from detecting moving olive fruit in each frame, a task which is accomplished during the first phase of the methodology adopted, motion-based tracking of olive fruit is also required. It consists in associating the detections that correspond to the same olive fruit in consecutive frames, over time. The aim of a multiple object tracking method is to assign a unique track identity for each object of interest in a sequence of frames, this assignment spanning over all the frames where the object appears. A review of recent advances and future research directions for multiple object tracking is contributed in [20]. Recently, deep learning techniques have also been used for multiple object tracking. A survey on works employing deep learning methods to solve the task of multiple object-tracking on single-camera videos is given in [21]. Likewise, deep learning has also been used for detection, as in [22], where a Faster Region-based Convolutional Neural Network (Faster R-CNN) is adapted for the task of fruit detection, using color Red-Green-Blue (RGB) and Near-Infrared (NIR) imagery.

The association of a new olive fruit detection in the current frame, to a track that was previously assigned and spanned several past frames, such that the track belongs to the same olive fruit, is performed, based on a combination of:

1. An Unscented Kalman Filter (UKF), an estimator of the motion of the tracks and
2. A minimization procedure of the total cost of all the potential assignments, performed by an algorithm solving a combinatorial optimization problem, called the assignment problem.

The locations, in the current frame, of all the already existing tracks displayed in the previous frame, are predicted by the UKF, based on a motion model of the tracks. A constant velocity or a constant acceleration model can be adopted. Moreover, new observation information is available by applying the detector in the current frame and consists in the locations of the centroids of the current detections. The assignment algorithm finds the association of predicted track locations to detector issued centroid locations that minimizes the sum of the Euclidean distances between predictions and observations, resulting in correct associations of detected olive fruit to tracks belonging to the same olive fruit. The observation information is also used in order to refine the predictions of the UKF, in its update phase. In the subsequent sections, we outline the recursive steps of the UKF algorithm and of the Kuhn-Munkres (or Hungarian) algorithm, that solves the assignment problem.

4 The Unscented Kalman Filter

Kalman filters are recursive nonlinear estimators [23–27]. The goal is to estimate the unobserved state $\boldsymbol{x}_k$ of a (in general) nonlinear dynamic system, at time instant k, given past observations up to time k. The determination of this a posteriori density is referred to as a Bayesian recursion, a problem that in general intractable, but can be greatly simplified under the assumption of Gaussian densities for the quantities involved.

The application of a Linear Kalman Filter (LKF) is limited to linear system dynamics [23], through which it analytically propagates the Gaussian state random variable, while an Extended Kalman Filter (EKF) handles non-linearities present in the system dynamics, by an analytical first-order linearization of the system dynamics, involving Jacobian matrices. However, the linear approximation can lead to sub-optimal performance, or even divergence of the filter, when the non-linearities in the system are severe, or when there exist significantly large initialization errors. The Unscented Kalman Filter [24–27] adopted in the present work provides a statistical alternative approach to the analytical linearization, based on the Unscented Transform (UT) and a small set of carefully selected sample points, as it will be evident hereafter. It is a generalization of the Kalman filter to nonlinear systems, that avoids the linearization required by the EKF, that is implemented easily, confers superior performance, deals also with non-Gaussian noise distributions and is computationally less demanding than other estimators, such as Particle Filters [28]. In [29], Kalman filtering is combined with convolutional neural networks and a data association strategy for vehicle detection and tracking in occlusion situations, while UKF filters are used for detection and tracking of multiple moving objects with occlusion, in nonlinear motion scenarios, in [30].

We consider a discrete-time nonlinear system with an evolution in time written is state space as:

$$x_k = f(x_{k-1}, u_k) + w_k$$
$$y_k = h(x_k, u_k) + v_k, \cdots k = 1, 2, \cdots \tag{7}$$

where x_k is the unobservable state vector at discrete time instant k, u_k a known exogenous input vector, and y_k a measurement observation vector, with dimensions n_x, n_u, and n_y, respectively. The initial state x_0 is considered to be known with uncertainty. The vector-valued functions f and h are assumed to be known. The system is subject to the additive process noise vector w_k (of dimension n_x) with Gaussian density $p(w_k) = \mathcal{N}(0, Q_k)$, with zero mean (expectation $\mathcal{E}(w_k) = 0$) and known covariance matrix $Q_k = \mathcal{E}(w_k {w_k}^T)$. The covariance matrix generalizes to multiple dimensions the notion of variance (how measurements are spread out from the average value) and is always a symmetric and positive semi-definite matrix.[5] The observed measured output y_k is also subject to an additive measurement noise source v_k (of dimension n_y) with Gaussian density $p(v_k) = \mathcal{N}(0, R_k)$, with zero mean (expectation $\mathcal{E}(v_k) = 0$) and known covariance matrix $R_k = \mathcal{E}(v_k {v_k}^T)$. Process and measurement noise are supposed to be uncorrelated, i.e., $\mathcal{E}(w_k {v_k}^T) = 0$. In the more general case, not considered here, the additivity of the noise sources can be dropped. An estimation $\widehat{x}_0 = \mathcal{E}(x_0)$ of the initial unknown state x_0 is also considered to be available, together with the covariance matrix $P_0 = \mathcal{E}\left[(x_0 - \widehat{x}_0)(x_0 - \widehat{x}_0)^T\right]$.

In order to track multiple olive fruit on a conveyor belt, it is necessary to compensate the eventual errors in the positions of their centroids, that are provided by the detector. An UKF can play an essential role in this aim. The movement is considered to take place in two dimensions. A constant velocity model has been chosen for the movement of the olive fruit and for M olive fruit to be tracked in the frame, the state vector x_k, at time instant k, comprises the positions and velocities of each object in the x and y axes, i.e., $\left\{\left(x_k^i, v_k^{xi}, y_k^i, v_k^{yi}\right)/i = 1, 2, \cdots M\right\}$ with a total of $4M$ components. Only the position coordinates $\left(x_k^i, y_k^i\right)$ of each moving object are measured by the detector and are given as the coordinates of its centroid, so that the output vector y_k comprises in total $2M$ components. The task of the UKF is to provide, for each frame captured, the best estimate of the position and velocity components of all the moving olive fruit in the frame, given as observations the positions provided by the detector.

Let us now derive the constant velocity process and measurement model for a particular olive fruit i, described by the state vector $\left[x_k^i, v_k^{xi}, y_k^i, v_k^{yi}\right]^T$. The constant velocity model assumes constant velocity of each olive fruit between sampling instants k and $k-1$, a reasonable assumption for a conveyor belt moving with a

[5] A symmetric nxn matrix A is positive semi-definite, if and only if $x^T A x \geq 0$ for all vectors $x \in \mathbb{R}^n$.

constant speed. However, for reasons that will be evident later on, we start our analysis with a constant acceleration model in continuous time, described by the following equations, concerning axis x:

$$\begin{aligned} \dot{x}^i(t) &= v^{xi}(t) \\ \dot{v}^{xi}(t) &= a^{xi}(t) \\ \dot{a}^{xi}(t) &= 0, \textit{that is}\, a^{xi}(t) = a^{xi} = \textit{constant} \end{aligned} \tag{8}$$

Analogous equations hold for the y axis, with a^{xi} and a^{yi}, the respective constant accelerations between sampling instants. By integrating the last equations between time instants $k-1$ and k corresponding to two consecutive frames with sampling period $t_k - t_{k-1} = T$ and taking into account the constant accelerations between sampling instants, one easily gets in discrete time:

$$\begin{aligned} x_k^i &= x_{k-1}^i + v_{k-1}^{xi}T + \frac{1}{2}a^{xi}T^2 \\ v_k^{xi} &= v_{k-1}^{xi} + a^{xi}T \\ a_k^{xi} &= a_{k-1}^{xi} = a^{xi} = \text{constant} \end{aligned} \tag{9}$$

and analogous equations hold for the y axis. In compact vector form we obtain:

$$\begin{bmatrix} x_k^i \\ v_k^{xi} \\ y_k^i \\ v_k^{yi} \end{bmatrix} = \begin{bmatrix} 1 & T & 0 & 0 \\ 0 & 1 & 0 & 0 \\ 0 & 0 & 1 & T \\ 0 & 0 & 0 & 1 \end{bmatrix} \begin{bmatrix} x_{k-1}^i \\ v_{k-1}^{xi} \\ y_{k-1}^i \\ v_{k-1}^{yi} \end{bmatrix} + \begin{bmatrix} \frac{1}{2}T^2a^{xi} \\ Ta^{xi} \\ \frac{1}{2}T^2a^{yi} \\ Ta^{yi} \end{bmatrix} \tag{10}$$

where we have omitted the equations for the constant accelerations. The constant velocity model is obtained from the constant acceleration model, by setting $a^{xi} = a^{yi} = 0$ in the last equation and is thus described by the state evolution equation:

$$\begin{bmatrix} x_k^i \\ v_k^{xi} \\ y_k^i \\ v_k^{yi} \end{bmatrix} = \begin{bmatrix} 1 & T & 0 & 0 \\ 0 & 1 & 0 & 0 \\ 0 & 0 & 1 & T \\ 0 & 0 & 0 & 1 \end{bmatrix} \begin{bmatrix} x_{k-1}^i \\ v_{k-1}^{xi} \\ y_{k-1}^i \\ v_{k-1}^{yi} \end{bmatrix} \tag{11}$$

and the observation equation:

$$\begin{bmatrix} x_k^i \\ y_k^i \end{bmatrix} = \begin{bmatrix} 1 & 0 & 0 & 0 \\ 0 & 0 & 1 & 0 \end{bmatrix} \begin{bmatrix} x_k^i \\ v_k^{xi} \\ y_k^i \\ v_k^{yi} \end{bmatrix} \tag{12}$$

However, the more general equation (10) can be useful in order to design an appropriate process noise covariance matrix Q_k. Indeed, by setting $w_k^i = \begin{bmatrix} \frac{1}{2}T^2a^{xi} \\ Ta^{xi} \\ \frac{1}{2}T^2a^{yi} \\ Ta^{yi} \end{bmatrix}$ and relaxing the unrealistic constant velocity assumption ($a^{xi} = a^{yi} = 0$) with zero mean Gaussian white process noise accelerations $a^{xi}\tilde{\mathcal{N}}(0, \sigma_{xi}^2)$ and $a^{yi}\tilde{\mathcal{N}}\left(0, \sigma_{yi}^2\right)$, it is easy to calculate that $\mathcal{E}\left(w_k^i\right) = 0$ and that the process noise covariance matrix Q_k can be selected as:

$$Q_k = \mathcal{E}\left(w_k^i w_k^{i\,T}\right) = \mathcal{E}\begin{bmatrix} \begin{bmatrix} \frac{1}{4}T^4 & \frac{1}{2}T^3 \\ \frac{1}{2}T^3 & T^2 \end{bmatrix}(a^{xi})^2 & \begin{bmatrix} \frac{1}{4}T^4 & \frac{1}{2}T^3 \\ \frac{1}{2}T^3 & T^2 \end{bmatrix}a^{xi}a^{yi} \\ \begin{bmatrix} \frac{1}{4}T^4 & \frac{1}{2}T^3 \\ \frac{1}{2}T^3 & T^2 \end{bmatrix}a^{xi}a^{yi} & \begin{bmatrix} \frac{1}{4}T^4 & \frac{1}{2}T^3 \\ \frac{1}{2}T^3 & T^2 \end{bmatrix}(a^{yi})^2 \end{bmatrix} \tag{13}$$

With uncorrelated accelerations in the two axes, i.e., $\mathcal{E}(a^{xi}a^{yi}) = 0$, one gets:

$$Q_k = \begin{bmatrix} \begin{bmatrix} \frac{1}{4}T^4 & \frac{1}{2}T^3 \\ \frac{1}{2}T^3 & T^2 \end{bmatrix}\sigma_{xi}^2 & \begin{bmatrix} 0 & 0 \\ 0 & 0 \end{bmatrix} \\ \begin{bmatrix} 0 & 0 \\ 0 & 0 \end{bmatrix} & \begin{bmatrix} \frac{1}{4}T^4 & \frac{1}{2}T^3 \\ \frac{1}{2}T^3 & T^2 \end{bmatrix}\sigma_{yi}^2 \end{bmatrix} \tag{14}$$

The UKF is applied recursively as two distinct phases, called prediction and update phase, that alternate. During the prediction phase, based on the estimate of the state from the previous step of the algorithm, a predicted state estimate and respective covariance matrix are produced, which do not include any observation information from the current time step. The predicted state estimate is termed a priori state estimate (before the measurement update). During the update phase, the state (and respective covariance matrix) estimate is refined, by incorporating the current observation information, resulting in an improved state estimate, termed a posteriori state estimate (after the measurement update). The update phase can be skipped, if an observation is not available during the present time step or can be repeated continuously in the presence of multiple independent observations. A graphical characterization of the implementation steps of an Unscented Kalman Filter is depicted in Fig. 8 and detailed hereafter.

4.1 Prediction Phase of the UKF

During the prediction phase, the UKF uses the so-called Unscented Transformation (UT), in order to calculate the statistics of the system state undergoing a nonlinear transformation, when propagated through the true nonlinear system. The UT uses the so-termed sigma points. This is a non-unique set of points, together with a non-unique set of weights, designed with a twofold purpose. On the one hand, they completely capture the mean value and covariance of the system state, when this state is supposed to be a Gaussian random variable. On the other hand, when a Gaussian random state is propagated through the state evolution equation of the nonlinear system, they capture the posterior mean and covariance (of the Gaussian random state after the transformation) accurately, up to at least the second order in a Taylor series expansion, in contrast to the EKF which can achieve only first-order accuracy and moreover demands for explicit Jacobian calculations.

Given the state $\widehat{x}_{k-1}$ and covariance P_{k-1} estimates of the UKF in the previous step $k-1$, a simple choice for the sigma points in the present step consists in using the following $2L + 1$ points, where $L = n_x$:

$$\mathcal{X}_{k-1} = \left[\widehat{x}_{k-1} \quad \widehat{x}_{k-1} + \left(\gamma\sqrt{P_{k-1}}\right)_i i = 1, \cdots, L \quad \widehat{x}_{k-1} - \left(\gamma\sqrt{P_{k-1}}\right)_i i = 1, \cdots, L\right. \tag{15}$$

with $\gamma = \sqrt{L + \lambda}$ and $\lambda = a^2(L + \kappa) - L$. Here, $\sqrt{P_{k-1}} = A_{k-1}$ is a symmetric lower triangular matrix A_{k-1}, called the square root of P_{k-1}, such that $P_{k-1} = A_{k-1}A_{k-1}^T$ and which can be computed efficiently by the so-called. Finally, Cholesky decomposition. In relation (15), $(\gamma\sqrt{P_{k-1}})_i$ denotes the ith column of the matrix $\gamma\sqrt{P_{k-1}}$ a and κ are scaling parameters. The parameter a determines the spread of the sigma points around $\widehat{\boldsymbol{x}}_{k-1}$ and typical values $1 \leq a \leq 1e - 4$ are used. The augmentation of a results in a wider spread and a better capture of the distribution and its nonlinearities. The value chosen for the parameter κ usually is set between 0 and $3 - L$. It can be proven that this particular choice of sigma points has the same statistics as $\widehat{\boldsymbol{x}}_{k-1}$, i.e., a mean value of $\widehat{\boldsymbol{x}}_{k-1}$ and a covariance matrix P_{k-1}. We now propagate the $2L + 1$ sigma points through the nonlinear function f, to obtain the following propagated sigma points:

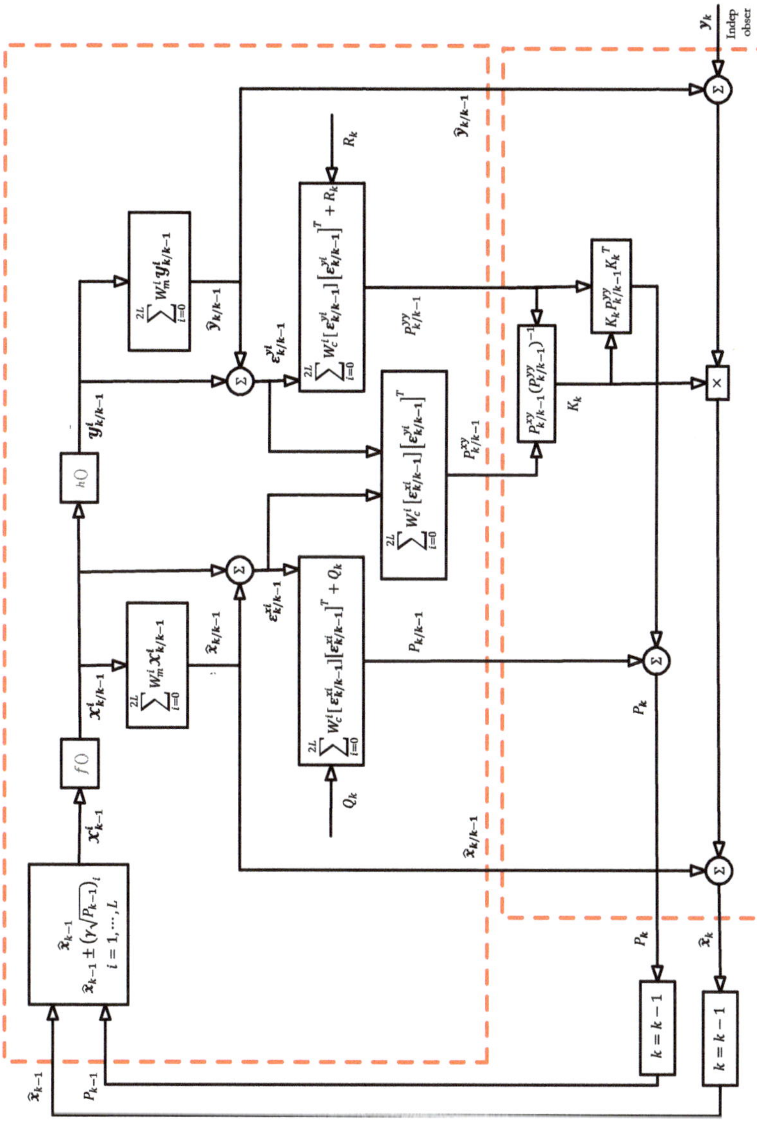

Fig. 8 UKF implementation steps

$$\mathcal{X}^i_{k/k-1} = f\left(\mathcal{X}^i_{k-1}, u_k\right)\ i = 0, 1, \cdots, 2L \tag{16}$$

where $\mathcal{X}^i_{k-1}$ is the ith sigma point, i.e., the ith column of the matrix $\mathcal{X}_{k-1}$ of sigma points and $\mathcal{X}^i_{k/k-1}$ is the respective i^{th} propagated sigma point (posterior sigma point). The subscript $k/k-1$ (k given $k-1$) characterizes a value at time step k, based on information from the previous step $k-1$. As it was already mentioned, the statistics of $f(\widehat{x}_{k-1}, u_k)$ can be approximated by calculating the a priori state estimate as a properly chosen weighted mean of the posterior sigma points:

$$\widehat{x}_{k/k-1} = \sum_{i=0}^{2L} W^i_m \mathcal{X}^i_{k/k-1} \tag{17}$$

and the covariance matrix of the predicted state as:

$$P_{k/k-1} = \sum_{i=0}^{2L} W^i_c \left[\mathcal{X}^i_{k/k-1} - \widehat{x}_{k/k-1}\right]\left[\mathcal{X}^i_{k/k-1} - \widehat{x}_{k/k-1}\right]^T + Q_k \tag{18}$$

The weights are chosen from the equations:

$$W^0_m = \frac{\lambda}{L+\lambda} \quad W^i_m = \frac{\lambda}{2(L+\lambda)}\ i = 1, 2, \cdots, 2L$$
$$W^0_c = \frac{\lambda}{L+\lambda} + 1 - a^2 + \beta \quad W^i_c = \frac{\lambda}{2(L+\lambda)}\ i = 1, 2, \cdots, 2L \tag{19}$$

Prior knowledge of the distribution of $\widehat{x}_{k-1}$ can be incorporated in the value of the parameter β. For example, in the case where $\widehat{x}_{k-1}$ is a Gaussian distribution, the choice $\beta = 2$ is optimal.

Relations (17) and (18) approximate the statistics of $f(\widehat{x}_{k-1}, u_k)$, i.e., of the propagated $\widehat{x}_{k-1}$, to the second order for a non-Gaussian input $\widehat{x}_{k-1}$, irrespective of the nonlinearity f, to the third order for a Gaussian input $\widehat{x}_{k-1}$ and the accuracy of the approximation can be extended to higher order moments with a proper choice for the parameters a and β. It is noted here that, unlike the linear case, it generally holds that a Gaussian random variable transformed through a nonlinear equation, does not result in a Gaussian random variable output.

Next, the transformed sigma points $\mathcal{X}^i_{k/k-1}$ are propagated through the observation function h to get the output sigma points:

$$\mathcal{Y}^i_{k/k-1} = h\left(\mathcal{X}^i_{k/k-1}\right)\ i = 0, 1, \cdots, 2L \tag{20}$$

which are used to calculate the statistics of the predicted output (mean value, covariance matrix, and cross-covariance matrix between predicted state and predicted output), as follows:

$$\widehat{y}_{k/k-1} = \sum_{i=0}^{2L} W_m^i \mathcal{Y}_{k/k-1}^i$$

$$P_{k/k-1}^{yy} = \sum_{i=0}^{2L} W_c^i \left[\mathcal{Y}_{k/k-1}^i - \widehat{y}_{k/k-1}\right]\left[\mathcal{Y}_{k/k-1}^i - \widehat{y}_{k/k-1}\right]^T + R_k$$

$$P_{k/k-1}^{xy} = \sum_{i=0}^{2L} W_c^i \left[X_{k/k-1}^i - \widehat{x}_{k/k-1}\right]\left[\mathcal{Y}_{k/k-1}^i - \widehat{y}_{k/k-1}\right]^T \tag{21}$$

4.2 *Update Phase of the UKF*

In the update phase, taking into account the new independent measurement vector y_k available at time k, the a posteriori (after the measurement update) state $\widehat{x}_k$ is calculated, as a correction to the a priori state $\widehat{x}_{k/k-1}$, by:

$$\widehat{x}_k = \widehat{x}_{k/k-1} + K_k\left(y_k - \widehat{y}_{k/k-1}\right) \tag{22}$$

with K_k the Kalman gain matrix:

$$K_k = P_{k/k-1}^{xy}\left(P_{k/k-1}^{yy}\right)^{-1} \tag{23}$$

The covariance matrix of $\widehat{x}_k$ is:

$$P_k = P_{k/k-1} - K_k P_{k/k-1}^{yy} K_k^T \tag{24}$$

The estimates $\widehat{x}_k$ and P_k are then used as initialization estimates for the next step of the UKF and the procedure is repeated a desired number of times recursively.

5 The Kuhn-Munkres (Hungarian) Algorithm

The association of current frame detections (observations) to existing tracks of the previous frame (track data) is a variant of the assignment problem, where agents (tracks in our case) are assigned to tasks (detections in our case), with a known cost, which depends on the specific agent-task assignment. Under the restrictions that, at most one agent is assigned to each task and that, at most one task is assigned to each agent, it is required to minimize the total cost of assignments.

If the number of agents is not equal to the number of tasks (unbalanced assignment), some agents or tasks may remain unassigned. Indeed, in our case, it is expected that some detections in the current frame will be assigned to some tracks of the previous frame, corresponding to olive fruit that continue their movement in the current frame, that some tracks will remain unassigned, corresponding, for example, to olive fruit that have left the field of view and finally that some new detections will also remain unassigned, because they correspond to new olive fruit entering the field of view and thus these detections will begin new tracks.

The problem can be represented by a matrix of all potential costs, where the element in row i and column j represents the cost of assigning the j-th task to the i-th agent (Euclidean distance between the centroid of the j-th detection and the i-th track location prediction provided by the UKF). The assignment problem can be stated as follows: find a set of independent matrix elements, i.e., elements such that no two of them can lie in the same row, or column of the matrix, such that the sum of these elements is minimum. Moreover, an extra cost of leaving a track, or a detection unassigned is also considered, with the choice of higher values for this cost prohibiting non-assignments.

While a brute-force approach for a balanced assignment with n agents and n tracks has at least an exponential runtime complexity ($n!$ operations are needed to find all possible permutations), there exists a polynomial $O(n^3)$ runtime complexity algorithm to solve the assignment problem, known as the Kuhn-Munkres (or Hungarian) algorithm, initially given by Kuhn and completed by Munkres. The Kuhn algorithm [31] is based on the following two remarks:

- The same optimal assignment solution is obtained, if a number is added, or subtracted from all the elements of any row, or any column of the cost matrix.
- If z is the maximum number of independent zero elements of the cost matrix, then there are z lines (rows or columns of the matrix) which contain all these zero elements.

The algorithm is outlined hereafter:

Step 1: Subtract the smallest element of each row, from every element of this row.
Step 2: Repeat an analogous procedure for all columns of the resulting matrix of Step 1.
Step 3: Find a minimal set of lines (rows, or columns) that contain all the zeros of the output matrix of Step 2. If their number equals n, there exists a set of n independent zeros and the elements of the initial cost matrix at these positions constitute the required solution. Else, proceed to Step 4.
Step 4: If the number of lines of Step 3 is less than n, find the smallest element of the output matrix of the previous step, not covered by any of these lines, subtract it from each uncovered row and add it to every covered column.

The procedure terminates after a finite number of steps. Munkres completed the algorithm, by giving a constructive procedure for carrying out Step 3 [32]. A solution to the extension of the assignment problem to rectangular $n \times m$ matrices was proposed in [33]. While the rectangular problem could be solved by appropriately augmenting the rectangular cost matrix to form a square one, with padding zeros and then solving the square assignment problem, the authors have developed a more effective direct solution to the rectangular assignment problem.

6 Results

6.1 Sample Collection

One olive variety (Kalamon) was considered for this study. 50 olive fruit were handpicked in olive orchards located in Agrinio, Greece (38.624723, 21.409865), in October 2019.

6.2 Simulation Design

For the purpose of capturing the images of the fruit, an image acquisition system was set up. The aim at this point was not to implement a final version of this device, integrated in a conveyor system, but a simulation mimicking its main features regarding image capture processing and tracking conditions. In order to create a video, we have shot images of olives. Since most of the sorters align the olives in channels, we aligned the olives in lines. An example of an image of olives is displayed in Fig. 9.

The olive fruit were placed on a semi-translucent plastic surface and were illuminated from below by a set of four fluorescent tubes, in order to avoid shadows, cast from the olive fruit. Additional lighting was provided by two sets of four fluorescent lamps each, placed in a perpendicular configuration and by the incorporated camera flash. The camera was placed perpendicularly to the imaging area and

Fig. 9 An example of an image of olives shot using a white background and aligned in two lines

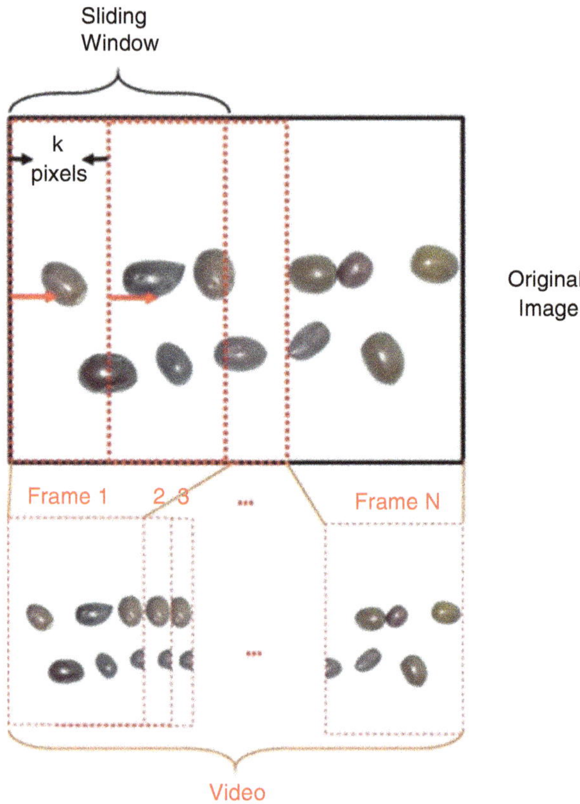

Fig. 10 An illustration of the video creation process

was set in manual mode with aperture f/5, shutter speed 1/500, focal length 17 mm, and ISO-200 sensitivity. The images were initially saved in a JPEG format with a resolution of 3264x2448 and 24 bits of color depth.

The video[6] was created using a frame window which was slided from the original frame and in this way additional frames were added to the video. A cyclical shift of the frames was applied, so that olive fruit exiting a frame, reappear in the currently viewed frame, some amount of time later. As one can easily observe, the critical parameter that controls the velocity of the hypothetical conveyor was the number k of pixels which were used to slide the window. An illustration of this process is shown in Fig. 10.

[6]An overview of the centroid extraction process can be accessed through https://www.youtube.com/watch?v=LgUJuib5yq0

6.3 Results Using Kalman Filtering[7]

Apart from the use of the Kuhn-Munkres algorithm in order to solve the assignment problem, we also followed a heuristic procedure described hereafter, which gave better results in the case of augmented noise in the predictions.

The first step, in order to associate the detections in the current frame to the tracks of the previous frame, consists in predicting the locations of the tracks in the current frame, by applying the prediction phase of the Kalman filter. Then, the prediction of the track i in the current frame is associated to the detection j of the current frame, only if the following conditions are both met:

1. The Euclidean distance from the predicted track i to j is the minimum of all the distances from track i to any of the detections of the current frame.
2. Reciprocally, it should also hold that the Euclidean distance from detection j in the current frame, to track i, is the minimum of all the distances from detection j to any of the predicted track locations.

This reciprocal condition is not met, if for example a predicted track remains in the current frame, while its corresponding detection is only partially inside the frame, having a bounding box with an edge outside of it. This partial detection is not considered as valid, since the centroid corresponding to its visible part is moving and is not steady with respect to the olive fruit, as it is exiting the frame. In such a case, the track is considered unassigned and deleted. From the other part, unassigned detections corresponding to olive fruit entering the frame create new tracks.

In Fig. 11, two consecutive frames of the tracking process are shown. Olive fruit are moving from right to left. In the upper, first frame, tracks 322 to 329 are identified. The green circles correspond to the centroids issued from the detection process, the green rectangles correspond to the detected bounding boxes and the yellow circles correspond to the centroids after adding Gaussian noise with a standard deviation of 20 pixels in the horizontal and five pixels in the vertical direction. These noisy centroids are considered to be available, as measurements, to the Kalman filter. The red and blue circles give the estimates of the prediction and update phases of the Kalman filter, respectively. It is evident that even touching olive fruit are separately detected and tracked by the algorithms. Two olive fruit are partially visible on the left edge of the frame. These olive fruit do not produce valid detections and their predictions, even if they remain totally in the frame, are considered as unassigned and are thus deleted. Similarly, two olive fruit are partially visible on the right edge, just entering the frame and thus are not yet detected. The detections signed 328 and 329 are presented with their nominal and noisy centroids and they have initiated new tracks, but they do not have associated Kalman prediction and update estimates yet. In the consecutive, lower, frame, olive fruit 322 to 324 have totally or partially exited the frame and thus do not have any valid detections. Olive fruit 328 and 329 follow their path through the frame, while new

[7] A video of the Kalman filtering results can be found in (https://youtu.be/q4T-poMhXlA)

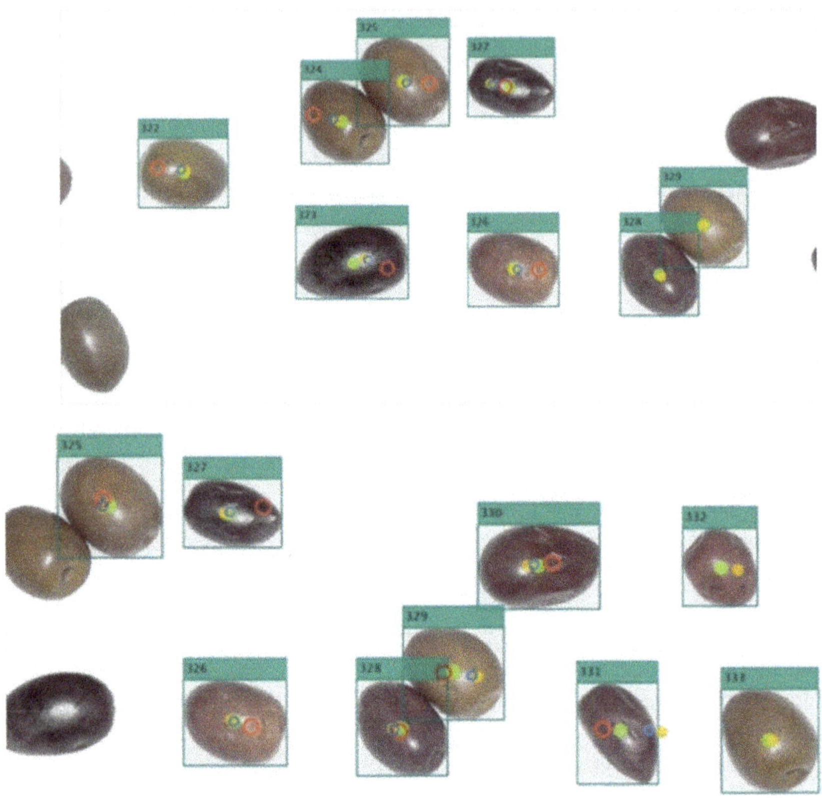

Fig. 11 Two consecutive frames of the tracking process

detections 332 and 333 entering the frame, have just initiated new tracks. The horizontal shift between the horizontal frames was 200 pixels, corresponding to a conveyor belt speed of 35 $cm\ s^{-1}$.

7 Evaluation of the Results

In order to evaluate the detection and tracking algorithms used, the ground truth olives were manually annotated and segmented (Fig. 12 shows an example), so that the localization of rectangular ground truth bounding boxes and ground truth centroids became possible. These annotations serve as a validation set in the sequel.

The bounding boxes issued from the detection algorithm were then compared to the ground truth bounding boxes, in each frame, by means of the IoU (Intersection over Union) measure. IoU is defined as the quotient of the intersection area of two bounding boxes, over their union area. Two bounding boxes in the exact same

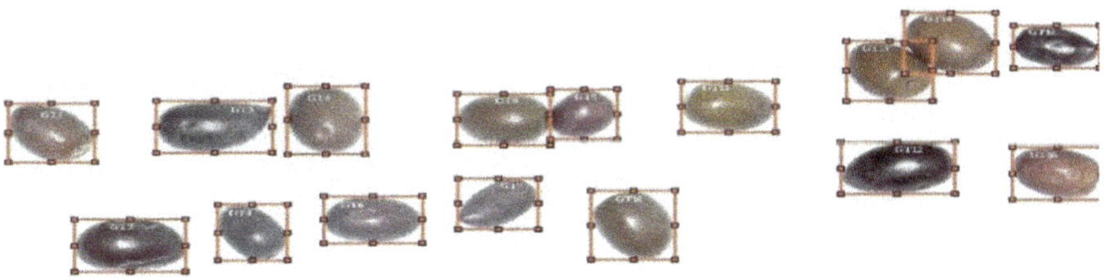

Fig. 12 Manual annotation of ground truth bounding boxes

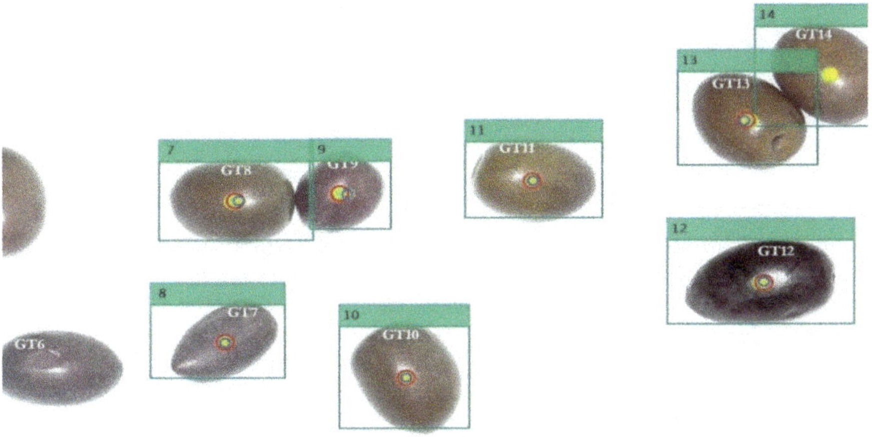

Fig. 13 Frame for which IoU values are reported

Table 1 IoU values when the detection is based on the frame of Fig. 13 (first column) and when the detection is based on the respective, blurred, frame of Fig. 14 (second column)

Ground truth id	IoU Detection without noise (Fig. 13)	IoU Detection under noise (Fig. 14)
GT7	0.9613	0.7211
GT8	0.9370 (0.0052)	0.7663
GT9	0.9349	0.6555 (0.0697)
GT10	0.9243	0.7836
GT11	0.9286	0.8232
GT12	0.9833	0.7841
GT13	0.9547 (0.0999)	0.7857 (0.1240)
GT14	0.9699 (0.1002)	0.7942 (0.1144)

position have an IoU equal to 1, while two non-intersecting bounding boxes have an IoU equal to 0. By using the Munkres algorithm and the IoU measure, ground truth objects are then associated with detections.

Example results, concerning the frame of Fig. 13, are reported in Table 1. For each ground truth olive fruit bounding box, the value of IoU with the bounding box

Fig. 14 Blurred version of the frame of Fig. 13

of the respective detection issued from the Munkres algorithm is given. In parentheses, values of IoU with other, than the associated, detections, if any, are also given. The first column concerns the application of the detection algorithm to the frame of Fig. 13, without considering any noise. For obtaining the values of the second column, the detection algorithm was applied to a blurred image of the frame that emulates Gaussian motion in both axes (see Fig. 14). As it was expected, the IoU values are lower in this case, but a one-to-one correspondence between ground truth instances and detections is still performed without errors (see detected bounding boxes in Fig. 13), an indication of the robustness of the detection algorithm to considerable levels of noise. Notice that the bounding boxes of Fig. 13 are not firmly surrounding the respective ground truth olive fruit, since they are calculated based on the blurred version of Fig. 13, depicted in Fig 14.

In object detection, a hit of a ground truth instance by a detection bounding box is achieved, if the corresponding IoU exceeds a predefined threshold. Thus, for different IoU thresholds, the number of correctly predicted ground truth instances (true positives), the number of missed, non-detected ground truth instances (false negatives), and the number of false-positive detections that do not correspond to ground truth instances can be calculated. Then, the standard metrics of precision, recall, and harmonic mean can be used for evaluation purposes. However, due to the high degree of accuracy obtained with the methodologies adopted in the present work, despite the high levels of noise, a further analysis with the synthetic data used cannot give reliable results, and a further analysis is therefore foreseen, with real data of olive fruit moving on a conveyor belt.

The tracking algorithm is evaluated based on a graph of the ground truth objects that appear in each frame, together with their respective track ids. Such a graph appears in Fig. 15, spanning a total of 54 frames. Ground truth objects appearing in a frame are seen in the columns, with numbers indicating the respective track they belong to. The appearance of a ground truth object in consecutive frames can be

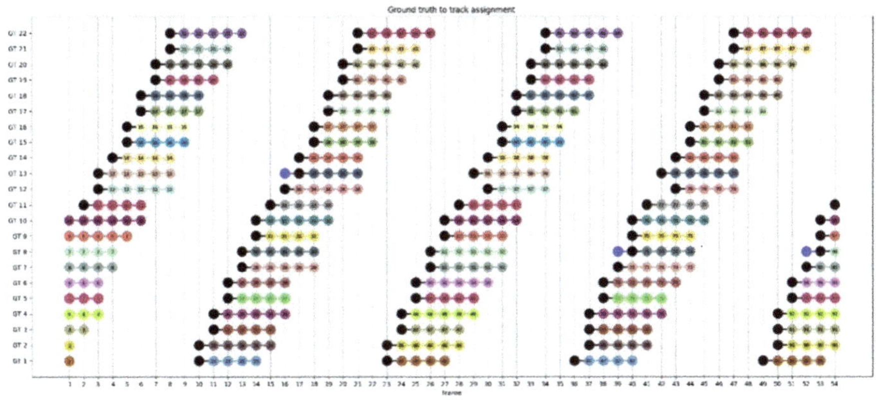

Fig. 15 Graph-based tracking spanning 54 consecutive frames. Frames can be seen in columns, while the appearances of each ground truth object can be followed horizontally

followed horizontally. A same color horizontally indicates that the ground truth object continues to belong to the same track. The re-appearance of a ground truth object occurs, since, due to the circular shift applied, an object exiting the current frame from its left side, re-enters the frame, after some time, from its right side. Black dots indicate that a ground truth object is detected, but not yet assigned to a track, since a Kalman prediction is not yet available for that object (it has just entered entirely the visible frame). Blue dots denote anomalies, like ground truth objects not detected, which happens very sporadically in our examples. A zoom-in of Fig. 15, around the frame with id 4, which is seen in Fig. 13, can be seen in Fig. 16. Modes of failure, like false-positive tracks not associated with ground truth objects, false-negative ground truths not associated with a track, multiple tracks/ground truths associated with the same ground truth/track have to be also further investigated, based on real and not synthetic data, as in the case of olive fruit moving on a conveyor belt.

8 Conclusions

A method for the detection and tracking of moving olive fruit was presented. The method is based on the application of the Watershed Transform to extract the centroid of the olive fruit areas. Then a Kalman filter together with an algorithm that solves the association problem between Kalman predictions and detections in the current frame is employed, in order to track the centroids in consecutive frames. A simulation experiment was designed and implemented to test our method with synthetic data. For all the experiments our method managed to track all the olive fruit with high accuracy even when we added severe noise. As for future work we plan to collect more images of olive fruit during the collecting season during October till

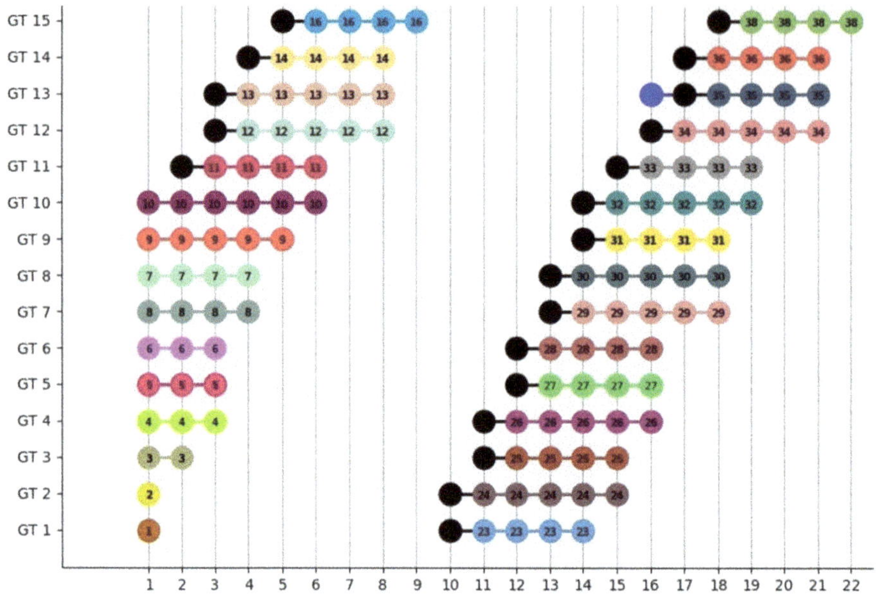

Fig. 16 Zoom-in of Fig. 15, around frame 4, which corresponds to Fig. 13

December and test our method with real data of olive fruit moving on a conveyor belt. Also, additional machine learning algorithms for tracking the centroids should be employed for comparative reasons.

References

1. Díaz, R. (2016). Classification and quality evaluation of table olives. In *Computer vision technology for food quality evaluation* (pp. 351–367). Academic Press.
2. Guzmán, E., Baeten, V., Pierna, J. A. F., & García-Mesa, J. A. (2015). Determination of the olive maturity index of intact fruit using image analysis. *Journal of Food Science and Technology, 52*(3), 1462–1470.
3. Puerto, D. A., Gila, D. M. M., García, J. G., & Ortega, J. G. (2015). Sorting olive batches for the milling process using image processing. *Sensors, 15*(7), 15738–15754.
4. Kavdır, İ., Büyükcan, M. B., & Kurtulmuş, F. (2018). Classification of olives using FT-NIR spectroscopy, neural networks and statistical classifiers. *Journal of Food Measurement and Characterization, 12*(4), 2493–2502.
5. Ponce, J. M., Aquino, A., & Andújar, J. M. (2019). Olive-fruit variety classification by means of image processing and convolutional neural networks. *IEEE Access, 7*, 147629–147641.
6. Krizhevsky, A., Sutskever, I., & Hinton, G. E. (2017). Imagenet classification with deep convolutional neural networks. *Communications of the ACM, 60*(6), 84–90.
7. Szegedy, C., Ioffe, S., Vanhoucke, V., & Alemi, A. (2016). *Inception-v4, inception-resnet and the impact of residual connections on learning*. arXiv:1602.07261 preprint.

8. Ponce, J. M., Aquino, A., Millan, B., & Andújar, J. M. (2019). Automatic counting and individual size and mass estimation of olive-fruit through computer vision techniques. *IEEE Access, 7*, 59451–59465.
9. Vincent, L., & Soille, P. (1991). Watershed in digital spaces: An efficient algorithm based on immersion simulations. *IEEE Transactions on Pattern Analysis and Machine Intelligence, 13*, 583–598.
10. Haris, K., Efstratiadis, S. N., Maglaveras, N., Pappas, C., Gourassas, J., & Louridas, G. (1999). Model-based morphological segmentation and labelling of coronary angiograms. *IEEE Transactions on Medical Imaging, 18*(10), 1003–1015.
11. Winter, M., Mankowski, W., Wait, E., De La Hoz, E. C., Aguinaldo, A., & Cohen, A. R. (2019). Separating touching cells using pixel replicated elliptical shape models. *IEEE Transactions on Medical Imaging, 38*(4), 883–893.
12. Lin, G., Adiga, U., Olson, K., Guzowski, J. F., Barnes, C. A., & Roysam, B. (2003). A hybrid 3D watershed algorithm incorporating gradient cues and object models for automatic segmentation of nuclei in confocal image stacks. *Cytometry Part A, 56*(1), 23–36.
13. Lu, N., & Ke, X. (2007). A segmentation method based on gray-scale mor-phological filter and watershed algorithm for touching objects image. In *Fourth international conference on fuzzy systems and knowledge discovery (FSKD 2007)* (pp. 474–478).
14. Otsu, N. (1979). A threshold selection method for gray-levels histograms. *IEEE Transactions on Pattern Analysis Machine Intelligent, 13*, 583–598.
15. Karvelis, P., Tzallas, A., Fotiadis, D., & Georgiou, I. (2008). A multichannel watershed-based segmentation method for multispectral chromosome classification. *IEEE Transactions on Medical Imaging, 2*(5), 697–708.
16. Gonzalez, R., & Woods, R. (2017). *Digital image processing* (4th ed.). Addison-Wesley Longman Publishing.
17. Vincent, L. (1993). Morphological grayscale reconstruction in image analysis: Applications and efficient algorithms. *IEEE Transactions on Image Processing, 2*, 176–201.
18. Soille, P. (1999). *Morphological image analysis: Principles and applications*. Springer-Verlag.
19. Karvelis, P., Likas, A., & Fotiadis, D. (2010). Identifying touching and overlapping chromosomes using the watershed transform and gradient paths. *Pattern Recognition Letters, 31*(16), 2474–2488.
20. Luo, W., Xing, J., Milan, A., Zhang, X., Liu, W., Zhao, X. & Kim T. (2014). *Multiple object tracking: A literature review*. arXiv: Computer Vision and Pattern Recognition.
21. Ciaparrone, G., Luque Sánchez, F., Tabik, S., Troiano, L., Tagliaferri, R., & Herrera, F. (2020). Deep learning in video multi-object tracking: A survey. *Neurocomputing, 381*, 61–88.
22. Sa, I., Ge, Z., Dayoub, F., Upcroft, B., Perez, T., & McCool, C. (2016). DeepFruit: A fruit detection system using deep neural networks. *Sensors, 16*, 1222.
23. Kalman, R. E. (1960). A new approach to linear filtering and prediction problems. *ASME-Journal of Basic Engineering, 382*(1), 35–45.
24. Julier, S. J. & Uhlmann, J. K. (1997). A new extension of the Kalman filter to nonlinear systems. In Proceedings of SPIE: The International Society for Optical Engineering, Orlando, FL, 3068, 182–193.
25. Julier, S. J., & Uhlmann, J. K. (2004). Unscented filtering and nonlinear estimation. *Proceedings of the IEEE, 92*(3), 401–422.
26. van der Merwe, R., Wan, E., & Julier, S. (2004). *Sigma-point Kalman filters for nonlinear estimation and sensor-fusion: Applications to integrated navigation*. AIAA Guidance, Navigation, and Control Conference, Providence.
27. Julier, S. J. (2002). *The scaled unscented transformation*. Proceedings of the 2002 American Control Conference, Anchorage, AK, USA, 6, 4555–4559.
28. Wang, X., Li, T., Sun, S., & Corchad, J. (2017). A survey of recent advances in particle filters and remaining challenges for multitarget tracking. *Sensors, 17*(12), 2707.
29. Guo, L., Ge, P., He, D., & Wang, D. (2019). Multi-vehicle detection and tracking based on Kalman filter and data association. In H. Yu, J. Liu, L. Liu, Z. Ju, Y. Liu, & D. Zhou (Eds.),

Intelligent robotics and applications, ICIRA 2019, lecture notes in computer science, 11744. Springer.

30. Chen, X., Wang, X. & Xuan, J. (2018). *Tracking multiple moving objects using Unscented Kalman Filtering techniques*. arXiv:1802.01235.
31. Kuhn, H. W. (1955). The Hungarian method for the assignment problem. *Naval Research Logistics Quarterly, 2*, 83–97.
32. Munkres, J. (1957). Algorithms for the assignment and transportation problems. *Journal of the Society for Industrial and Applied Mathematics, 5*(1), 32–38.
33. Bourgeois, F., & Lassalle, J. (1971). An extension of the Munkres algorithm for the assignment problem to rectangular matrices. *Communications of the ACM, 14*(12), 802.

Integrating Spatial with Qualitative Data to Monitor Land Use Intensity: Evidence from Arable Land – Animal Husbandry Systems

Thymios Dimopoulos, Christos Vasilakos, and Thanasis Kizos

1 Introduction

1.1 Land Use Intensity and Farming Systems

Agriculture was intensified in the twentieth century with extensive mechanization and industrialization to cover the dietary needs on the growing global population [1], however, with severe impact upon resources [2], nutrient cycles [3], ecosystem services [4], and biodiversity [5–7]. The quantification and mapping of this intensity remains a challenge [8]. Some of the most important challenges are related to how land cover/land use is defined and how we can incorporate actual practices of farming systems practiced at the individual field level and project them at a landscape level.

Land cover is usually defined as *"the observed (bio)physical cover on the earth's surface"* [9]. It is typically used to describe vegetation and man-made features, but also areas where the surface consists of bare rock or bare soil and water surfaces. Although land cover appears as a straightforward concept, it is strongly influenced by the scale and the sources that are used for the definition of the classes that are used. Scale dependence implies that the classes used are influenced by the level of detail within each cover and between different covers, while source dependence means that the classes are related to the means used to collect information (satellite imagery, aerial photography, field survey, or their combinations) [10].

T. Dimopoulos (✉)
Mediterranean Institute for Nature and Anthropos – MedINA, Athens, Greece
e-mail: thymiodimo@med-ina.org

C. Vasilakos · T. Kizos
Department of Geography, University of the Aegean, Mytilene, Greece

D. D. Bochtis et al. (eds.), *Information and Communication Technologies for Agriculture—Theme II: Data*, Springer Optimization and Its Applications 183,
https://doi.org/10.1007/978-3-030-84148-5_7

Land use, on the other hand, is "*characterized by the arrangements, activities and inputs people undertake in a certain land cover type to produce, change or maintain it*" [9]. Land use therefore establishes a direct link between land cover and how people utilize the land and socio-economic activity. Despite the differences in their content, the two terms are being used interchangeably [10]. Moreover, land cover often substitutes insufficient knowledge for land use [11], since satellite imagery allows rapid classifications compared to more elaborate needs for characterizing land use. The same is true for landscape change as well, where many studies claiming to analyze landscape change, quite often worked with mapped information showing land cover and at best certain aspects of land use (see e.g., [12]).

The quantification of land use intensity has been pursued through i) inputs, ii) outputs, and iii) changes of system properties [8, 13]. Inputs that are used include the size of cultivated land, technological capital, mechanization level, plant protection products (expressed as quantity per ha), and human capital (labor) [8, 14]. Outputs that are used include harvests per area and time, and change in system properties – that are affected by inputs and in a smaller degree by outputs – are expressed with concepts such as biodiversity, complexity, primary biomass production, water and nutrient availability in the soil [13, 15], with different spatial metrics proposed. Most of these metrics are related to data from satellite images, sometimes combined with field measurements since land cover and other spectral data cannot cover all aspects of intensity [8, 16–18]. Outputs can be calculated with the use of spatial modeling [19–21] and some of the aspects of system properties can also be estimated or calculated with time series and landscape metrics. But, other system properties require extended field measurements, such as biodiversity [22–25], ecosystem services [26, 27], or qualitative research data [28–32]. Inputs are also elusive, as different levels of intensification of similar practices can have very similar or even identical spectral signatures [33, 34] and again statistical, field, or qualitative data have to be utilized [35, 36].

Farming systems represent efforts to classify together individual farms in a location. Even if all farms are inherently different in size, number of plots, family circumstances, resource availability, structural characteristics (equipment), etc., we can group together farms or crops/animals within farms, under a "farming system", that have *"broadly similar resource bases, enterprise patterns, household livelihoods and constraints, and for which similar development strategies and interventions would be appropriate"* ([37], p.2). Therefore, within them, *"farm households with similar characteristics and constraints"* ([37], p. 2) are grouped, although inevitably within such systems there will be a considerable degree of heterogeneity. Dixon, Gulliver and Gibbon ([37], p.2) propose a number of key factors to group farms into farming systems, including: *"(i) the available natural resource base; (ii) the dominant pattern of farm activities and household livelihoods, including relationship to markets; and (iii) the intensity of production activities"*. The use of cultivation practices, including inputs and outputs, can also be used for a classification and this links farming systems with land use intensity. In this sense, farming systems are groups of farms of the same crop/animal type that use similar inputs in type, magnitude, and intensity, and produce similar outputs and similar impacts to natural resources, biodiversity, and landscape.

The understanding of changes in land use intensity in mixed crop-livestock systems requires an understanding of the level of integration of crops and livestock, as interchanges between them affect both inputs and outputs. In order to quantify changes in land use intensity one has to take into consideration the interaction between crop and livestock production (see [38, 39] for more insights on crop-livestock interactions).

1.2 Land Use/Land Cover (LULC) Extraction

Remote sensing can assist planning and application of agricultural practices by providing information on crop status at different spatial and temporal scales for different end-users [40, 41]. Weiss et al. [42] report research developments in agricultural applications including crop breeding, agricultural land use monitoring, crop yield forecasting, as well as ecosystem services in relation to soil and water resources or biodiversity loss. One widely used application of remote sensing in agriculture is the mapping and monitoring of agricultural land abandonment. The space and time of agricultural land abandonment are essential to understand the environmental and social consequences of these transitions and at the same time to distinguish them from other transitions such as fallow land [43]. Different data sources have been used including very high spatial resolution orthophotomaps [44], Landsat [45] to 250-m MODIS [46] and combinations such as hyperspectral images and Light Detection and Ranging (LiDAR) data [47].

LULC extraction from remote sensing data is based on the spectral and spatial properties of targets in a study area. The growth of computational approaches, the evolution of sensors' characteristics, and the availability of satellite data have fueled the development of different approaches in image classification. The most widely used methods are the supervised ones, including the traditional approaches of maximum likelihood and minimum distance, as well as more recently, the modern machine learning classifiers, especially as pixel-based classifiers [48–52].

Opposing to pixel-based classifiers, the object-based classification of remote sensing data has been extensively explored not for classifying single pixels, but groups of pixels that represent already existing objects in a GIS database [53]. This approach is well known as object-based image analysis (OBIA) and is applied in high and very high-resolution images. OBIA approaches have shown outstanding performance in spatial planning and monitoring [54]. Li et al. [55] demonstrated very high accuracy in a cropland mapping based on OBIA segmentation with a time series of Landsat-MODIS dataset. A research comparing pixel-based and OBIA-based classification in a complex agricultural landscape showed that the simultaneous usage of both spectral and spatial/geometric features of image objects improved the land use classification, oppose to the independent usage of spectral or spatial/geometric feature [56]. Csillik et al. [57] compared multiple single-band

and multi-band object-based time-constrained Dynamic Time Warping (DTW) classifications for crop mapping based on Sentinel-2 time series of vegetation indices. Sentinel-2 has been also used in mapping the extent and the location of cultivated fields based on OBIA and the results revealed significantly higher field boundary accuracies than the supervised per-pixel approach [58]. Finally, OBIA method combined with Random Forest classifier and spatiotemporal rules applied in Spot 6/7, Sentinel-2, and Landsat-8 data improved the crop type mapping in a complex agricultural landscape, characterized by small field size, fragmented landscape, and highly variable cropping practices [15, 59].

Today, very high-resolution multispectral images are available for high precision land use mapping. However, multitemporal classification often is based on aerial black and white (B&W) panchromatic images. There is an extended archive of such images with some imagery dating back to the 1930s for some part of the world [60], oppose to multispectral images of Landsat with lower resolution dating back to 1970s. One of the main drawbacks of these images is the limited spectral information. The land use extraction is based mainly on targets' features, i.e., shape and texture thus their usage is limited to few studies [61].

This chapter investigates changes in land use intensity in a crop-livestock farming system. A mixed methodology is presented that combines data of LULC types extracted with remote sensing techniques from B&W aerial images, statistics, and qualitative data. From these data, changes in land use intensity are assessed through changes in land use management and agricultural land area. A synthesis of the results of these different tools demonstrates how land and capital was managed on Lemnos from 1960 to 2000 through its mixed crop-livestock farming system.

2 Methodology

2.1 Study Area

Lemnos is a Greek island (478km^2, 260 km coastline), partly volcanic, whose gentle geomorphology is in contrast with most of the Greek islands (Fig. 1). Vegetation on the island is mainly phrygana and low maquis along with extensive areas of arable fields and vineyards. The climate is Mediterranean semi-arid, with mild winters and hot dry summers, strong winds, and annual precipitation of 500 mm [62].

The adaptability of local farmers to the environmental and socio-cultural conditions of Lemnos has formed a particular land use system, the '*mandra'* system, which shapes the landscape of the island. This mixed crop livestock farm system is mainly characterized by, and organized around, the element of the '*mandra'* – a traditional stone wall structure with an animal shed, the farmer's/shepherd's hut, and a barn. On the periphery of the '*mandra'* the pastures and/or arable land with cereal and leguminous crops destined for fodder and/or human consumption (Fig. 2).

Fakos study area is located at the southwest part of Lemnos, covers 54.50 km^2, and includes the communities of Kontias, Tsimandria, and Portianou. The crop-

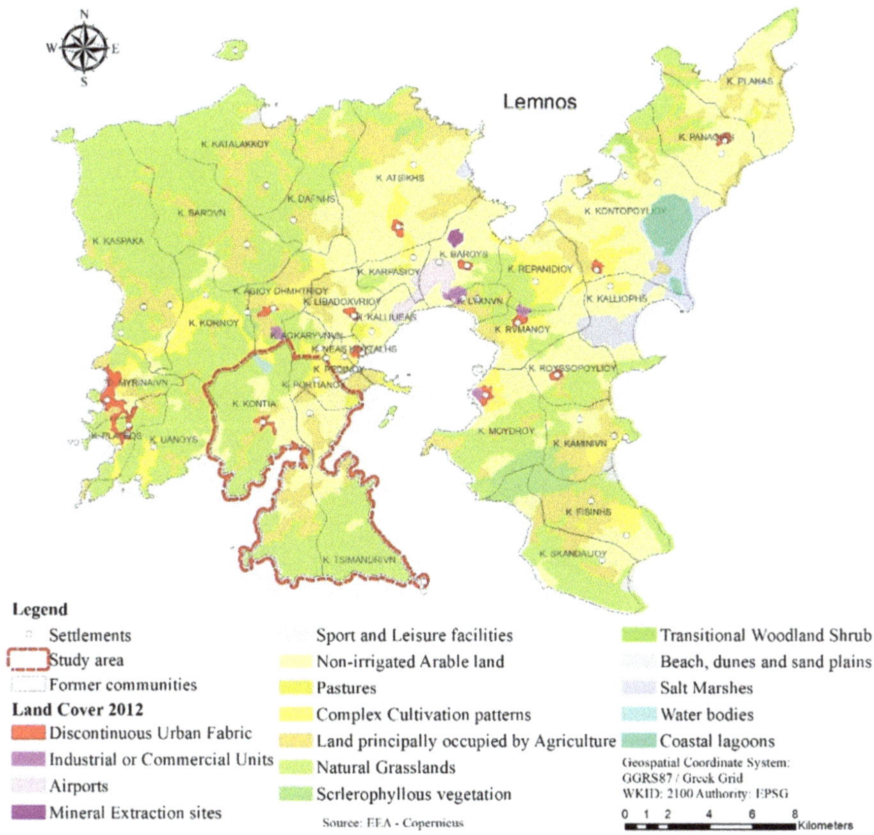

Fig. 1 The study area of Lemnos Island, Greece

livestock '*mandra*' system in Fakos has presented elements of intensification in the post-war period. The number of farms has declined by 70% in the period 1961–2000 (from 578 to 179 farms), but the Utilized Agricultural Area (UAA) has increased by 5.4% according to agricultural census by the Hellenic Statistics Authority [63] (Table 1). At the same time the number of parcels has decreased by 55% and their average size has increased indicating aggregation of land to fewer and larger farms. This intensification trend is also present in terms of livestock breeding, as the number of sheep per farm increased from 30 to 100.

In a published survey conducted in 2018 in Lemnos which focused on recording agricultural practices [64], three types of farm types were deduced from the analysis of farm characteristics:

- *Group A* includes small- and medium-sized mixed farms which present a balanced portion of cultivated land and rent more than half of their UAA, while the ratio of animals per UAA is high;

Fig. 2 The 'mandra' system is a crop-livestock system that produces a mosaic of agropastoral land uses

- *Group B* represents big mixed farms with many animals, in which a large part of the land (almost 50%) belongs to the farmers;
- *Group C* includes large land farming holdings, which rent almost all their farmland.

In the Fakos area, farms were dominantly type A (over 70%) and type B. The complexity of agricultural practices was analyzed with three composite indexes:

- *Diversification*, more related to crop rotation, number of crop types, fallowing, and intercropping;
- *Complementarity*, expressing the relations between cultivation and grazing patterns;
- *Intensification*, related to pesticide use and green manure.

Fakos area presents a relatively high score on diversification, a relatively high complementarity score, and a rather moderate intensification score. This leads to diverse farms in terms of crops and their practices, with less important than other localities but still significant complementarity with grazing and less intensification than other localities.

Table 1 Agricultural statistics of Fakos study areas in 1960 and 2000

	Population	Number of holdings	Utilized agricultural area (ha)	Arable land (% UAA)	Grazing land (% UAA)	Grazing/Arable	Irrigated land	Parcels	Parcel size	Number of sheep	Sheep per farm
1960	2188	578	4162.3	39.05	57.88	1.48	168.70	4148	1.003	9236	27.9
2000	1248	179	4388.66	33.69	65.58	1,94	125.80	1885	2.328	13,652	100.7
Change (%)	−42.96	−69.03	5.44	−13.71	13.30	39.06	−25.43	−54.56	132.10	47.81	260.82

2.2 *Materials and Methods*

In an effort to measure and understand changes in land use intensity in Fakos area, we have combined quantitative and qualitative data that assess different insights of these changes. This mixed method included the following steps: first, the timeline of historical changes was compiled based on secondary literature, statistical information, and information from key local informants. Then, remote sensing was used to assess land cover changes and conversions that represent changes in intensity. This information was complemented by participatory mapping with local farmers. A comparison and synthesis of results was used in the end.

Timeline of Changes

Statistical data on agriculture provided by the Hellenic Statistics Authority along with various sources was used to create a basic timeline of the agricultural history of Lemnos. An open interview with two key informants of the regional department of agricultural economy in Lemnos was conducted to explore the broader time frame of changes in agriculture. The questions focused primarily on historical changes of the agricultural history which brought changes in production, practices, and characteristics of agricultural holdings. This information was used to build a narrative of events and understand changes in intensity of land use, land expansion, and abandonment.

Remote Sensing Data

Black and white panchromatic air-photographs were obtained for 1960 (1:30,000), 1980 (1:35,000), and 2002 (1:40,000) and a time series of three photo-mosaics for each study site was developed by applying the Structure-from-Motion workflow [65, 66] within Agisoft Photoscan software [67]. The photo-mosaics produced, overlapped 97% of the area covering 5316.4 ha. Prior to analysis, the photo mosaics were georeferenced and rectified to EGSA87 Greek Grid projection with the same resolution in order to make comparison between images possible.

Land cover maps for 1960, 1980, and 2002 were produced by aerial photographs using additional texture features, through object-oriented image analysis (OBIA). Seven classes were selected based on ground truth data obtained through a survey conducted in 2018 [64], as well as on data from vegetation maps of 1996 from the Ministry of Agriculture. The main distinction that was used to assess land use intensity changes was between grasslands and crops, while other classes were used supplementary to describe change from agricultural land to other (Urban, Shrublands, Bare, Waterbodies, and Wetlands).

Segmentation is a very important procedure of OBIA and the overall accuracy of classification mainly depends on the extracted features during this process

[68, 69]. In image segmentation pixels are aggregated into spectrally homogenous image objects using segmentation algorithms [70]. In multispectral image classification, the spectral signature of each image pixel in the multidimensional feature space is used to extract information during the segmentation process. However, in panchromatic images such as B&W aerial photographs, a single-dimensional space provides limited information and makes segmentation more difficult. In order to produce image objects that represent broad categories such as grasslands or urban, some processing before the segmentation and classification process has been applied.

Texture features were used to enlarge signature space and provide information on the pixel level for the segmentation process. Haralick et al. [71] have provided the most accredited and well-known texture measures based on the Grey Level Co-occurrence Matrix (GLCM). Caridade et al. [72] have tested the Haralick features in pixel-based classification, suggesting that the method has proven useful with good results. Halounová [73] uses GLCM in OBIA in order to find solutions for automatic information extraction from B&W aerial orthophotographs.

From several GLCM texture features, three were selected (mean, homogeneity, contrast) to enlarge the original image. Additional information was selected to contribute to OBIA analysis and especially to the segmentation process, based on the results of the discussion with key informants, as slope seems to contribute in distinguishing crops from grassland. As machinery was used to work the land, high slope became an obstacle for modern agriculture. In this way selecting slope as a layer that contributes to OBIA was considered important. Vogels et al. [61] have selected slope as a covariate to enhance the discrimination between 'cropland' and 'other land cover' in the classification procedure.

In image segmentation there are several methods and algorithms (for more information see [74]). Here, a multiresolution segmentation in a three-level procedure was applied. This is done in order to overcome the problem of under-segmentation as pointed out by Liu and Xia [70]. The scene is segmented at a scale of 400, classified, then specific classes are segmented again at a scale of 200, reclassified, and then the same procedure is done again for a scale of 75. This procedure has the advantage of refining the results in order to get more precise classification and more meaningful features, avoiding under-segmentation.

After the multiresolution segmentation, supervised classification based on nearest neighborhood classification was applied. The classifier used for this study is based on fuzzy membership functions [75]. Supervised classification is based on training algorithms with characteristics of sampling areas that have been chosen for each land cover class. Through this procedure, spectral signatures are created for each land cover based on the statistics, texture, form and mutual relations of image objects that are used to specify each object's class definition. In order to build these sampling areas, ground truth data, or prior knowledge is required. Sample selection for each class was selected through interpretation of the photomosaics as no ground information was available on those dates. The basic distinction for samples was defined by homogeneity of brightness, regularity of shapes, size, similarity to other bordering image objects, and direction.

After training the classifier for each class of the classification date, multiple class descriptions were selected in order to build the fuzzy expression which allows the evaluation of each feature for classification [76]. A threshold for the required membership degree is defined. This threshold takes values from 0.0 to 1.0, where 1.0 expresses probability (a complete assignment) to a class and 0.0 expresses absolutely improbability. For this study it was set to 0.2. If this threshold is not reached, the object remains "unclassified'. The class description parameters selected for each date are presented in Table 2.

The classification was followed by the accuracy assessment. The commonly used accuracy assessment elements in pixel-based change detection include overall accuracy, producer's accuracy, user's accuracy, and the kappa coefficient [77]. However, a basic difference with object oriented-based analysis is that study units can be either pixels or image objects. Khorram et al. [78] argued that pixel-based accuracy assessments tend to underestimate object-based map accuracy. In this study, a polygon-wise validation is used with an error matrix that was provided from E-Cognition software based on image objects. Conducting a thorough classification accuracy assessment requires validation data to compare to classification predictions. As this was not possible, interpretation of the aerial photos was needed in order to assess classification accuracy.

To produce the error matrix with a Test and Training Area (TTA) mask in E-cognition, two scenes were used. Classification as produced from the classifier and a revised version that misclassification errors were corrected based on interpretation of the aerial photos. From this comparison three accuracy measures were produced: Producer's, User's, and Overall Accuracy. Producer's accuracy indicates how well training set image objects of the given cover type are classified. User's accuracy indicates the probability that an image object classified into a given category actually represents that category on the ground [76]. Kappa statistics indicate how much the classifier omits errors, compared to a randomly assigned classification.

Post classification comparison of the land cover results provided changes that occurred between these periods, while cross tabulation was used to assess land conversions. These land conversions were synthesized in overarching processes in an effort to find evidence of changes in intensity of land use, grasping both agricultural expansion or abandonment and management of agricultural land (increase/decrease in labor capital and inputs) (Table 3).

2.3 Participatory Workshop

A focus group workshop targeting local farmers of an elder age (over 50 years) was held to mark changes in the same period, from 1960 to 1980 and 2002. The workshop took place on July 10, 2019 in one of the villages of the area, Portianou. Participants were nine mostly aged (older than 60) still active farmers with mixed farms, covering a range of farm sizes. A participatory mapping approach was used to

Table 2 Classification parameters used for each year classification

	Original	Max difference	Area	Length/Width	Asymmetry	Border index	Elliptic fit	Main direction	Radius of largest enclosed ellipse	Radius of smallest enclosing ellipse	Rectangular Fit	Roundness	Number of Edges
1960	✓	✓			✓	✓	✓	✓	✓	✓		✓	✓
1980		✓	✓	✓	✓	✓	✓	✓	✓	✓	✓		✓
2002		✓			✓	✓	✓	✓	✓				✓

Table 3 Characterization of land use changes

		To						
		Grasslands	Crops	Urban	Vegetation	Bare	Water	Wetlands
From	Grasslands	–	Intensification	Other	Abandonment	Abandonment	Other	Abandonment
	Crops	Extensification	–	Other	Abandonment	Abandonment	Other	Abandonment
	Urban	–	–	–	–	–	–	–
	Vegetation	Other	Agricultural expansion	Other	–	Other	Other	Other
	Bare	Other	Agricultural expansion	Other	Other	–	Other	Other
	Water	Agricultural expansion	Agricultural expansion	Other	Other	Other	–	Other
	Wetlands	Agricultural expansion	Agricultural expansion	Other	Other	Other	Other	–

record what land use changes they have witnessed, as farmers were asked to point out on maps with the help of place names and other landmarks such as *'mandras'* (livestock pens).

3 Results and Discussion

3.1 Image Processing

Table 4 presents the results of mosaicking and georeferencing of the tiles for each year. For each year, the root mean square error (RMS) of the co-registration was quite low. Figure 3 presents the homogeneity, the contrast, the variance, and the second moment based on the 1960 images. Both homogeneity and contrast capture well the distinction between crops and urban or shrublands, providing extra information for enlargement of the picture's signature space.

The multiresolution segmentation in three steps is presented in Fig. 4. In every step an evaluation of the classification and manual correction of some results was necessary in order to avoid the replication of errors in lower levels. The

Table 4 Co-registration of the images

Year	Scale	Tiles	Pixel size (m)	Points of reference	RMS (m)
1960	1:30000	10	0.635	305	5.25
1980	1:35000	12	0.741	265	5.16
2002	1:40000	14	0.847	250	5.19

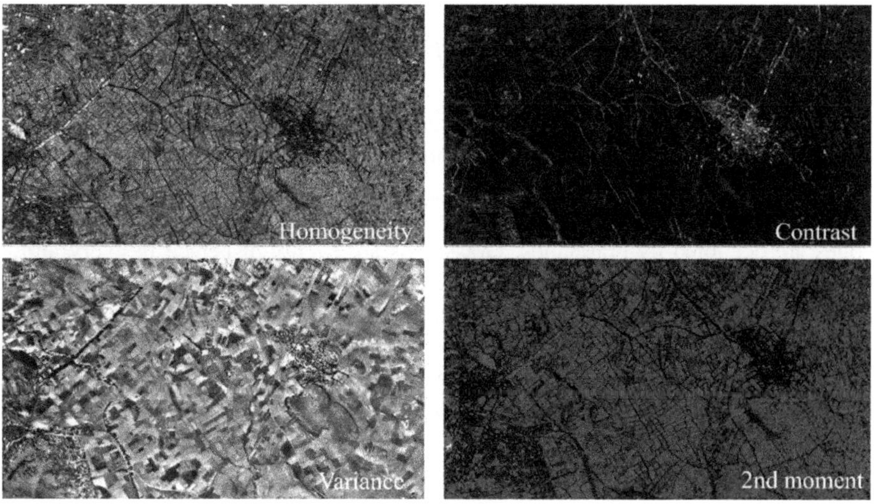

Fig. 3 GLCM textural features for 1960 photomosaic

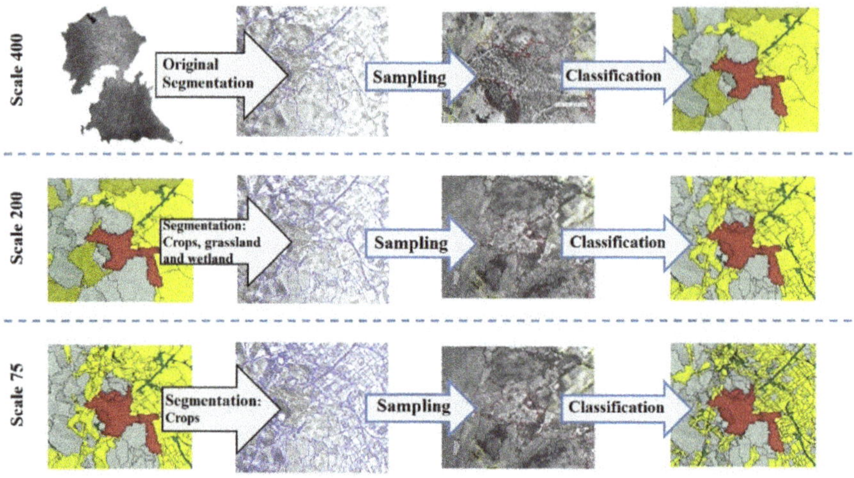

Fig. 4 Segmentation and classification three-step procedure

multiresolution segmentation algorithm identifies single image objects of one pixel in size and merges them with their neighbors, based on relative homogeneity criteria. This homogeneity criterion is a combination of spectral and shape criteria [79]. In order to create meaningful image objects, the following parameters were used for each year orthophoto. The use of different texture features has resulted in separate adequate combination for each year. Table 5 summarizes the settings that were used during segmentation.

3.2 *Land Cover Type Extraction and Change Detection*

The overall classification accuracy for each date is given in Table 6. Producer's Accuracy for Grasslands and Crops is high (over 0.87) while other classes have lower scores.

Crops, as expected, are prominent in lowlands whereas the higher altitudinal zones are occupied by grasslands. Grasslands are evenly dispersed in every altitudinal zone as localized geomorphological factors can limit crop cultivation even in the lowlands, leaving space for grasslands. Bare land can be found mainly in the area of Fakos as the peninsula has steeper slopes, both at its shores as well as inland. Moreover, linear features such as roads were also categorized under this class. Bare land was in some cases misclassified under grasslands. The appearance of water class in years 1980 and 2002 is a result of the construction of Aghios Dimitrios irrigation dam. Shrubland vegetation at this scale has been identified as ravines and linear features such as tree lines alongside main roads. This has contributed to misclassification errors that are represented in the low producer's accuracy. Urban

Table 5 Segmentation parameters used for each year

Date	Shape	Compactness	Band combination				
			Original	Contrast	Homogeneity	Slope	Mean
1st level: scale 400							
1960	0.2	0.8	1	1	0	0.7	0
1980	0.2	0.7	1	1	0	0.5	0
2002	0.2	0.8	1	1	1	1	0
2nd level: Scale 200							
1960	0.2	0.6	1	0	1	0.7	0
1980	0.2	0.6	1	1	0	0	0
2002	0.2	0.8	1	1	1	0	0
3rd level: Scale 75							
1960	0.6	0.3	0	0	1	0	1
1980	0.2	0.6	1	1	1	0	0
2002	0.2	0.6	1	1	1	0	0

areas occupy the edge of the main plains at the feet of hills over-looking the plain and are evenly dispersed across the mainland. There are two areas that appear constantly as wetlands north and south of Diapori, as well as two smaller areas east and west of Nevgatis which appear in the classification of 1960 and 2002 but not in 1980.

Crops and grassland constitute over 90% of the area in all three dates and in overall there seems to be an increase (4%) in agricultural land from 1960 to 2002, a fact validated by the agricultural statistics (Table 1). The other land cover classes are marginal as only bare land has a presence near to 4% and all others account for less than 2%. Grassland increased from 50% in 1960 to 60% in 1980 and decreased to 55% in 2002 presenting an increase of 13.5% compared to 1960, which is also documented in ELSTAT statistics (Table 1). Crops represented 41% in 1960, shrinking to 32% in 1980 and increased to 37% in 2002, presenting an overall loss of −8%, slightly different from −14% that is given in ELSTAT statistics.

Change represents 22.18% of the area in the period of 40 years, while 12.06% has changed more than once. The most important change seems to be the interchange of land between the two dominant classes. The small area size of other classes can lead to erroneous conclusions as spectral and resolution differences may cause unexpected changes due to false classification. Particularly changes from vegetation or bare to grasslands were considered erroneous as in steep slopes, such as ravines or cliffs, there were misclassification issues between these three classes. Also change from urban areas to other uses has been ignored as it is highly unlikely for such a transformation to take place.

Table 6 Classification results

		Grasslands	Crops	Urban	Shrublands	Bare	Waterbodies	Wetlands
Error matrix 1960	Producer's Accuracy	0.885	0.874	1	0.711	0.871	-	0.692
	User's Accuracy	0.885	0.882	0.982	0.808	0.583	-	0.777
	Kappa Index Per Class	0.788	0.764	1	0.706	0.868	-	0.688
	Overall Accuracy: 0.875							
	Kappa Index Overall 0.781							
Error matrix 1980	Producer's Accuracy	0.937	0.969	0.823	0.699	0.468	1	0.259
	User's Accuracy	0.971	0.873	1	0.979	1	1	0.891
	Kappa Index Per Class	0.879	0.944	0.82	0.695	0.461	1	0.256
	Overall Accuracy: 0.924							
	Kappa Index Overall: 0.866							
Error matrix 2002	Producer's Accuracy	0.997	0.972	0.879	0.634	0.493	1	0.795
	User's Accuracy	0.823	0.864	0.985	0.786	1	1	0.623
	Kappa Index Per Class	0.996	0.941	0.8	0.63	0.484	1	0.792
	Overall Accuracy: 0.909							
	Kappa Index Overall: 0.842							

3.3 *Land Conversions*

Land use trajectories revealed that extensification is the basic trend as cropland transformed to grasslands represents 38.7% of all changes from 1960 to 2002. Intensification coexists in at the same time, representing 25%, as grasslands are converting to crops. Agricultural expansion represents 8.7% as different land uses are turned to agricultural land. Cumulatively intensification and agricultural expansion account for 33.7% of the changes. Abandonment, which represents a variety of transformations from agricultural land to bare land, shrubland vegetation, and wetlands, stands for 12.45% of total land transformations in the period of 1960–2002.

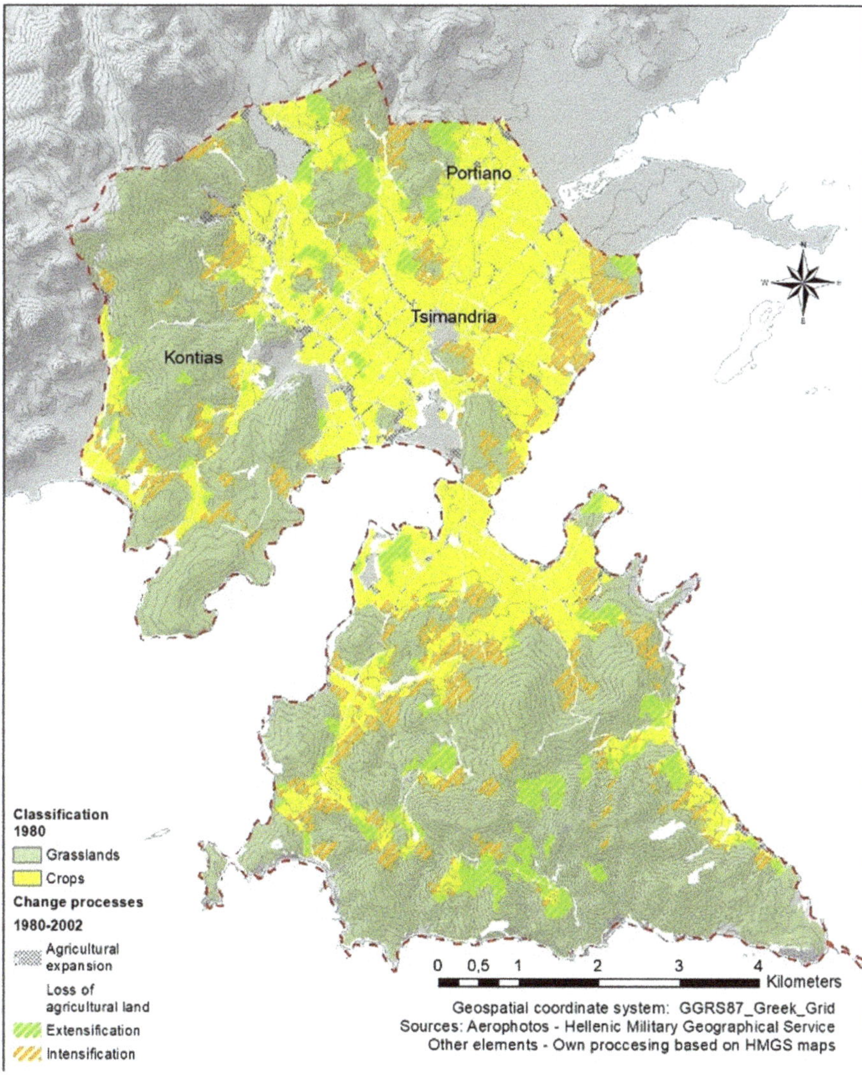

Fig. 5 Land cover types of 1980 and changes in 1980–2002

Looking at the two periods separately, although the main processes coexist there seems to be a distinct pattern between them (Fig. 5 and Fig. 6). Extensification seems to be the main process of change from 1960 to 1980 affecting mainly the hilly uplands as more remote and marginal areas are being converted from crops to grasslands or are abandoned, whereas intensification is the main trend for the next 20 years mainly in the lowlands as modernized agriculture (irrigation fields, aggregation of land, use of machinery) replaces more extensive land uses and traditional landscape elements such as tree hedges.

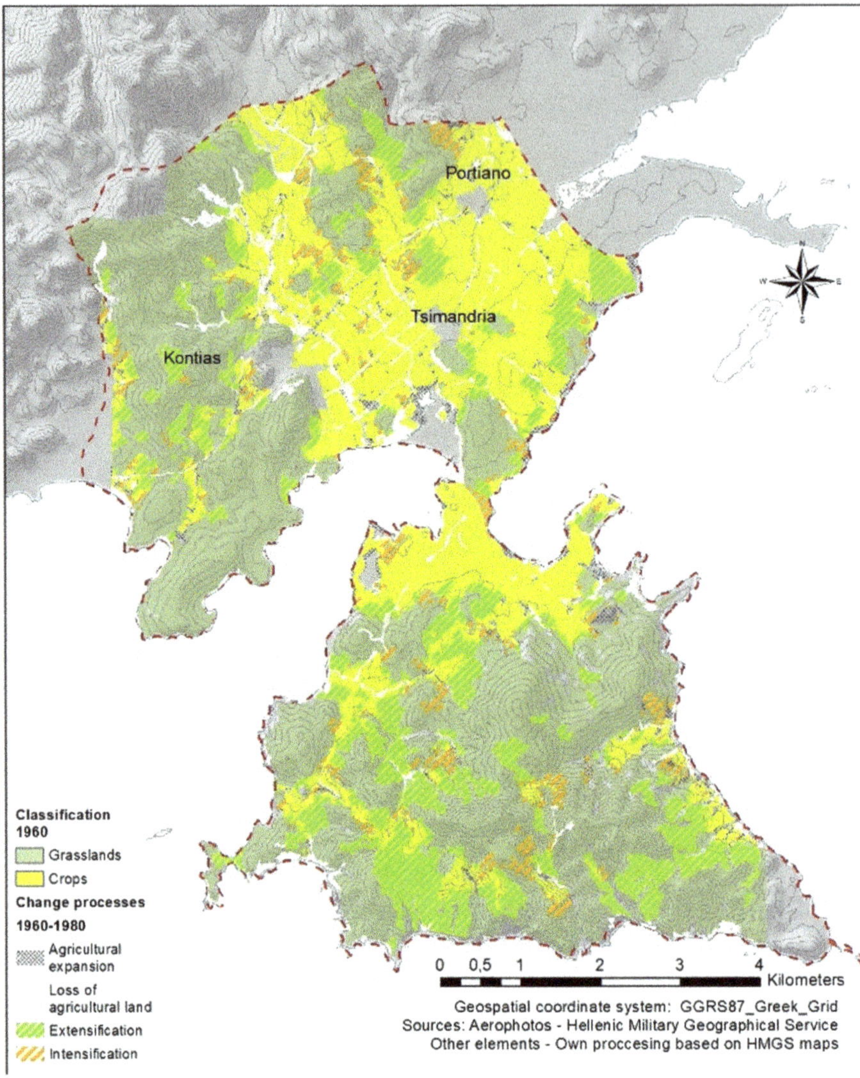

Fig. 6 Land cover types of 1960 and changes in 1960–1980

3.4 Results from Qualitative Methods

Results from the participatory mapping were grouped into broader narratives that correspond to the selected processes of change and persistence. These narratives were enriched, and cross-checked with information from key informants.

- *Cotton to cereals*: Cultivation of cotton was gradually abandoned during the 1960s and changed to cereal crops. This process is characterized as extensification as it has been a change between irrigated crops to rain-fed crops and thus a decrease in technological capital and inputs. It has been related to the abandonment of wells that opened in the 1950s for cotton cultivation.

- *Rainfed to irrigated*: Thc construction of the dam and the irrigation network in the plains of Kontias and Tsimandria in the mid-1970s contributed to the change from cereals to clover, corn, and vegetables.
- *Cropland to grassland*: This process of extensification was directly connected with the use of machinery and access to roads. In other parts of the island this has been connected with the abandonment of terraces.
- *Grasslands to cropland*: Participants recognized some areas where grasslands have turned to cropland, after 1980. This intensification process was a result of agricultural machinery used to plow low slope areas.
- *Abandonment*: In the case of southeastern Fakos peninsula participants described the area as abandoned, i.e., neither cropland nor grasslands. This process was directly related to abandoned holdings (mandras).
- *Persistent areas*: When it came to defining what has remained unchanged, participants tended to refer to specific holdings (mandras) where land is still cultivated inside areas that had undergone some kind of change, usually extensification or abandonment.

3.5 Comparison and Synthesis of Results

The overarching processes derived from remote sensing results on land cover change conversions describe changes mostly related to agricultural abandonment and land use, from grassland to cropland and vice versa. Qualitative data describe a wider range of conversions. The recorded conversions from both remote sensing and quantitative methods describe the changes in land use intensity as shown in Fig. 8. Technological capital and inputs increase as rain-fed crops are turning to irrigated grasslands or other land covers are turning to crops. Labor and inputs decrease as irrigated crops are turning to rain-fed, crops are turning to grassland, and agricultural land turns to other land covers (is being abandoned). This wide range of land use intensity changes cannot be captured by remote sensing techniques alone.

Furthermore, results from remote sensing describe many areas to remain unchanged as 65.76% of total land cover has not been transformed at all in both periods. Comparison with workshop results reveals that although these areas have not witnessed land cover changes, changes in management and land use intensity are evident (Fig. 7). Especially the plains of Kontias and Tsimandria have undergone a double process of extensification and intensification, during the period of 1960 to 1980 that was captured only through qualitative methods. Evidence from remote sensing indicated that there has been a change in the structure of the fields during that period that can be related to the construction of the irrigated network as a regular road network appears and some linear tree hedges disappear. This structural change is reflected also in statistical data for the period, with parcel size increasing by 132%.

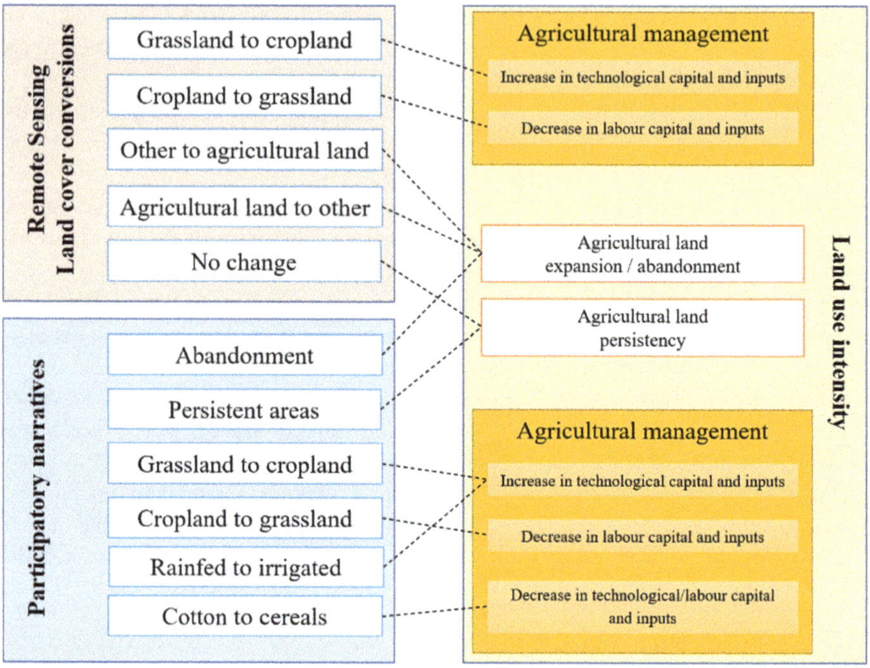

Fig. 8 Land use conversions from remote sensing and participatory approaches and relation to land use intensity

Results from remote sensing on land abandonment and agricultural expansion are marginal and were not confirmed by farmers. On the contrary, the area of Kastria and the southeastern part of 'wild' Fakos, that was described through remote sensing as unchanged, was marked as abandoned in 2002 by the farmers.

3.6 Farming Systems and Land Use Intensity

To understand the changes in land use intensity a reference to the characteristics of the *mandra* farming system is required. The characteristics of the agricultural practices of holdings of Fakos today, as presented in Dimopoulos et al. [64], show that complementarity plays a big role in the management of land in Lemnos, along with diversification, while intensification (represented as pesticides and manure inputs) is significantly less important.

Following the timeline, the role of complementarity throughout the period is evident with alternate strategies developed by farmers. To overcome the loss of income by the decline of cotton production farmers invested in livestock, making use of their land either for production of commercial crops or for fodder (or forage) or turning crops to grasslands (Fig. 9). Outputs were sold in markets directly or

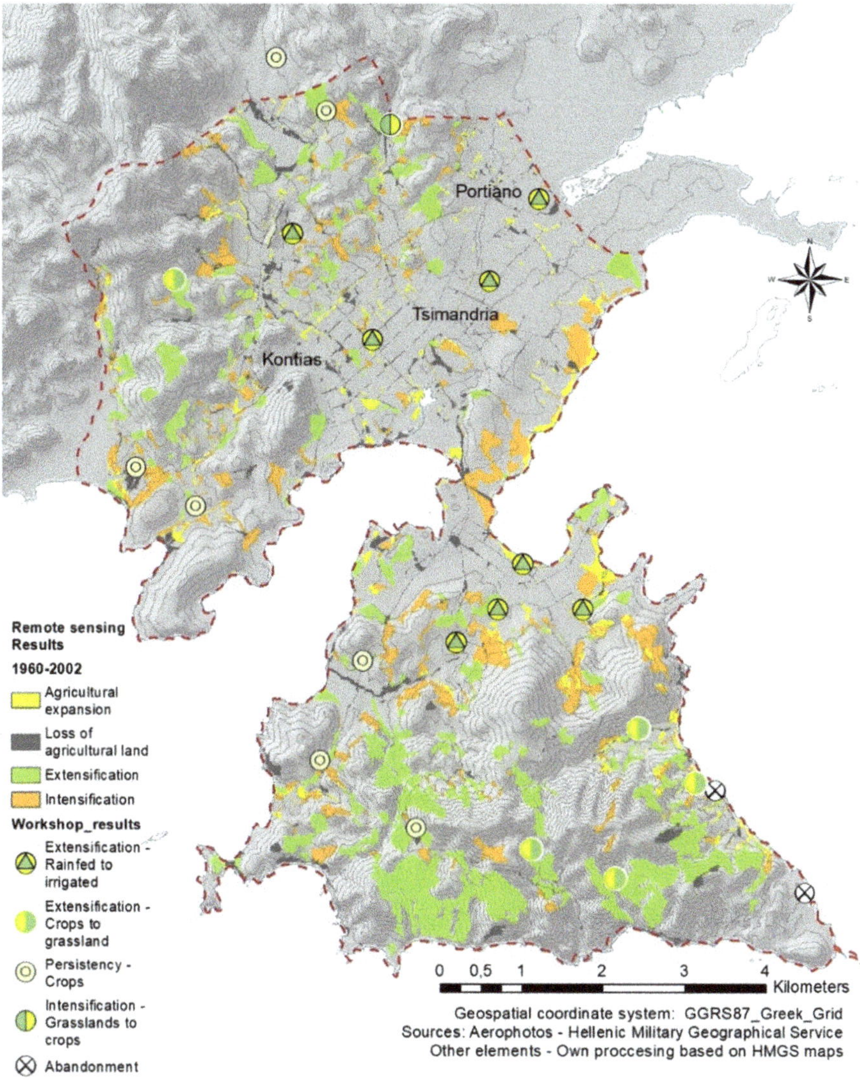

Fig. 7 Comparison of participatory mapping workshop and remote sensing results on land use intensity changes

processed to added value products such as cheese and other dairy products. This narrative is supported by the fact that farms grew in size (both in terms of land and livestock) but retained their mixed character as the ratio of grazing to arable land shows (Table 1). In comparison, other islands of the Aegean during the same period experienced a dichotomy between crop cultivation and livestock breeding (Spilanis and Kizos, 2015). This indicates that farmers in Lemnos have adapted their farming system to the restricted capabilities of an island and the economic reality of

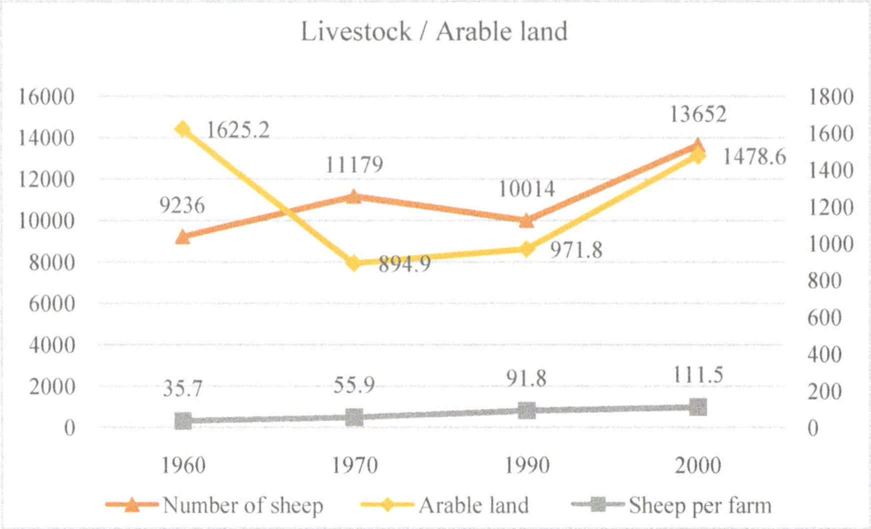

Fig. 9 Livestock and arable land

industrialized agriculture and market economy. This farming system has integrated changes in intensity within each own structure. While in other islands farmers in order to cope with integration to the market economy either had to intensify through specialization (of livestock breeding or olive orchards) or abandon, in Lemnos intensification came through diversification in production, investment of capital in livestock, complementarity, and aggregation of land.

4 Conclusion: Ways Forward in Integrating Qualitative Data in Land Use Intensity

The benefits of crop-livestock integration in mixed farming systems have been highlighted by several authors [80, 81]. However, a decline of integrated crop-livestock systems in commercial agriculture has occurred from 1960 to 2000 due to major structural landscape changes [82]. Thus, changes in land use intensity in a crop-livestock farming system should be thoroughly analyzed, mapped, and interpreted. The overall goal of this chapter was to investigate changes in land use intensity in a crop-livestock farming system through time using a combination of remote sensing techniques and qualitative methods.

Methodological plurality is a key to represent complex land use intensity changes. Within this research the Structure-from-Motion processing approach was applied to B&W aerial images in order to create an orthophoto map for each study period. This approach is user-friendly to non-experts users with no or limited knowledge of analytic photogrammetry [83]. Moreover, results demonstrated that remote sensing techniques such as OBIA can successfully extract information on land cover conversions from panchromatic aerial photographs. These land conversions can represent changes in land use management or agricultural expansion and abandonment. There are other land use management changes such as irrigated to rain-fed crops on the plains of Kontias and Tsimandria that were only recorded through qualitative methods and could not be captured with remote sensing. Only through comparison and synthesis of the results from both methodological tools can one retrieve a more accurate picture of changes in labor and capital invested in agricultural activities throughout the years. In this case study capturing these changes proved more relative than agricultural expansion or abandonment as the landscape seemingly remained unchanged (see [84]).

This case study shows that in some farming systems land use intensity changes cannot be represented through simple dichotomist differentiations. Changes from crops to grasslands and vice versa cannot represent holistically changes in land use intensity in crop-livestock farming systems, as investment in capital can take other forms such as livestock. In crop-livestock systems, the level of integration and complementarity might change in order to adapt to external drivers of change, redirecting labor and capital inputs within the same farming system. Changes in land use intensity in such cases should capture all the aspects of the farming system such as agricultural area, livestock, and the relation between the two.

Acknowledgments This paper is a result of the project "Translation of OAP activities into acknowledged landscape approaches (M6) - (17071)" funded by the MAVA Foundation for Nature.

References

1. Davis, K. F., Gephart, J. A., Emery, K. A., et al. (2016). Meeting future food demand with current agricultural resources. *Global Environmental Change*. https://doi.org/10.1016/j.gloenvcha.2016.05.004
2. Smith, P., House JI, Bustamante, M., et al. (2016). Global change pressures on soils from land use and management. *Global Change Biology*. https://doi.org/10.1111/gcb.13068
3. Foley, J. A. (2011). Can we feed the world & sustain the planet? *Scientific American*. https://doi.org/10.1038/scientificamerican1111-60
4. Schröter, D., Cramer, W., Leemans, R., et al. (2005). Ecology: Ecosystem service supply and vulnerability to global change in Europe. *Science, 80*. https://doi.org/10.1126/science.1115233
5. Herzog, F., Steiner, B., Bailey, D., et al. (2006). Assessing the intensity of temperate European agriculture at the landscape scale. *European Journal of Agronomy*. https://doi.org/10.1016/j.eja.2005.07.006
6. Stoate, C., Báldi, A., Beja, P., et al. (2009). Ecological impacts of early 21st century agricultural change in Europe - a review. *Journal of Environmental Management*.
7. Tilman, D., Cassman, K. G., Matson, P. A., et al. (2002). Agricultural sustainability and intensive production practices. *Nature*.
8. Kuemmerle, T., Erb, K., Meyfroidt, P., et al. (2013). Challenges and opportunities in mapping land use intensity globally. *Current Opinion in Environment Sustainability*.
9. Di Gregorio, A., & Jansen, L. J. M. (2000). Land cover classification system (LCCS): Classification concepts and user manual. *FAO*. https://doi.org/10.1017/CBO9781107415324.004

10. Lambin, E. F., Geist, H., & Rindfuss, R. R. (2008). Introduction: Local processes with global impacts. In *Land-use and land-cover change*.
11. Fisher, P., & Unwin, D. (2005). Land use and land cover: Contradiction or complement. In *Re-presenting GIS*.
12. Bürgi, M., Bieling, C., von Hackwitz, K., et al. (2017). Processes and driving forces in changing cultural landscapes across Europe. *Landscape Ecology*. https://doi.org/10.1007/s10980-017-0513-z
13. Erb, K. H., Haberl, H., Jepsen, M. R., et al. (2013). A conceptual framework for analysing and measuring land-use intensity. *Current Opinion in Environment Sustainability*.
14. Binswanger, H. P., & Rosenzweig, M. R. (1986). Behavioural and material determinants of production relations in agriculture. *Journal of Development Studies*. https://doi.org/10.1080/00220388608421994
15. Matson, P. A. (1997). Agricultural intensification and ecosystem properties. *Science, 277*(80), 504–509. https://doi.org/10.1126/science.277.5325.504
16. Beeson, P. C., Daughtry, C. S. T., Hunt, E. R., et al. (2016). Multispectral satellite mapping of crop residue cover and tillage intensity in Iowa. *Journal of Soil and Water Conservation*. https://doi.org/10.2489/jswc.71.5.385
17. Estel, S., Kuemmerle, T., Levers, C., et al. (2016). Mapping cropland-use intensity across Europe using MODIS NDVI time series. *Environmental Research Letters*. https://doi.org/10.1088/1748-9326/11/2/024015
18. Temme, A. J. A. M., & Verburg, P. H. (2011). Mapping and modelling of changes in agricultural intensity in Europe. *Agriculture, Ecosystems and Environment*. https://doi.org/10.1016/j.agee.2010.11.010
19. Doraiswamy, P. C., Moulin, S., Cook, P. W., & Stern, A. (2003). Crop yield assessment from remote sensing. *Photogramm. Eng. Remote Sensing*.
20. Fang, H., Liang, S., & Hoogenboom, G. (2011). Integration of MODIS LAI and vegetation index products with the CSM-CERES-maize model for corn yield estimation. *International Journal of Remote Sensing*. https://doi.org/10.1080/01431160903505310
21. Prasad, A. K., Chai, L., Singh, R. P., & Kafatos, M. (2006). Crop yield estimation model for Iowa using remote sensing and surface parameters. *International Journal of Applied Earth Observation and Geoinformation*. https://doi.org/10.1016/j.jag.2005.06.002
22. Bubová, T., Vrabec, V., Kulma, M., & Nowicki, P. (2015). Land management impacts on European butterflies of conservation concern: A review. *Journal of Insect Conservation*.
23. De Palma, A., Kuhlmann, M., Roberts, S. P. M., et al. (2015). Ecological traits affect the sensitivity of bees to land-use pressures in European agricultural landscapes. *Journal of Applied Ecology*. https://doi.org/10.1111/1365-2664.12524
24. Frenzel, M., Everaars, J., & Schweiger, O. (2016). Bird communities in agricultural landscapes: What are the current drivers of temporal trends? *Ecological Indicators*. https://doi.org/10.1016/j.ecolind.2015.11.020
25. Liu, Y., Rothenwöhrer, C., Scherber, C., et al. (2014). Functional beetle diversity in managed grasslands: Effects of region, landscape context and land use intensity. *Landscape Ecology, 29*, 529–540. https://doi.org/10.1007/s10980-014-9987-0
26. de Groot, R. S., Alkemade, R., Braat, L., et al. (2010). Challenges in integrating the concept of ecosystem services and values in landscape planning, management and decision making. *Ecological Complexity*. https://doi.org/10.1016/j.ecocom.2009.10.006
27. Rega, C., & Spaziante, A. (2013). Linking ecosystem services to Agri-environmental schemes through SEA: A case study from northern Italy. *Environmental Impact Assessment Review*. https://doi.org/10.1016/j.eiar.2012.09.002
28. Chen, N., Li, H., & Wang, L. (2009). A GIS-based approach for mapping direct use value of ecosystem services at a county scale: Management implications. *Ecological Economics*. https://doi.org/10.1016/j.ecolecon.2008.12.001

29. Maes, J., Egoh, B., Willemen, L., et al. (2012). Mapping ecosystem services for policy support and decision making in the European Union. *Ecosystem Services, 1*, 31–39. https://doi.org/10.1016/j.ecoser.2012.06.004
30. Mouchet, M. A., Paracchini, M. L., Schulp, C. J. E., et al. (2017). Bundles of ecosystem (dis)services and multifunctionality across European landscapes. *Ecological Indicators*. https://doi.org/10.1016/j.ecolind.2016.09.026
31. Plieninger, T., Dijks, S., Oteros-Rozas, E., & Bieling, C. (2013). Assessing, mapping, and quantifying cultural ecosystem services at community level. *Land Use Policy*. https://doi.org/10.1016/j.landusepol.2012.12.013
32. Sherrouse, B. C., Clement, J. M., & Semmens, D. J. (2011). A GIS application for assessing, mapping, and quantifying the social values of ecosystem services. *Applied Geography*. https://doi.org/10.1016/j.apgeog.2010.08.002
33. Alcantara, C., Kuemmerle, T., Prishchepov, A. V., & Radeloff, V. C. (2012). Mapping abandoned agriculture with multi-temporal MODIS satellite data. *Remote Sensing of Environment, 124*, 334–347. https://doi.org/10.1016/j.rse.2012.05.019
34. Redo, D. J., & Millington, A. C. (2011). A hybrid approach to mapping land-use modification and land-cover transition from MODIS time-series data: A case study from the Bolivian seasonal tropics. *Remote Sensing of Environment*. https://doi.org/10.1016/j.rse.2010.09.007
35. Godinot, O., Leterme, P., Vertès, F., & Carof, M. (2016). Indicators to evaluate agricultural nitrogen efficiency of the 27 member states of the European Union. *Ecological Indicators*. https://doi.org/10.1016/j.ecolind.2016.02.007
36. Kerr, J. T., & Cihlar, J. (2003). Land use and cover with intensity of agriculture for Canada from satellite and census data. *Global Ecology and Biogeography*. https://doi.org/10.1046/j.1466-822X.2003.00017.x
37. Dixon, J., Gulliver, A., & Gibbon, D. (2001). Farming systems and poverty: Improving farmers' livelihoods in a changing world.
38. Bell, L. W., & Moore, A. D. (2012). Integrated crop–livestock systems in Australian agriculture: Trends, drivers and implications. *Agricultural Systems, 111*, 1–12. https://doi.org/10.1016/j.agsy.2012.04.003
39. Moraine, M., Duru, M., Nicholas, P., et al. (2014). Farming system design for innovative crop-livestock integration in Europe. *Animal*. https://doi.org/10.1017/S1751731114001189
40. Bégué, A., Arvor, D., Bellon, B., et al. (2018). Remote sensing and cropping practices: A review. *Remote Sensing*.
41. Defourny, P., Bontemps, S., Bellemans, N., et al. (2019). Near real-time agriculture monitoring at national scale at parcel resolution: Performance assessment of the Sen2-Agri automated system in various cropping systems around the world. *Remote Sensing of Environment*. https://doi.org/10.1016/j.rse.2018.11.007
42. Weiss, M., Jacob, F., & Duveiller, G. (2020). Remote sensing for agricultural applications: A meta-review. *Remote Sensing of Environment*. https://doi.org/10.1016/j.rse.2019.111402
43. Yin, H., Prishchepov, A. V., Kuemmerle, T., et al. (2018). Mapping agricultural land abandonment from spatial and temporal segmentation of Landsat time series. *Remote Sensing of Environment*. https://doi.org/10.1016/j.rse.2018.02.050
44. Kolecka, N. (2018). Height of successional vegetation indicates moment of agricultural land abandonment. *Remote Sensing*. https://doi.org/10.3390/rs10101568
45. Prishchepov, A. V., Radeloff, V. C., Dubinin, M., & Alcantara, C. (2012). The effect of Landsat ETM/ETM+ image acquisition dates on the detection of agricultural land abandonment in Eastern Europe. *Remote Sensing of Environment*. https://doi.org/10.1016/j.rse.2012.08.017
46. Dubovyk, O., Menz, G., Conrad, C., et al. (2013). Spatio-temporal analyses of cropland degradation in the irrigated lowlands of Uzbekistan using remote-sensing and logistic regression modeling. *Environmental Monitoring and Assessment*. https://doi.org/10.1007/s10661-012-2904-6

47. Liu, X., & Bo, Y. (2015). Object-based crop species classification based on the combination of airborne hyperspectral images and LiDAR data. *Remote Sensing, 7*, 922–950. https://doi.org/10.3390/rs70100922
48. Foody, G. M., & Mathur, A. (2004). A relative evaluation of multiclass image classification by support vector machines. *IEEE Transactions on Geoscience and Remote Sensing, 42*, 1335–1343. https://doi.org/10.1109/TGRS.2004.827257
49. Phiri, D., & Morgenroth, J. (2017). Developments in landsat land cover classification methods: A review. *Remote Sensing, 9*, 967. https://doi.org/10.3390/rs9090967
50. Sohn, Y., & Rebello, N. S. (2002). Supervised and unsupervised spectral angle classifiers. *Photogrammetric Engineering & Remote Sensing, 68*, 1271–1280.
51. Strahler, A. H. (1980). The use of prior probabilities in maximum likelihood classification of remotely sensed data. *Remote Sensing of Environment, 10*, 135–163. https://doi.org/10.1016/0034-4257(80)90011-5
52. Vasilakos, C., Kavroudakis, D., & Georganta, A. (2020). Machine learning classification ensemble of multitemporal Sentinel-2 images: The case of a mixed mediterranean ecosystem. *Remote Sensing*. https://doi.org/10.3390/rs12122005
53. Walter, V. (2004). Object-based classification of remote sensing data for change detection. *ISPRS Journal of Photogrammetry and Remote Sensing, 58*, 225–238. https://doi.org/10.1016/j.isprsjprs.2003.09.007
54. Blaschke, T. (2010). Object based image analysis for remote sensing. *ISPRS Journal of Photogrammetry and Remote Sensing*.
55. Li, Q., Wang, C., Zhang, B., & Lu, L. (2015). Object-based crop classification with landsat-MODIS enhanced time-series data. *Remote Sensing, 7*, 16091–16107. https://doi.org/10.3390/rs71215820
56. Goodin, D. G., Anibas, K. L., & Bezymennyi, M. (2015). Mapping land cover and land use from object-based classification: An example from a complex agricultural landscape. *International Journal of Remote Sensing, 36*, 4702–4723. https://doi.org/10.1080/01431161.2015.1088674
57. Csillik, O., Belgiu, M., Asner, G. P., & Kelly, M. (2019). Object-based time-constrained dynamic time warping classification of crops using Sentinel-2. *Remote Sensing*. https://doi.org/10.3390/rs11101257
58. Watkins, B., & Van Niekerk, A. (2019). Automating field boundary delineation with multi-temporal Sentinel-2 imagery. *Computers and Electronics in Agriculture*. https://doi.org/10.1016/j.compag.2019.105078
59. Crespin-Boucaud, A., Lebourgeois, V., Lo Seen, D., et al. (2020). Agriculturally consistent mapping of smallholder farming systems using remote sensing and spatial modelling. *ISPRS - Int Arch Photogramm Remote Sens Spat Inf Sci, XLII-3*(W11), 35–42. https://doi.org/10.5194/isprs-archives-XLII-3-W11-35-2020
60. Morgan, J. L., Gergel, S. E., & Coops, N. C. (2010). Aerial photography: A rapidly evolving tool for ecological management. *Bioscience*. https://doi.org/10.1525/bio.2010.60.1.9
61. Vogels, M. F. A., de Jong, S. M., Sterk, G., & Addink, E. A. (2017). Agricultural cropland mapping using black-and-white aerial photography, object-based image analysis and random forests. *International Journal of Applied Earth Observation and Geoinformation, 54*, 114–123. https://doi.org/10.1016/j.jag.2016.09.003
62. Panitsa, M., Snogerup, B., Snogerup, S., & Tzanoudakis, D. (2003). Floristic investigation of Lemnos island (NE Aegean area, Greece). *Willdenowia, 33*, 79–105. https://doi.org/10.3372/wi.33.33108
63. ELSTAT. (2019). Hellenic Statistical Authority. https://www.statistics.gr/en/home/. Accessed 29 Mar 2019.
64. Dimopoulos, T., Dimitropoulos, G., & Georgiadis, N. (2018). The land use systems of Lemnos island. Terra Lemnia project: Recording of land use systems & practices. Strategy 1.1, action

1.1.1. https://terra-lemnia.net/wp-content/uploads/2019/04/Terra-Lemnia-1.1.1-Land-Use-Systems-of-Lemnos-Dec-2018.pdf. Accessed 18 Apr 2020.
65. Clapuyt, F., Vanacker, V., & Van Oost, K. (2016). Reproducibility of UAV-based earth topography reconstructions based on structure-from-motion algorithms. *Geomorphology, 260*, 4–15. https://doi.org/10.1016/j.geomorph.2015.05.011
66. Westoby, M. J., Brasington, J., Glasser, N. F., et al. (2012). 'Structure-from-motion' photogrammetry: A low-cost, effective tool for geoscience applications. *Geomorphology, 179*, 300–314. https://doi.org/10.1016/j.geomorph.2012.08.021
67. Agisoft LLC (2016) Agisoft PhotoScan User Manual: Professional Edition, Version 1.2. In: User Manuals.
68. Blaschke, T., Lang, S., & Hay, G. J. (2008). Object-based image analysis: Spatial concepts for knowledge-driven remote sensing applications. *LibTuDelftNet*.
69. Su, T., & Zhang, S. (2017). Local and global evaluation for remote sensing image segmentation. *ISPRS Journal of Photogrammetry and Remote Sensing*. https://doi.org/10.1016/j.isprsjprs.2017.06.003
70. Liu, D., & Xia, F. (2010). Assessing object-based classification: Advantages and limitations. *Remote Sensing Letters*. https://doi.org/10.1080/01431161003743173
71. Haralick, R. M., Shanmugam, K., & Dinstein, I. (1973). Textural features for image classification. *IEEE Transactions on Systems, Man, and Cybernetics, SMC-3*, 610–621. https://doi.org/10.1109/TSMC.1973.4309314
72. Caridade, C. M. R., Marçal, A. R. S., & Mendonça, T. (2008). The use of texture for image classification of black; white air photographs. *International Journal of Remote Sensing, 29*, 593–607. https://doi.org/10.1080/01431160701281015
73. Halounová, L. (2005). Automatic classification of B&W aerial orthophotos. In M. Oluic (Ed.), *New strategies for European remote sensing* (p. 768).
74. Hossain, M. D., & Chen, D. (2019). Segmentation for object-based image analysis (OBIA): A review of algorithms and challenges from remote sensing perspective. *ISPRS Journal of Photogrammetry and Remote Sensing, 150*, 115–134. https://doi.org/10.1016/j.isprsjprs.2019.02.009
75. Benz, U. C., Hofmann, P., Willhauck, G., et al. (2004). Multi-resolution, object-oriented fuzzy analysis of remote sensing data for GIS-ready information. *ISPRS Journal of Photogrammetry and Remote Sensing*. https://doi.org/10.1016/j.isprsjprs.2003.10.002
76. Rahman, M. R., & Saha, S. K. (2008). Multi-resolution segmentation for object-based classification and accuracy assessment of land use/land cover classification using remotely sensed data. *Journal of the Indian Society of Remote Sensing*. https://doi.org/10.1007/s12524-008-0020-4
77. Lu, D., Mausel, P., Brondízio, E., & Moran, E. (2004). Change detection techniques. *International Journal of Remote Sensing*. https://doi.org/10.1080/0143116031000139863
78. Khorram, S. K., Biging, G. S., Chrisman, N. R., et al. (1999). *Accuracy assessment of remote sensing-derived change detection.*
79. Darwish A, Leukert K, Reinhardt W (2003) IMage segmentation for the purpose of object-based classification. In: IGARSS 2003. 2003 IEEE international geoscience and remote sensing symposium. Proceedings (IEEE cat. No.03CH37477). IEEE, pp. 2039–2041.
80. Ryschawy, J., Choisis, N., Choisis, J. P., et al. (2012). Mixed crop-livestock systems: An economic and environmental-friendly way of farming? *Animal*. https://doi.org/10.1017/S1751731112000675

81. Stark, F., González-García, E., Navegantes, L., et al. (2018). Crop-livestock integration determines the agroecological performance of mixed farming systems in Latino-Caribbean farms. *Agronomy for Sustainable Development, 38*, 4. https://doi.org/10.1007/s13593-017-0479-x
82. Garrett, R. D., Ryschawy, J., Bell, L. W., et al. (2020). Drivers of decoupling and recoupling of crop and livestock systems at farm and territorial scales. *Ecology and Society*. https://doi.org/10.5751/ES-11412-250124
83. Vasilakos, C., Chatzistamatis, S., Roussou, O., & Soulakellis, N. (2019). Comparison of terrestrial photogrammetry and terrestrial laser scanning for earthquake response management. In *Lecture notes in geoinformation and cartography* (pp. 33–57).
84. Dimopoulos, T., & Kizos, T. (2020). Mapping change in the agricultural landscape of Lemnos. *Landscape and Urban Planning, 203*, 103894. https://doi.org/10.1016/j.landurbplan.2020.103894

Air drill Seeder Distributor Head Evaluation: A Comparison between Laboratory Tests and Computational Fluid Dynamics Simulations

Ignacio Rubio Scola, Sebastián Rossi, and Gastón Bourges

1 Introduction

The adaptation of Agriculture 4.0 technologies in the agricultural machinery produced by Argentinean industries is a big challenge. Sowing systems have different levels of technological adaptation linked to intelligent agriculture, and one pending issue is their self-regulation in order to maintain performance indicators at optimal levels in case of changes in sowing conditions [1]. According to Bragachini [2], the application of precision agriculture technologies and techniques in crop management allows to increase the average yield of the main crops by 10 to 15% by means of efficient input management, which has a direct impact on the gross margin of production processes in intensive and extensive crops [2]. Today, extensive agriculture is seeking greater productivity, and for that purpose, among other means, larger working widths are combined with increases in sowing speeds. In this regard, Nielsen [3] showed that increasing the speed of sowing increases the variability of plant spacing, as well as the percentage of losses [3]. Liu et al. [4] observed that the standard deviation increases as planting speed increases [4].

I. Rubio Scola (✉)
CIFASIS, French Argentine International Center for Information and Systems Sciences, National Scientific and Technical Research Council (CONICET) and the National University of Rosario (UNR), Rosario, Argentina
e-mail: irubio@fceia.unr.edu.ar

S. Rossi
National Institute of Industrial Technology (INTI), Rosario, Argentina
e-mail: srossi@inti.gob.ar

G. Bourges
National Institute of Industrial Technology (INTI) and the National University of Rosario (UNR), Rosario, Argentina
e-mail: gbourges@inti.gob.ar

D. D. Bochtis et al. (eds.), *Information and Communication Technologies for Agriculture—Theme II: Data*, Springer Optimization and Its Applications 183,
https://doi.org/10.1007/978-3-030-84148-5_8

Regarding the increase in work capacity, equipments such as air drill seeders present important technical benefits due to their high autonomy, maneuverability, productivity, easy loading, and adjustment, but with the following deficiencies: low precision of transversal distribution, seed damage, frequent risks of obstruction, and high energy consumption [5]. Air drill seeders are defined in ASAE/ASABE S506 as the type of machinery that uses a centralized hopper and a volumetric metering mechanism to contain and meter the seeds, respectively [6]. Seeds are conveyed pneumatically through the system ducts, from the tank to each sowing body. Among the different options available in the market, there may be compartmentalized single-hoppers, with the possibility of containing fertilizer and seeds, or two-hopper systems, each containing seeds and fertilizer independently. A great capacity of work is a very interesting condition for the production in large-scale applications.

Usually, these systems show a lack of uniformity in seeds distribution between sowing lines, which corresponds to the disparity of air-seed mixture flows between the distributor head's outlet ducts. Specific works related to air drill seeders can be mentioned in the study, on a distribution head, of Allam y Wiens [7], which describes experiments on nine different air drill seeders [7]. In Allen RR [8], three types of no tillage wheat seeders were tested [8]. In Kumar et al. [9], authors developed an air drill seeding machine for small grains as wheat, oats, barley, and sorghum [9]. The work of Kumar and Durairaj [10] focuses on the influence of the distributor head and the feeding tube in the trajectory of seeds in air drill seeders [10]. McCartney et al. [11] studied, designed, and evaluated modifications in the metering and agitation systems for pasture seeding using three different types of air planters [11].

Yatskul and Lemiere [12] carried out an experimental and theoretical study of pneumatic transport in an air seeder [12]. This work proposes a method to measure flow concentration and air velocity to optimize an existing seeder or to design a new one. The work presented in [5] studied the influence of divider head geometry and working conditions on the seed's distribution accuracy, while in [13] the influence of an air-stream loading system type on the total energy consumption and seeds-metering precision was studied. In Yatskul and Lemiere [13], the authors determined a minimal air velocity and flow concentration per type of seeds relative to the pipe diameter [14].

There exists an extended literature where the behavior of the air seeds flow in components of the air drill transport system is analyzed using computational methods (Computational Fluid Dynamics – CFD). Those carried out by [15–17] are mentioned. From these simulations, the proportions of seed mass in each outlet and the residence times of each seed in the distributor head area were analyzed. Additionally, in Lei et al. [17] numerical results were compared with tests, in a test bench, where the mass of each outlet of the distributor head was measured [17]. In works like [18–20], numerical evaluations of the behavior of the air-seed mixture in distributor heads were performed. Other interesting works include [21, 22], where the authors justified the design and technological parameters of a horizontal seed, by means of laboratory and field experiments. Among other results, they found that seed placement quality depends on the splitter position and the slope angle of its plates. In another work, Mudarisov et al. [23] proposed a technique of substantiating parameters of the two-phase "air-seeds" flow to apply it on the numerical implementation of a mathematical model, describing the technological process of the grain seeder distribution [23]. A work related to particle trajectories is [24], where the influence of conical directing agents on the speed and trajectory of particles movement in the vertical pneumatic conduit is studied. In Bourges [25], numerical air-seeds flow simulations on a commercial distributor head of an air drill seeder are described [25].

From previous works, in order to look for more efficient air drill seed transport and distribution system designs, it is necessary to understand the behavior of the transported seeds in air flows. Furthermore, with the implementation of agriculture 4.0 and ICT technologies in the sowing processes, it is necessary to control and regulate this type of seeder machines in real time. In order to understand the behavior of air-seed flow in air drill systems, in the present work, results obtained in controlled laboratory tests are analyzed and contrasted with the numerical results from [25].

2 Materials and Methods

2.1 Tested Model Description

The tested distributor head is described in Fig. 1. Air-seeds mixture flow goes through a horizontal tube, of inner diameter $D_i = 0.058$m, then subsequently up to an elbow and later by a vertical tube. Thereafter, the airflow enters in a cylindrical region, called the expansion zone, where it is split between each of the outlet tubes. This configuration is based on an actual distributor head, obtained from the company E. Gherardi e Hijos SA, located in Casilda (Santa Fe province, Argentina).

2.2 Description of Distributor Head's Test Bench

Experiments are carried out in a distributor head test bench of air drill seeders, located in facilities of the School of Mechanical Engineering of the National University of Rosario. The objective of this test bench is to carry out controlled experiments to evaluate the incidence of the different factors, in pneumatic transport of seeds in distributor heads. It seeks to replicate the operation of the pneumatic transport system of an air drill seeder, without its complexity and adapted to available facilities. A test bench is designed according to the operating scheme of Fig. 2, to evaluate a single distributor head per test. It has a centrifugal fan, which can provide up to $12m^3\ s^{-1}$ at a relative pressure of 1400 *Pa*, a seed hopper with an approximate capacity of 4 *kg* of soybean (Glycine max), a roller-type meter coupled through a chain to a gear motor controlled by a Variable Frequency Drive, and a seed distributor head, provided by the company E. Gherardi e Hijos SA (Fig. 1). Outlet

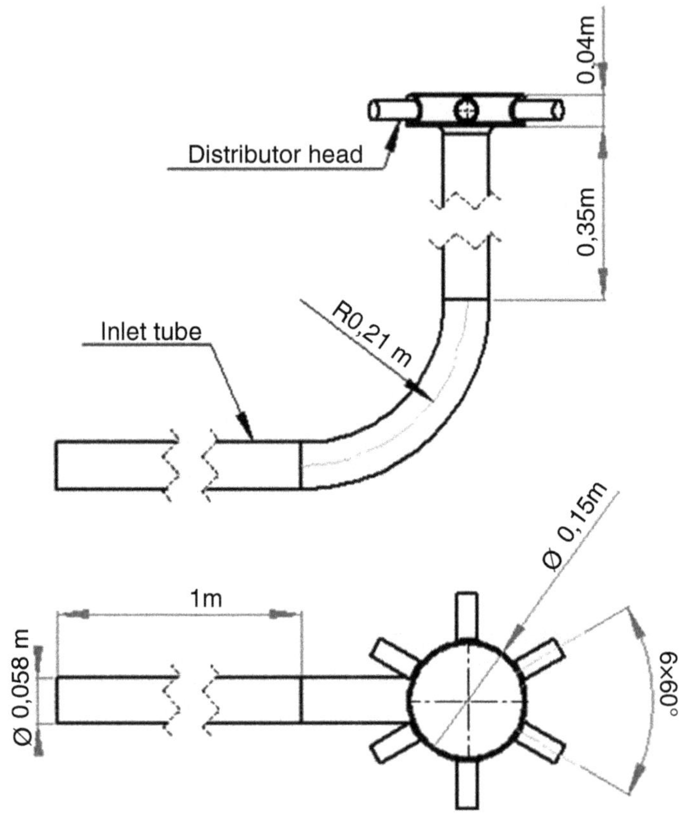

Fig. 1 Geometry of original model. Top and side views. Dimensions in meters [m]

ducts are attached to the distributor by corresponding connections, and discharge seeds in plastic containers. The test bench is based on experiments carried out by [7].

2.3 Experiment Design

Tests were designed in six blocks, with six repetitions each. Each block corresponds to a different location of each outlet orifice of the distributor, with respect to the outlet hoses. In other words, the distributor head is rotated in such a way that each block corresponds to a different "exit hole" for each hose. This allowed to distinguish the effect of the "exit orifice" and the effect of the "hose". The "hose" is the position where the hoses are located, and it is invariant throughout the test. The "exit orifice" is an orifice belonging to the distributor head, which rotates along with the distributor in each test the equivalent to the angular distance between each tube, that is, 60 degrees. During the experiments, soybean seeds (Glycine max) are used, being

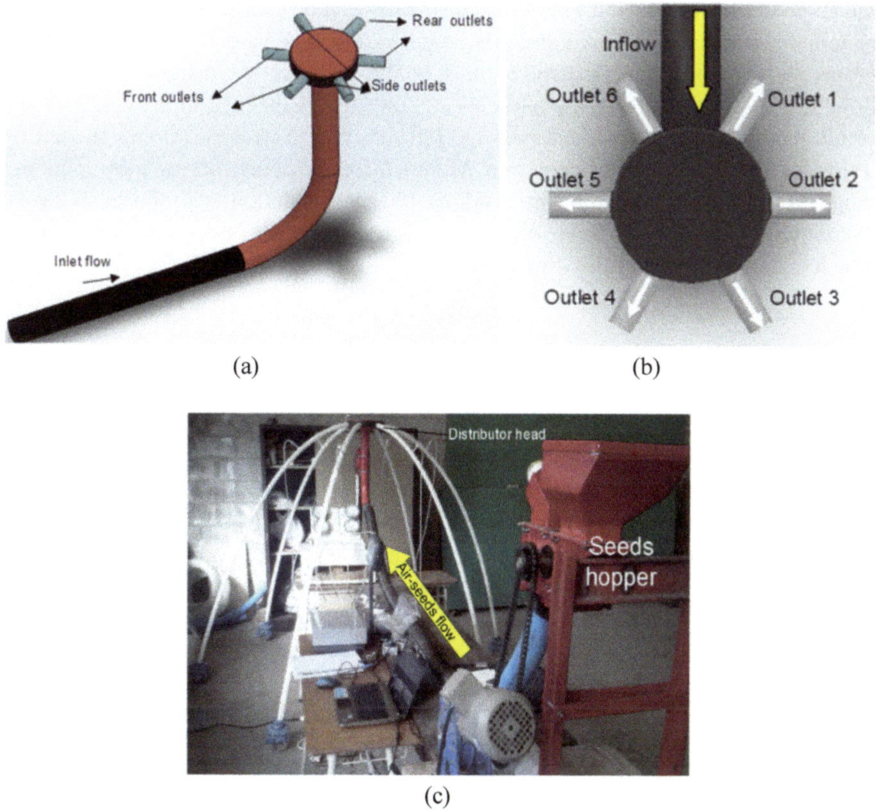

Fig. 2 (**a**) Flow configuration. (**b**) Distributor flow scheme. (**c**) Test bench of distributor heads

the mass of 1000 seeds 0.147 *kg*. In each test, 2 *kg* of seeds are deposited in the metering hopper. These are then injected into the air stream, which has a speed of 20 $m\ s^{-1}$, in the duct leading to the distributor. The duration of each test is 4 min. In summary, the volumetric flow of air is $Q_{va} \approx 0.53\ m^3\ s^{-1}$ while the volumetric flow of seed, for a soybean bulk density $\rho_b = 750\ kg\ m^3$, is $Q_{vs} \approx 1.110^{-5}\ m^3\ s^{-1}$.

2.4 *Numerical Simulations*

Air-Seeds Mixture Flow

Air-seed mixture flow modeling is a case of multiphase flow [26], where the continuous phase corresponds to the air's model and the discrete phase is associated with seeds (particles). The two-phase gas-solid mixture is considered as a gas with solid particles in suspension. In pneumatic transport, the flow regimes vary according to the gas flow rate, the characteristics of the particles transported, and the geometry of the transport system. These rates can be classified as dense or dilute pneumatic transport [27]. In the first regime, the gas velocity and the particle properties form a regime in which most of the particles are dispersed in the gas flow. In the second regime, the particles are not fully suspended in the gas, and as a result, they are transported as a package [28]. Particle trajectories are evaluated by following the movement of each one. Modeled forces, acting on each particle, are: drag, lift, and momentum. In the work of [29] a distinction is proposed between dense and scattered flows based on the volumetric fraction of the scattered phase α_p, which is defined by the following equation:

$$\alpha_p = \lim_{\delta V \to V^0} \frac{\delta V_p}{\delta V} \tag{1}$$

where δV_p is the volume of the dispersed phase, contained in the volume δV. The author suggests that flows with $\alpha_p < 0.001$ can be considered as diluted, while when $\alpha_p > 0.001$ they are dense. Other authors, e.g [30, 31], propose a volumetric fraction limit to define the dispersed phase much higher, i.e. $\alpha_p < 0.1$. In the present work, since the volumetric fraction of the intervening phases is inferior to 0.001, the air-seed mixture can be considered as dispersed.

Air Flow

Fluid flow equations are solved using the Reynolds Average Navier-Stokes technique [32], with the Realizable $k - \varepsilon$ turbulence model [33]. The algorithm is stationary and is solved with ANSYS Fluent software. A pressure-based solver is used, with coupling between velocity and pressure. Regarding the domain discretization, an unstructured mesh of 434,215 tetrahedral elements and 87,318 nodes is used. Boundary conditions, as described in Fig. 3 and Table 1, are: normal inlet velocity, null outlet relative pressure, and logarithmic wall velocity gradients on internal walls. In this model, the geometry of the inlet pipe and a portion of the outlet hoses are considered. Due to computational resources limitations, their full length is not modeled, although its internal diameter is held.

Particles Trajectory

Trajectories of particles are calculated with ANSYS Fluent software, using discrete phase module (DPM) [34]. The considered forces on the particle are the gravity force and the aerodynamic drag force. The forces balance per unit mass on each particle is given by:

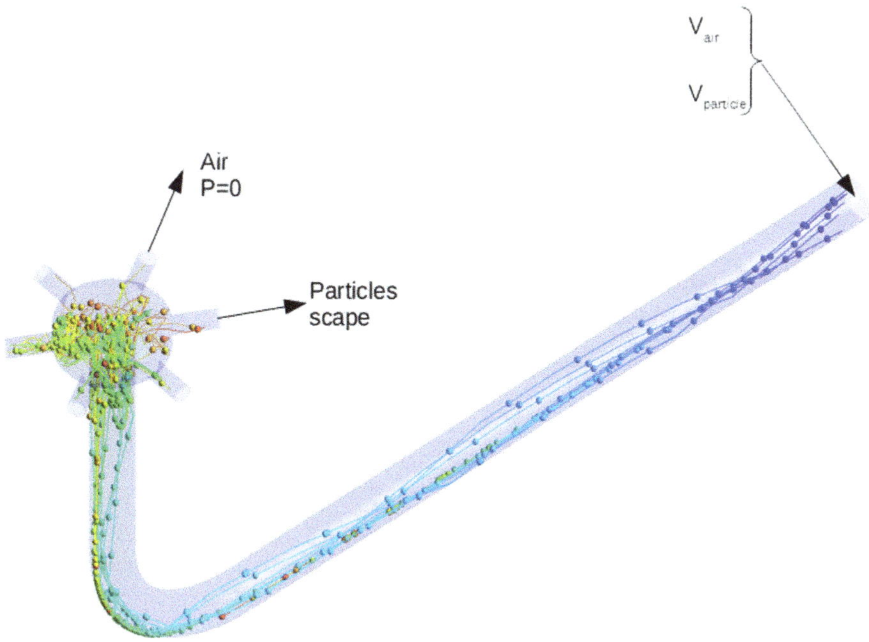

Fig. 3 Boundary conditions of the studied model

Table 1 Boundary conditions description

Phase	Inlet	Internal walls	Outlet
Continuum	Normal velocity $v_i = 20\ m\ s^{-1}$	Log. wall velocity Gradient	Null pressure
Discrete	Normal velocity $v_{ip} = 1\ m\ s^{-1}$	Specular reflection	Escape

$$\frac{du_p}{dt} = F_d + \frac{g(\rho_p - \rho)}{\rho_p} \tag{2}$$

where F_d is the aerodynamic drag force per unit mass of the particle and can be computed as:

$$F_d = \frac{18\mu C_D R_{ep}}{24\rho_p D_p^2}(u - u_p) \tag{3}$$

where u is the continuous phase velocity, up particle velocity, ρ and $v_i = 20\ m\ s^{-1}$ are density and dynamic viscosity of the fluid, ρ_p and D_p are particle's density and diameter. Reynolds particle number R_{ep} is obtained from the function:

$$R_{ep} = \frac{\rho D_p |u_p - u|}{\mu} \tag{4}$$

and drag coefficient C_D is computed as:

$$C_D = a_1 + \frac{a_2}{R_{ep}} + \frac{a_3}{R_{ep}^2} \tag{5}$$

where a_1, a_2, and a_3 are constants dependent of R_{ep} values [35]. Particle rebounds against system's boundaries are considered as elastic collisions with specular reflection. Regarding the rebound model, momentum change is thoughtful as a restitution coefficient

$$e_n = \frac{V_{s,n}}{V_{i,n}} \tag{6}$$

where e_n is the normal coefficient of restitution, and $V_{s,\,n}$ and $V_{i,\,n}$ refer to the relative velocity after and before collision, respectively. Normal coefficient defines absorbed momentum on each bounce. Similarly, the coefficient of tangential restitution e_t relates the tangential velocities post and pre impact. In this work, normal and tangential coefficients of restitution are, both, equal to 0.8 [36]. Collisions between particles are not considered.

Stokes number is the ratio between the characteristic stopping time of a particle and the characteristic flow time [37]. Then, Stokes number of air-seeds flow mixture is:

$$S_t = \frac{\tau_p}{t_f} \tag{7}$$

where τ_p is the characteristic time of the particle:

$$\tau_p = \frac{\rho_p D_p^2}{18\mu} \tag{8}$$

and t_f is the characteristic time of the fluid:

$$t_f = \frac{L_f}{V_f} \tag{9}$$

Let L_f be the length of the input pipe of the flow, and V_f be the mean velocity of the fluid inlet [38]. If S_t is much less than one ($S_t \ll 1$), response time of the flow particles is much lower than characteristic time associated with the flow field. That is, particles have time to respond to changes in velocity of flow field. On the other hand, if S_t is much larger than one ($S_t \gg 1$) the response time of the particles is

Table 2 Seeds and fluid parameters in the computation of the Stokes number

Particle		
$\rho_p[kg\ m^{-3}]$		700
$D_p[m]$		0.004
$\tau_p[s]$		34.1
Fluid		
$L_f[m]$		1.86
$V_f\ [m\ s^{-1}]$		20
$t_f[s]$		0.093
Stokes number		
S_t		367

greater than the time of the characteristic flow field, which prevents it from responding to the speed changes that occur in the latter [26]. Table 2 describes particles and fluids characteristics for Stokes number calculation.

From Table 2, soybean Stokes number is greater than one, which means that particles response time is higher than flow field ones. Therefore, it is expected that the behavior of the particles does not coincide with the behavior of the air flow.

Discrete Phase Model Setup

Trajectories of particles are calculated using a two-way coupling method between continuous and discrete phases model simulations, where the presence of particles affect flow behavior and vice versa. These effects are transmitted to the flow as DPM sources. The DPM solution and the flow solution should reach a converged and self-consistent solution [34]. This interaction between phases is updated every 10 iterations of continuous phase. Discrete phase tracking is set as unsteady, where the particles time step is set as 0.03 *s*.

3 Results

3.1 Experimental Results

From the experiments, it can be concluded that the "hose" effect has no relevant incidence in results, whereas on the contrary, the effect of the "exit orifice" affects results. Figure 4 shows the box plot of seeds masses obtained on each outlet. Horizontal lines in boxes represent the medians while circles represent the means of mass measurement in all tests, at each outlet. Notice that outlets one and six have larger mean, while outputs three and four have lower values. That is, frontal outlets send more seeds than rear ones. This is confirmed in the analysis of means, in Fig. 5, where the null hypothesis is rejected, in other words, means are equal and the difference between them is put back into evidence.

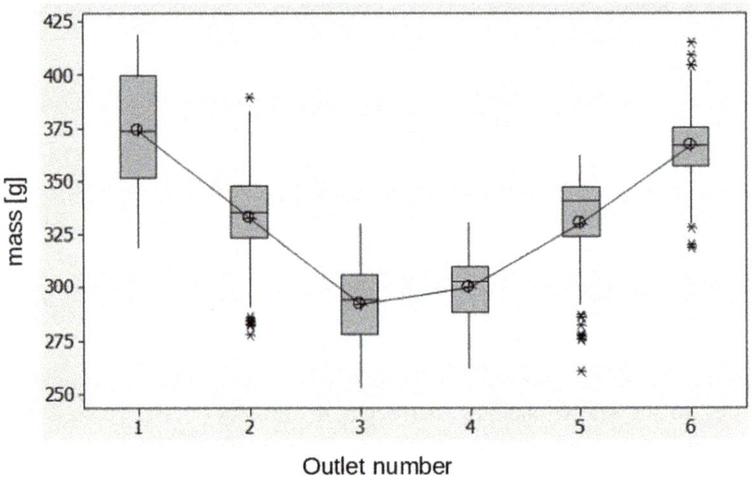

Fig. 4 Box plot: seed mass [*g*] vs outlet number

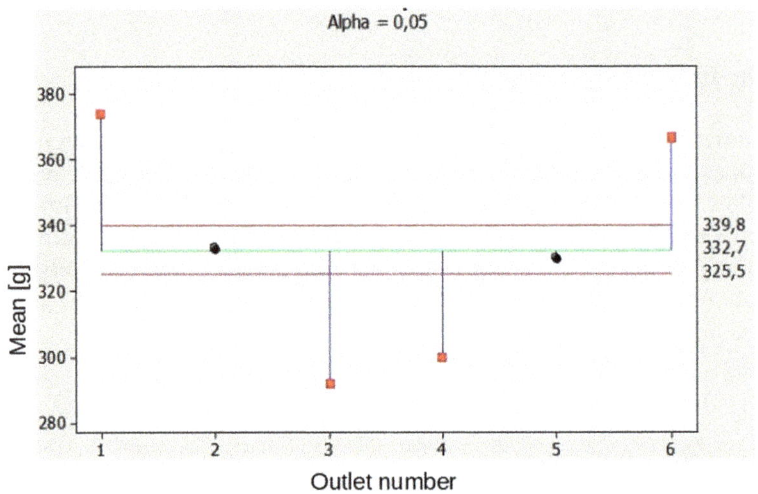

Fig. 5 Analysis of means: seed mass [*g*] vs outlet number

The Shapiro-Wilk test [39] confirms the normality of obtained data for the six outlets. Figure 6 shows the histogram of the standardized residuals on which the Shapiro-Wilk test was applied.

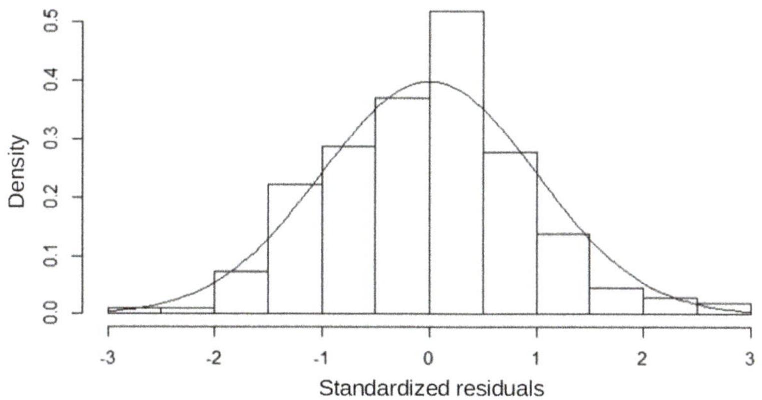

Fig. 6 Histogram of standardized residuals on each outlet

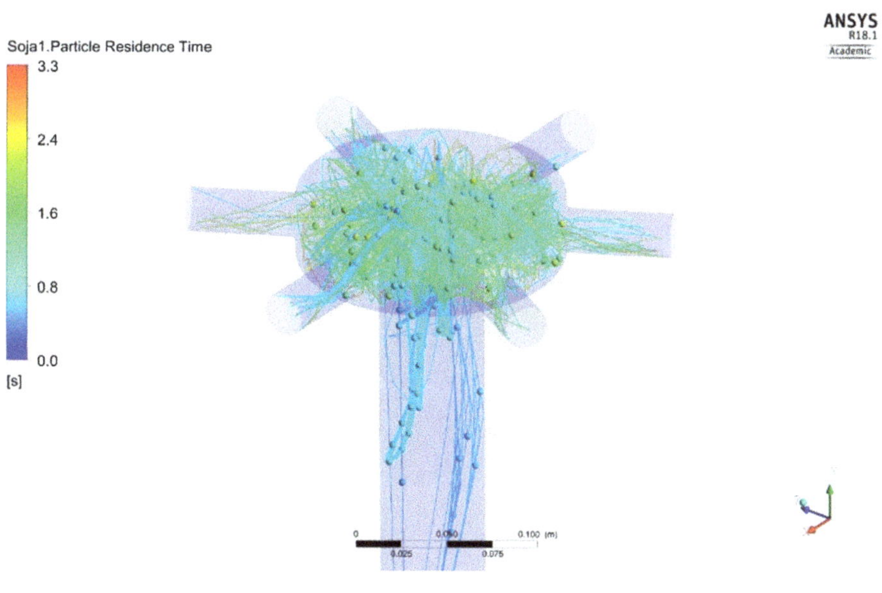

Fig. 7 Trajectory of particles. Residence time [s]

3.2 *Numerical Results*

From numerical simulations, a difficulty of the particles to escape through outlet ducts is observed. The cause of this difficulty could be due to the high number of rebounds of the particles inside the expansion area of the distributor head. Figure 7 illustrates the particles' trajectories which appear partially on the vertical tube and the expansion zone and where the particles' residence times reach up to 3.3 *s*. In this sector the particles remain for a longer period of time, while in the vertical tube

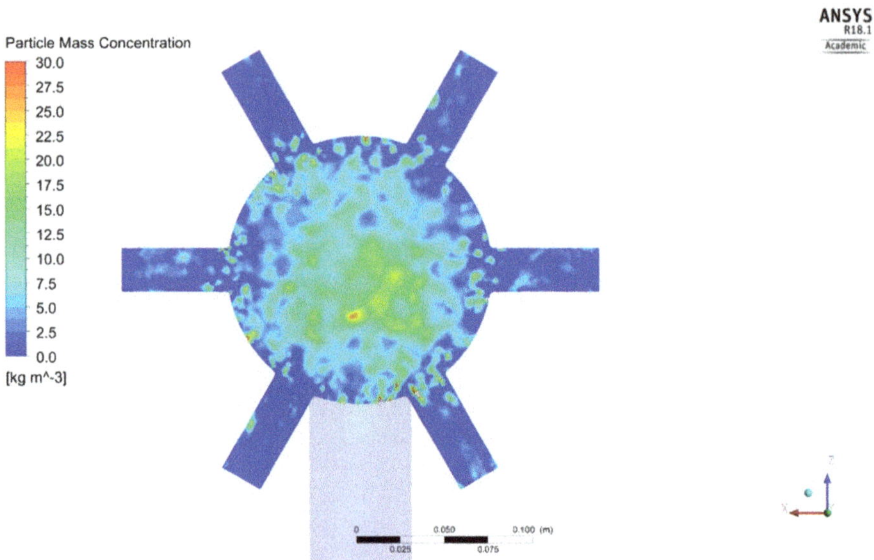

Fig. 8 Particle mass concentration of particles inside the distributor head, in [kg m^{-3}]

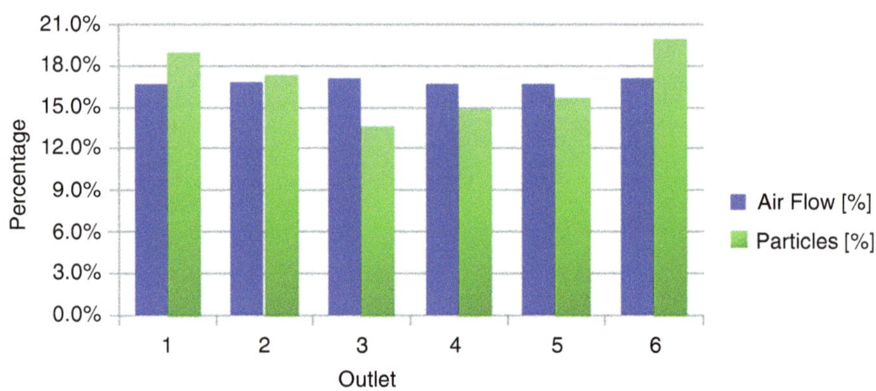

Fig. 9 Bar charts percentage of particles and air flux (abscissa) vs. output number (ordinate)

particle's residence time is lower than 0.8 *s*, that is, almost a quarter of the time the particles stay in the expansion zone.

This is confirmed in Fig. 8, where particle mass concentration is shown in the fluid field. Note that the highest concentration of particles occurs in the central zone of the expansion sector, and particularly, close to frontal outlets one and six, in accordance with Fig. 2.

Figure 9 shows the percentage of particles and air flux expelled on each outlet. Regarding particles distribution on outlet tubes, frontal tubes have a greater flow of particles compared with back ones, especially outlets one and six, being the particles that leaves them 39% of the total. This distribution between outlets can be considered, approximately, symmetric with respect to a vertical plane that divides the domain into two equal parts, having on one side the outputs one to three, and on the other side the outputs four to six.

From Fig. 9 it can be inferred that airflow in each outlet (in percentages of the total air injected) is approximately the same, and therefore, its distribution between outlets is homogeneous, as oppose to seeds flow distribution. This lack of equivalence between seed and air flows distribution between outlets tubes asserts the idea that these types of particles, and due to their high inertia, behave independently of the continuous phase. The latter states the concept of Stokes number, defined in eq. (7). If Stokes number is small, ($\ll 1$), it means that particles motion is tightly coupled to fluid motion. If Stokes number is large, particles are not influenced by the fluid. Their response time is longer than the time fluid must act on it (the fluid time scale may be the rotation time of a characteristic eddy) and so particles will pass through the flow without much deflection in its initial trajectory [26]. According to Table 2, Soybean S_t is greater than one, which means that they have an important inertia with respect to the flow field.

3.3 Validation of the Numerical Model

Numerical results are compared with experimental tests. Seed's proportions are used, and the statistical term χ is defined as the percentage difference between the average of the experimental results, obtained for each of the outlet hoses x_m, and the results obtained in numerical simulations x_s. Student's t-test is applied [40], using a significance level $a = 0.05$. Null hypothesis is defined as the difference between the population means H_0: $\mu_m - \mu_s = 0$, while the alternative hypothesis is defined as H_a: $\mu_m - \mu_s \neq 0$.

Laboratory experiment yields a total of $n = 36$ samples of seed mass, for each output duct relieved. These samples are divided by the total amount of seed mass relieved in each experiment, and the percentage of mass in each outlet tube is obtained, with respect to the total extracted. On the other hand, from simulations, a quantity of particles (seeds) that come out of each of the distributor's outlets is obtained. These amounts of particles are divided by the total of outgoing ones (in all evaluated outlets), and in this way, their percentage in each output are obtained.

From Table 3 it is observed that, in outlets one and six, the p-value is lower than the level of significance a, which implies that there is significant evidence to reject the null hypothesis H_0 in favor of the alternative hypothesis H_a. That is, in outlets one and six the simulation does not approximate well the results of the tests, overestimating the value of these outputs. On the other hand, at the rest of the outlets, the simulation correctly approximates the experimental results. In both cases, the distribution of seeds between outlets presents a similar behavior, that is, the distribution curve has maximum flow rates at the front outlets one and six, and minimum at rear outlets three and four (Figs. 4 and 9).

Table 3 Calculation of statistical χ, Student's t and p-value, for a significance level $\alpha = 0.05$

Sample	S1	S2	S3	S4	S5	S6
Simul. Results [%]	18.90	17.30	13.50	14.70	15.60	20.00
Sample Means [%]	16.71	15.94	14.31	15.08	17.36	17.83
χ [%]	−2.19 ±0.021	−1.36 ±0.012	0.81 ±0.007	0.38 ±0.003	1.76 ±0.018	−2.17 ±0.023
t-student	−2.24	−1.493	0.940	0.438	1.754	−2.042
p-value	$0.0318 < \alpha$	$0.145 > \alpha$	$0.350 > \alpha$	$0.664 > \alpha$	$0.089 > \alpha$	$0.049 < \alpha$

4 Conclusions and Perspectives

Numerical and laboratory experiments were carried out in a seed's distributor head. Obtained trajectories through numerical simulations gave results in consistency with those obtained in controlled trials. In both cases, the highest flow rates of seeds are produced in frontal outlets, while rear outlets present the lowest flow. This suggests an important inertia influence on seeds trajectories, which is confirmed through the Stokes number of soybeans ($S_t > 1$). These results coincide with observations of manufacturers of this type of agricultural machinery. This way, the lack of uniformity of seeds distribution between the sowing line is linked to the disparity of mixing flows between the outputs of the distributor head.

A possible solution to this lack of homogeneity between the outputs flows of the head is to modify the geometric distribution of the output tubes, looking to reduce the difference of seed flows between them. These modifications have to consider that each one of the outlet ducts should have the same resistance to the passage of fluid. As future work, laboratory tests and numerical simulations will be carried out on modified distribution heads, with the aim of reducing this lack of homogeneity between each output tubes. In this sense, a path has already been started proposing modified head geometries, which are being evaluated on laboratory tests. Some of these modified distributor heads designs, proposed by the working group, are described in [25]. Finally, it is important to highlight that these results are useful to understand the physical process of seed transport and distribution. Through the combination of controlled laboratory tests and numerical models, more information on these phenomena can be obtained, and in this way, proposing automatic control and regulation systems to improve performance of air drill seeding systems.

Acknowledgments The authors thank the Faculty of Exact Sciences, Engineering and Surveying (FCEIA) of the National University of Rosario (UNR), the National Institute of Industrial Technology (INTI), and National Scientific and Technical Research Council (CONICET) for making possible the development of this research.

References

1. Saikai, Y., Patel, V., & Mitchell, P. D. (2020). Machine learning for optimizing complex site–specific management. *Computers and Electronics in Agriculture, 174*, 105381. https://doi.org/10.1016/j.compag.2020.105381
2. Bragachini, M. (2019). *Adopción de tecnología en el sector agropecuario argentino en los últimos 28 años. Technical report - INTA Manfredi.* http://www.agriculturadeprecision.org/descargaItem.asp?item=%2Farticulos%2Fmaquinaria%2Dagricola%2FAdopcion%2DTecnologia%2DSector%2DAgrop%2DArgentino%2Dultimos%2D28%2Danios%2Epdf. Accessed 10 Feb 2020.
3. Nielsen, R. (1995). Planting speed effects on stand establishment and grain yield of corn. *Journal of Production Agriculture, 8*(3), 391–393. https://doi.org/10.2134/jpa1995.0391
4. Liu, W., Tollenaar, M., Stewart, G., & Deen, W. (2004). Within–row plant spacing variability does not affect corn yield. *Agronomy Journal, 96*(1), 275–280. https://doi.org/10.2134/agronj2004.2750
5. Yatskul, A., Lemiere, J. P., & Cointault, F. (2017). Influence of the divider head functioning conditions and geometry on the seed's distribution accuracy of the air-seeder. *Biosystems Engineering, 161*, 120–134. https://doi.org/10.1016/j.biosystemseng.2017.06.015
6. ASABE S506OCT2010 (R2014) Terminology and Definitions for Planters, Drills and Seeders (2010) American society of agricultural and biological engineers' standards: St. .
7. Allam, R. K., & Wiens, H. (1983). *An investigation of air seeder component characteristics*. Tech. Rep., Prairie Agricultural Machinery Institute, Lethbridge, Alberta.
8. Allen, R. R. (1988). Performance of three wheat seeders in conservation tillage residue. *Applied Engineering in Agriculture, 4*(3), 191–196.
9. Kumar, V. F., Durairaj, C. D., & Balasurbramanian, M. (1999). Air assisted drill for small seeds. *Journal of Agricultural and Engineering Research, 8*(4), 259–265.
10. Kumar, V., & Durairaj, C. (2001). Influence of distributor head on the seed trajectory within the feeder plenum of an air drill. *International Agricultural Engineering Journal, 10*(3-4), 255–267.
11. McCartney, D., Boyden, A., & Stevenson, C. (2005). Development of agitators for seeding forages using air delivery systems. *Journal of Rangeland Ecology and Management, 58*(2), 199–203.
12. Yatskul, A., & Lemiere, J. (2014). Experimental determination of flow concentration for pneumatic conveying systems of air-seeders. *INMATEH - Agricultural Engineering, 44*(3), 19–26.
13. Yatskul, A., Lemiere, J. P., & Cointault, F. (2018). Comparative energy study of the air-stream loading systems of air-seeders. *Engineering in Agriculture, Environment and Food, 11*(1), 30–37. https://doi.org/10.1016/j.eaef.2017.09.003
14. Yatskul, A., & Lemiere, J. P. (2018). Establishing the conveying parameters required for the air-seeders. *Biosystems Engineering, 166*, 1–12. https://doi.org/10.1016/j.biosystemseng.2017.11.001
15. Bayati, M., & Johnston, C. (2017). *CFD-DEM investigation of seed clustering in an air seeder with the immersed boundary method*. Grande Prairie.
16. Lei, X., Liao, Y., Wang, L., Wang, D., Yao, L., & Liao, Q. (2017). Simulation of gas- solid two-phase flow and parameter optimization of pressurized tube of air- assisted centralized metering device for rapeseed and wheat. *Transactions of the Chinese Society of Agricultural Engineering, 33*(19), 67–75.
17. Lei, X., Liao, Y., Zhang, Q., Wang, L., & Liao, Q. (2018). Numerical simulation of seed motion characteristics of distribution head for rapeseed and wheat. *Computers and Electronics in Agriculture, 150*, 98–109. https://doi.org/10.1016/j.compag.2018.04.009
18. Dai, Y., Luo, X., Wang, Z., Zeng, S., Zang, Y., Yang, W., Zhang, M., Wang, B., & Xing, H. (2016). Design and experiment of rice pneumatic centralized seed distributor. *Nongye*

Gongcheng Xuebao /Transactions of the Chinese Society of Agricultural Engineering, 32, 3642. https://doi.org/10.11975/j.issn.10026819.2016.24.005
19. Pyataev, M., & Zyryanov, A. (2018). Defining rational parameters of a dividing head in a pneumatic grain seeding machine. *The bulletin of the KrasGAU, 3*, 88–94.
20. Konovalov, V., Kravtsov, A., Zaitsev, V., Petrov, A., & Petrova, S. (2019). Pneumatic system of a seeder with pneumatic sowing. *IOP Conference Series: Earth and Environmental Science, 403*, 012,131. https://doi.org/10.1088/1755-1315/403/1/012131
21. Mudarisov, S., & Sharafutdinov, A. (2019). Justification of constructive and technological parameters of the horizontal distributor of the pneumatic seeder. *Vestnik of the Bashkir State Agrarian University, 49*(1), 133–140. https://doi.org/10.31563/1684-7628-2019-49-1-133-140
22. Mudarisov, S., Sharafutdinov, A., & Farkhutdinov, I. (2019). More precise seed placement with a horizontal seed spreader of an air planter. *Vestnik of the Bashkir State Agrarian University, 50* (2), 131–136. https://doi.org/10.31563/1684-7628-2019-50-2-131-136
23. Mudarisov, S., Badretdinov, I., Rakhimov, Z., Lukmanov, R., & Nurullin, E. (2020). Numerical simulation of two-phase "air-seed" flow in the distribution system of the grain seeder. *Computers and Electronics in Agriculture, 168*, 105,151. https://doi.org/10.1016/j.compag.2019.105151
24. Kravtsov, A. V., Konovalov, V. V., Zaitsev, V. Y., Petrov, A. M., & Petrova, S. (2020). Simulation of the pneumatic system of a seed drill with a vertical flow direction. *KnE Life Sciences*. https://doi.org/10.18502/kls.v5i1.6060
25. Bourges, G. (2019). *Sistemas de siembra neumáticos*. Editorial Académica Española.
26. Crowe, C. T. (2005). *Multiphase flow handbook*. CRC Press. https://doi.org/10.1201/9781420040470
27. Lim, E. W. C., Wang, C. H., & Yu, A. B. (2006). Discrete element simulation for pneumatic conveying of granular material. *AICHE Journal, 52*(2), 496–509. https://doi.org/10.1002/aic.10645
28. Mohammadreza, E. (2014). *Cfd-dem modelling of two-phase pneumatic conveying with experimental validation. PhD thesis*, The University of Edinburgh, http://hdl.hadle.net/1842/9693.
29. Elghobashi, S. (1994). On predicting particle-laden turbulent flows. *Applied Scientific Research, 52*(4), 309–329. https://doi.org/10.1007/bf00936835
30. Lun, C. K. K., & Bent, A. A. (1994). Numerical simulation of inelastic frictional spheres in simple shear flow. *Journal of Fluid Mechanics, 258*, 335–353. https://doi.org/10.1017/s0022112094003356
31. Lun, C. K. K., Savage, S. B., Jeffrey, D. J., & Chepurniy, N. (1984). Kinetic theories for granular flow: Inelastic particles in Couette flow and slightly inelastic particles in a general flow field. *Journal of Fluid Mechanics, 140*, 223–256. https://doi.org/10.1017/s0022112084000586
32. Wilcox, D. (2006). *Turbulence Modeling for CFD*. DCW Industries.
33. Shih, T. H., Liou, W. W., Shabbir, A., Yang, Z., & Zhu, J. (1995). A new k-ε eddy viscosity model for high Reynolds number turbulent flows. *Computers & Fluids, 24*(3), 227–238. https://doi.org/10.1016/00457930(94)00032-t
34. ANSYS Fluent 160. (2014). Fluent theory guide.
35. Morsi, S. A., & Alexander, A. J. (1972). An investigation of particle trajectories in two-phase flow systems. *Journal of Fluid Mechanics, 55*(02), 193. https://doi.org/10.1017/s0022112072001806
36. Zhang, X., & Vu-Quoc, L. (2000). Simulation of chute flow of soybeans using an improved tangential force displacement model. *Mechanics of Materials, 32*(2), 115–129. https://doi.org/10.1016/s01676636(99)00043-5

37. Israel, R., & Rosner, D. E. (1982). Use of a generalized stokes number to deter- mine the aerodynamic capture efficiency of non-stokesian particles from a compressible gas flow. *Aerosol Science and Technology, 2*(1), 45–51. https://doi.org/10.1080/02786828308958612
38. Rao, A., Curtis, J., Hancock, B., & Wassgren, C. (2011). Numerical simulation of di- lute turbulent gas-particle flow with turbulence modulation. *AICHE Journal, 58*(5), 1381–1396. https://doi.org/10.1002/aic.12673
39. Shapiro, S. S., & Wilk, M. B. (1965). An analysis of variance test for normality (complete samples). *Biometrika, 52*(3/4), 591. https://doi.org/10.2307/2333709
40. Mendenhall, W., Beaver, R. J., & Beaver, B. M. (2007). *Introduccióonn a la probabilidad y estadística Ed. 12*. Thomson Learning.

Part III
Value Chain

Data-Based Agricultural Business Continuity Management Policies

Athanasios Podaras

1 Introduction

The modern industry is undoubtedly highly complex [1] and utterly dependent on information and communication technologies, briefly ICTs. The majority, if not all, of the business processes and decisions are ICT-supported and fully data-driven. Thus, any operational failure of information systems or any business inefficiencies caused by technological unavailability can seriously harm a modern enterprise in many different negative ways [2].

Business continuity creates an *"environment that enables the organization to continue its mission, provide services for its customers and maintain its viability in the event of unplanned events, weather, or emergency conditions by effectively planning and responding to disruptions caused by these events"* [3]. One of the core objectives of the Business Continuity Management, briefly BCM, is the timely restoration of the most critical business functions. In a detailed study conducted by Engemann and Henderson [4]: *"the objective of the business continuity management is to limit the impact of major disruptions at an acceptable level. This involves determining the criticality of organization functions, operations and processes, identifying interrelationships and dependencies, identifying critical timeframes and determining exposures over time. Data collection is an important activity throughout the BCM development process"* [4]. In this way, the researchers successfully indicate the importance of applying data-based BCM policies.

The value of data in combination with modern sophisticated data manipulation technologies is also important in the field of agriculture. The importance of

A. Podaras (✉)
Technical University of Liberec, Faculty of Economics, Department of Informatics, Liberec, Czech Republic
e-mail: athanasios.podaras@tul.cz

D. D. Bochtis et al. (eds.), *Information and Communication Technologies for Agriculture—Theme II: Data*, Springer Optimization and Its Applications 183,
https://doi.org/10.1007/978-3-030-84148-5_9

data-supported agricultural decision making has been underlined by recent studies. For instance, a research conducted by [5] highlights the important role of computerized systems as well as the data collection and processing tools in agriculture. In this study, the research team considers data and data processing tools as the basis for boosting agricultural productivity. As a consequence, effective BCM policies in the agrarian sector should be also dependent on quality data and sophisticated data processing solutions in every step of the business continuity management lifecycle.

In recent years, experts from the academic sphere have managed to derive significant results, both theoretical models [6] as well as technical research outcomes [7–9] regarding the business continuity management field. Moreover, the existence of global standards such as the BS 25999 [10] and ISO 22301 [11] in combination with an important number of developed software tools, the vendors of which has been included in a recent list presented by [12], have significantly ameliorated the quality in formulating business continuity management strategies within the agricultural domain.

Nevertheless, multiple experts in practice either skip serious BCM steps or apply them according to their personal aspect and experience. The agricultural field is not an exception. In the Czech Republic for instance, as stated by [13], "*only 5% of all organizations having a BC standard are from the agricultural sector*". Additional evidence with respect to the lack of standard and unified BCM policies within the agriculture industry is provided by specific business continuity reports regarding the dairy farms [14] and the cattle feed yard [15] industries. From these business continuity guidelines, it can be concluded that BCM policies differ from one agricultural industry to another.

According to the previous statements, two main elements should be taken into account when attempting to introduce effective agricultural BCM policies. The first prerequisite is the collection of sufficient, high-quality, and trustworthy data. Data-based decisions require adequate data volumes and business intelligence systems which can be utilized in order to host and process the collected data and conduct descriptive and predictive data analysis. Secondly, a standard BCM policy should be charted and followed. Arbitrary BCM solutions lead to poorly organized BCM strategies. Business intelligence multidimensional models can facilitate knowledge discovery based on standard facts, dimensions, and hierarchies for a fully data-driven agricultural BCM decision making and strategy implementation.

The present chapter aims to propose such a standard data-based approach for utilizing available agricultural business continuity management data in order to develop efficient business continuity multidimensional data models as key drivers for efficient decision making in the given field. The aforementioned goal can be fulfilled via the accomplishment of the following objectives:

(a) Based on the available BCM data from two different agricultural industries, two novel multidimensional data warehouse schemas are proposed as the conceptual static representation of pattern-oriented agricultural BCM data storage and analysis tools.

(b) Complementary business process modeling and notation (BPMN) [16] diagrams are used in order to represent the dynamic form of data-based business continuity decisions which stem from the available data that is organized in the above-mentioned data warehouse solution.
(c) Finally, a representative example of agricultural BCM classification is presented based on decision trees. The specific activity provides evidence of the parallel predictive capability of the data which is used to generate the proposed multidimensional schemas.

In order to achieve the aforementioned goal as well as the target objectives of the current research the rest of the chapter is organized as follows:

Motivation: further to the introductory background analysis, a motivation section is devoted to underline the importance of adopting business intelligence tools in the business continuity management scientific discipline. Moreover, the need for combining business intelligence and data mining techniques toward facing the big data challenges in the modern era is highlighted. The section is completed by mentioning the big data challenges in the field of agriculture.

Tools and Methods: a reference to the theoretical background of the present chapter is included in this section. Apart from the sources used to create the utilized datasets, the core business intelligence concepts regarding the multidimensional data warehouse theory are analyzed. Moreover, BPMN and machine learning preliminaries are introduced for the better comprehension of the target research outcomes.

Results: the results section is composed of the novel multidimensional models including the corresponding conceptual data warehouse schemas. The dimensions, facts, measures, and hierarchies of the proposed data models are thoroughly explained. Representative online analytical processing operations, briefly OLAP, in the form of the structured query language (briefly SQL) queries are also demonstrated. Additionally, example of business process modeling notation (BPMN) decision-making processes are designed to support agricultural BCM policies according to the OLAP query tasks. Moreover, a machine learning prediction rule entitled *risk exposure classifier* is induced based on standard data mining classification algorithms, namely decision trees. The classifier is boosted via the utilization of advanced classification techniques, that is the 10-folds cross validation and the random forests.

Discussion: the discussion section addresses all the possible limitations of the current research. Moreover, a deeper analysis of additional research activities required to increase the value of the current contribution is included.

Conclusions/Implications for Future Research: the chapter is finalized by summarizing the current data-based agricultural business continuity research achievements, the brief reference to the research approach including the utilized tools and methods, and by stating the possible future research activities that will be conducted to complete the present achievements.

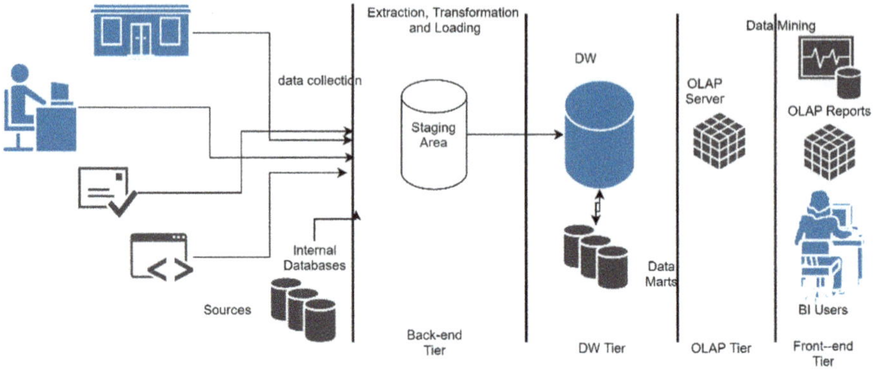

Fig. 1 The involved tiers of a full business intelligence solution (Source: own work inspired by [18])

2 Motivation

2.1 Business Intelligence Tools as Business Continuity Solutions in the Modern Era

Apart from the presence of various hazards which threaten the modern business environment, the exchange, reproduction, storage, analysis, and processing of large data volumes is another challenge for modern private companies and public organizations. In every sector, business units utilize sophisticated software solutions in order to face the so-called big data challenges, generate beneficiary knowledge and become more competitive.

Business intelligence systems [17] are considered to be an ideal solution to the modern big data challenges. Such systems are composed of different tiers in order to collect data from multiple sources existing in diverse formats and transform them into formats which permit further processing and ensure data integrity within the organizations. These data are further loaded into subject-oriented, time-variant multidimensional databases, namely data warehouses [18] which facilitate descriptive and predictive corporate decisions based on analytical reports and datasets (Fig. 1).

For the business continuity management domain which deals with organizational resilience, the incorporation of a complete business intelligence solution is crucial. Business intelligence systems are nowadays considered as a profitable investment for ensuring organizational resilience [19]. A combined business intelligence and data mining approach, similar to the current contribution, can facilitate the charting of a standard business continuity organizational policy.

2.2 *Business Continuity and Big Data Challenges in the Agricultural Domain*

Agriculture is a domain which faces various challenges from both the business continuity as well as the data manipulation standpoints. From the viewpoint of business continuity management, agricultural productivity is continuously threatened by natural, man-made, biological, and technological hazards. Multiple BCM researchers have highlighted the agricultural vulnerability to a diversity of threats including draughts and wildfires [20] while other researchers indicate the necessity to use modern information and big data technologies to mitigate or eliminate the major negative consequences of such threats with respect to an effective agricultural business continuity management strategy [21].

A recent study conducted by [5] underlines the importance of extracting and analyzing available data from various sources (e.g., geospatial data, farm data, sensors' data, and more) in order to enhance the agricultural productivity. However, an additional agricultural business continuity challenge is the lack of available data sources due to the confidential character of BCM data. The specific limitation is somehow justified by another study [22] where the value of small and limited datasets is considered the path to knowledge discovery and the better comprehension of vast datasets.

Another challenging issue that highly increased the motivation to conduct the current research, which is related to agri-business continuity management, is the influence of epidemics on the global agricultural industry. The recent COVID-19 unexpected worldwide spread has highlighted the necessity to apply effective BCM policies in agribusiness. The Agriculture Organization of the United Nations [23] claimed that "*the pandemic is impacting global food systems, disrupting regional agricultural value chains, and posing risks to household food security*". From the available business continuity guidelines [14, 15] which are the exclusive data sources of the current contribution, the epidemics are recorded as one of the most important hazards which should be taken into account when charting recovery strategies in agribusiness.

Furthermore, the study of the BCM data that are used in the current contribution revealed a lack of standard policy in agricultural BCM. As a consequence, and bearing in mind other BCM [13] and business intelligence [24, 25] studies in the agrarian sector, the conceptualization of a mixed business intelligence and data mining approach for standardizing the policies in resuming critical business functions in agriculture and dealing effectively with the numerous existing hazards in agribusiness has been considered indispensable. Data mining techniques have been also suggested for agricultural business continuity and disaster recovery predictions [26].

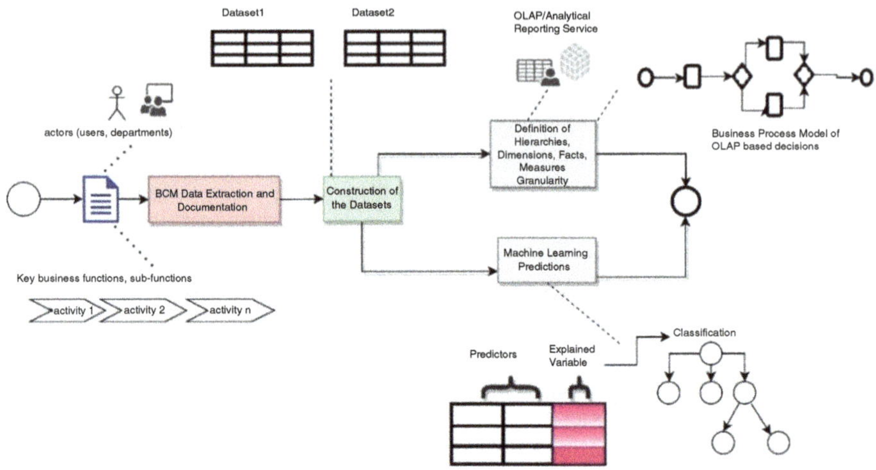

Fig. 2 The summarized research activities

2.3 Research Steps

In order to conduct the current research, the following activities have been implemented (Fig. 2):

- Data gathering and dataset construction: two datasets have been constructed based on the available data. The first dataset includes information about key business functions, their sub-functions, their department, and their level of criticality with respect to their recovery priority in both industries.
- Design of two different multidimensional models based on the aforementioned organized dataset and conceptualization of their data warehouse schemas.
- Illustration of the dynamic business process aspect of potential representative agricultural business continuity decisions based on online analytical processing operations is also included. Decision-making scenarios based on descriptive data are modeled with the help of BPMN diagrams.
- Utilization of machine learning classification techniques to conduct predictive rule-based agricultural business continuity decisions with respect to representative labeled output class variables. A practical example is also delineated.

3 Tools and Methods

3.1 Formulation of Datasets

The business continuity reports from two different agrarian industries are used for constructing the utilized datasets. The reports are very recent and can be easily

Table 1 Sample of selected records of dataset1 (key business functions, sub-functions, department, and criticality levels dataset)

Agrarian Industry	Key Business Function	Sub-Function	Department	Criticality Level (Recovery Priority)
CF	Cattle movement and holding	Cattle tracking/tagging	Feed yard	Mostly critical
CF	Facility maintenance	Water supply	Feedmill manager	Mostly critical
CF	Business administration	Payroll	Business manager	Moderately critical
CF	Information technology	Maintenance of software	IT manager	Least critical
DF	Parlor operations	Milk transport	Parlor manager	Mostly critical
DF	Calf rearing	Records management	Herd manager	Moderately critical
DF	Animal health	Treatment and vaccinations	Environmental manager	Moderately critical
DF	Information technology	On-site network and data systems management	IT manager	Least critical

Source: own work based on BCM data by [14, 15]

downloaded from the official website of the Texas Panhandle Regional Planning Commission [14, 15].

The first dataset (hereinafter dataset1) includes data regarding a number of critical business functions in the *dairy farms* and the *cattle feed yard* industries, while the second dataset (hereinafter dataset2) includes data regarding the natural hazards, the key business function(s) they influence as well as their risk exposure level. The data has been organized in spreadsheets using Microsoft Excel 2013 where the columns represent the observed variables. Samples of the constructed datasets are below illustrated (Tables 1 and 2).[1] The illustrated parts of the constructed datasets include the variables (columns) which are utilized for the OLAP analysis and the machine learning classification results.

3.2 Business Intelligence Multidimensional Data Models – Preliminary Concepts

Every business intelligence system is utilized in order to collect data from various sources, transform them to a specified format and organize them into a target

[1]The full datasets are also available and can be provided by the author upon request.

Table 2 Sample of dataset2 (selected records) – Hazards, causes, and risk exposure

Agrarian Industry	Hazard	Cause	Key Business Functions Impacted	Risk Exposure	Average Recovery Time (Hours)
CF	Weather	Natural	2	High	48
CF	Utility malfunction	Technological	8	Medium	90
CF	Cyber-attack	Man-made	2	Medium	48
CF	Flu pandemic	Biological	8	Low	90
DF	Pump failure	Technological	1	High	24
DF	Inadequate surveillance	Man-made	2	Medium	48
DF	Employee loss	Man-made	1	Low	72
DF	Liner leakage	Technological	1	High	24

CF Cattle Feed yards, *DF* Dairy Farms

multidimensional, integrated, subject-oriented, time-variant, and non-volatile database solution which is known as a *data warehouse* [27].

As stated by data researchers who focus on the multidimensional database systems, *"multidimensional databases are typically used for business intelligence (BI), especially for online analytical processing (OLAP) and data mining (DM). The advantages of using multidimensional databases for OLAP and DM rather than a relational database such as a dimensional data store (DSS) are that they use less disk space and have better performance"* [28].

In the multidimensional schema, *"facts correspond to events which are usually associated to numeric values known as measures and are referenced using the dimension elements"* [29]. Moreover, *"dimensions are modeled as hierarchies of elements, where each element belongs to a category. The categories are also organized into a hierarchy called hierarchy schema"* [29]. A standard multidimensional hierarchy schema is illustrated in Fig. 3.

A final concept of major importance in multidimensional data models is the information *granularity* (else called *grain*) which acts as an element that elevates aggregate models to higher abstraction levels and enables quantification of different results in individual levels in order to ensure enhanced knowledge-based and data-oriented decision support [30]. A typical multidimensional data model is usually represented in the form of a cube [31] (Fig. 3).

The above-mentioned multidimensional model is considered to be the basis of designing the target data warehouse (database) schema. One of the most commonly used database design approaches for developing a data warehouse solution is the relational database design approach [18, 31]. As in the classical operational databases, the data warehouse schema is composed of *entity types* representing the dimensions and facts, *attributes* which express the hierarchy levels that exist in

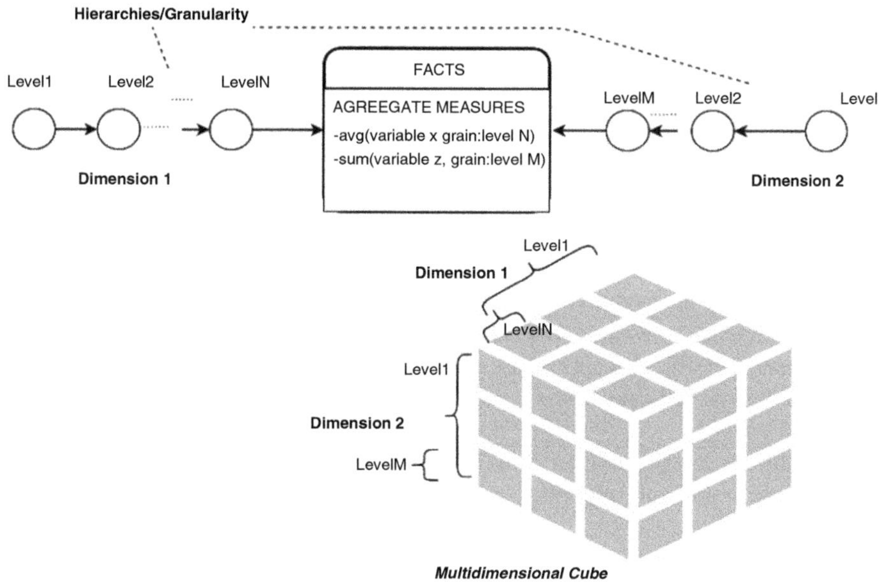

Fig. 3 A typical representation of a multidimensional data model

each dimension, the *measures* included in the facts entity type, as well as the *relationships* between the dimensions and the facts. The simplest form of such a schema is known as the *star schema*, where multiple dimensions are related to a unique facts' entity. A more complex form of such a schema is known as *the snowflake schema* where the dimension entity types are normalized. Additionally, it is possible to create data warehouse schemas where some dimensions are normalized while some others not (*starflake schema*), and other where each dimension can be related to more than one facts entity types (*constellation schema*) [18]. The currently proposed business continuity multidimensional models are illustrated in the form of relational data models via the star schema of a data warehouse. The proposed database helps conducting OLAP-based descriptive analytics based on the Structured Query Language (SQL).

3.3 Business Process Modelling Notation (BPMN) for Supporting Business Decisions Based on Multidimensional Data

While database schemas are used to design a static view of a decision support system, business process modeling tools are utilized for the dynamic representation of the processes and decisions which heavily rely on the aforementioned static data models [32]. The interpretation of the business process which is supported by a business intelligence system is of paramount importance. Researchers [31] thoroughly analyze the value of graph-based business process modeling tools in the context of business intelligence, due to the fact that the graphs include semantics and symbols that can illustrate relations between process objects. Thus, supported BI activities with the help of BPMN [16] and Petri nets [33] are often illustrated, due to their broad acceptance as representative graph-based business process modeling tools.

In the present study, the BPMN approach is selected since it *"offers a range of modeling elements with a set of icons representing these elements"* [31]. In this way, the process compliance with the proposed business intelligence agricultural business continuity management multidimensional data schemas can be ensured.

The core BPMN concepts and symbols such as the *inclusive* and *exclusive getaways* which indicate crucial decision points, the *sequence flows, message flows* and *activities,* and other like *pools* and *lanes* that delineate the core process participants are described in detail in [34]. In the present study, two selected OLAP-based decision-making processes are illustrated with the help of the BPMN diagrams.

3.4 A Robust Machine Learning Agricultural Business Continuity Classifier

Apart from the thorough and detailed analytical reporting services related to past and current business trends which are known as *descriptive analytics* that can be provided by a business intelligence multidimensional data warehouse solution, an additional remarkable and important BI activity is related to the formulation of business decisions based on *predictive analytics*. Prediction rules related to crucial business facts are a modern data-based decision-making trend nowadays and they are considered to be an integral part of the knowledge discovery process.

Prediction rules are mainly based on machine learning techniques such as classification, regression, and clustering. Classification based on *decision trees* [35] is one of the most commonly used data mining approaches to model prediction rules based on available datasets, when the *predicted* (else *output* or *explained*) variable is labeled as a class (i.e., 0/1, YES/NO, Green/Yellow/Red and more). In the present study, the classification based on the CART (classification and regression trees) [36] algorithm is used as a broadly accepted, standard, and adequately tested data mining classification approach. The specific algorithm has been also proposed for implementing machine learning tasks in the information security domain which is closely related to business continuity management [37].

In the current study, the second dataset (dataset2) (Table 2) is utilized as the basis to conduct rules-based classification regarding the level of *risk exposure* (high, medium, low) of key business functions in both agrarian industries. The .csv file including the full dataset is imported to and processed in the R-Package [38] for the induction of the target decision tree. The necessary R-Package libraries and scripts have been utilized in order to infer the appropriate classifier. The classifier is further boosted with the 10-fold cross-validation [39] procedure as well as the random forest ensemble classification techniques. Each model is measured for its accuracy based on the standard evaluation metrics used in classification process, namely the confusion matrix. Data mining methods such as random forests and 10-fold cross-validation are used by agricultural researchers in making predictions related to the agricultural production [40].

4 Results

4.1 The Multidimensional Data Models for Supporting Agricultural Business Continuity Management Decisions

Model 1: The Criticality Levels Multidimensional Model

The specific model relies on the dataset1 (Table 1). The model is composed of the following derived dimensions (Fig. 4):

- Dimension 1: Agricultural Operational Unit.

The specific dimension includes the hierarchies related to the operational units in both the dairy farms and the cattle feed yard agrarian sectors. The dimension is composed of the following levels of hierarchy:

Agrarian Industry (Level 1). Since the suggested data model can be expanded in order to host agricultural business continuity data from more sectors and industries, the specific hierarchy (or dimensional attribute) should be taken into consideration.

Key Business Function (Level 2). Using the drill-down OLAP operation the domain experts may analyze either dairy or cattle feed yard BCM data according to the

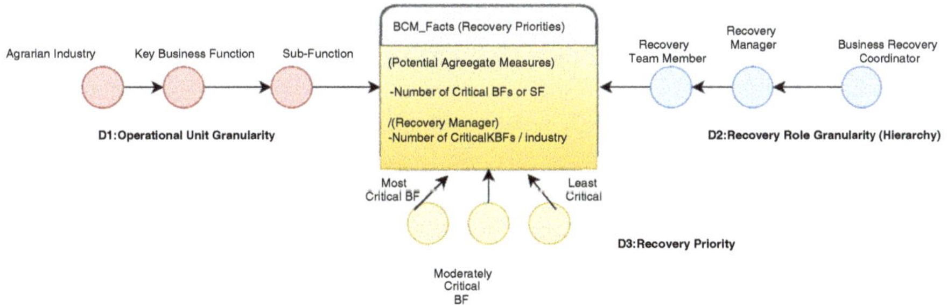

Fig. 4 The criticality levels multidimensional model

next level of hierarchy, namely the key business function (KBF). The relationship between the Agrarian Industry and the Key Business Functions is one-to-many.

Sub-function (Level 3): Each business function is related to multiple sub-functions. As a result, the information granularity can be further expanded. Again, the relationship between the Key Business Function and the Sub-Functions is one-to-many.

Dimension 2: Recovery Role.

The specific dimension refers to the data regarding the recovery team operational units in both the dairy farms and the cattle feed yard industries. The dimension is composed of the following levels of hierarchy:

Business Recovery Coordinator (Level 1). According to the dairy farms and cattle feed yard business continuity guides, the recovery teams and all the activities regarding the resumption of the critical business functions are guided and supervised by the recovery coordinator. Thus, the rest of the involved human actors should report to the specific leaders with regard to the implementation of any BCM tasks.

Recovery Manager (Level 2). At the second level of hierarchy with respect to the recovery team, the Recovery Managers are observed. Each recovery coordinator supervises many (two) recovery managers. In the case of the dairy farms the specific level is composed by the *dairy manager* and the *office manager.* Respectively, in the cattle feed yard industry the recovery managers are the *office* and *feed yard managers.* It can be realized that the relationship between the first and the second levels of recovery team hierarchies is one-to-many.

Recovery Team Members (Level 3): The lowest level of the recovery team information granularity includes the recovery team members. The team members are the responsible or representatives for one or more business functions and they should refer directly to the corresponding recovery managers. In general, the relationship between the recovery manager and the team members is one-to-many. Exceptionally, in both industries, the Animal Health/Hospital Administration and Environmental Management KBF representatives are supervised by the office managers and the feed yard/dairy managers. Based on this fact, the final form of the specific relationship in this case is defined as many-to-many.

- Dimension 3: Recovery Priority (or Criticality Level).

The specific dimension refers to the importance of every key business function (KBF) in both industries. The specific dimension includes attributes which share the same level of hierarchy and information granularity. The dimension refers to the criticality of the key business functions. Every KBF in both agrarian industries is regarded as *mostly critical, moderately critical,* or *least critical.*

Potential facts for which aggregate functions (i.e., SUM, AVERAGE, MEAN, COUNT, and more) might be applied in order to conduct analytical reports based on OLAP operations (drill-down, rollup, drill across, slicing, dicing, and pivot analysis) are the *number of critical business functions or sub-functions per industry, the mean*

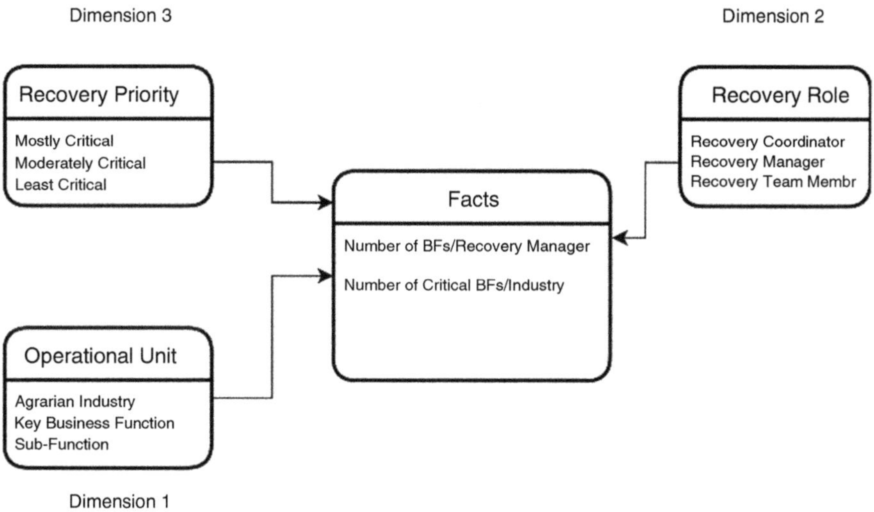

Fig. 5 The conceptual Star schema of the proposed multidimensional model (model 1)

value of moderately critical functions per industry or the average number of sub-functions controlled by each recovery manager (level 2). The corresponding star schema of the above illustrated multidimensional model is also included (Fig. 5).

Example of OLAP-based decisions in the dairy farms industry

In order to further facilitate the comprehension of the value of the multidimensional OLAP-based decision-making process in the agricultural business continuity management, the following representative decision-making scenario is considered.

It is assumed that the BCM coordinator in the dairy sector, needs to verify the ability of the recovery teams to bring back to normal the mostly critical key business functions in the sector. The corresponding recovery exercises are requested by the coordinator and supervised by the recovery manager. If the teams exceed the necessary recovery time as proposed by BCM experts [41] (according to which mostly critical business functions should be recovered within 24 hours) then the exercises should be repeated till the demanded recovery time requirement is fulfilled. In such case, the input data required by the recovery coordinator is based on the following possible OLAP query:

-SQL OLAP Query regarding Number (COUNT) of Mostly Critical Key Business Functions per Recovery Manager in the Dairy Farms industry:

```
SELECT COUNT(F.Key Business Functions) AS NumberofKeyFunctions, D2.
    RecoveryManager, D3.Mostly Critical
FROM RecoveryRoles D2
INNER JOIN.
(Recovery Priorities D3 INNER JOIN (Operational Unit D1 INNER JOIN Facts
    F ON D1.OperationalUnitID = F.OperationalUnitID)
```

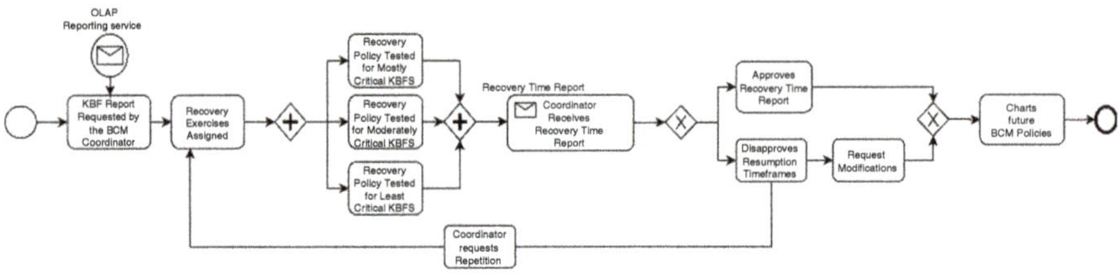

Fig. 6 An OLAP-based decision regarding the resumption of key business function in a specific agricultural industry

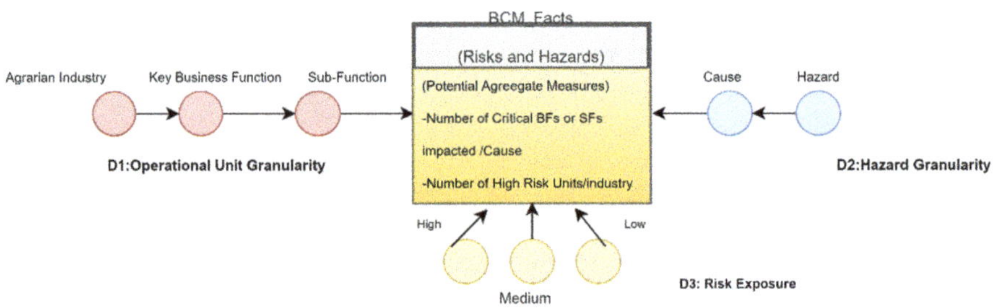

Fig. 7 The proposed risks/hazards multidimensional model

ON D3.RecoveryPriorityID = F.RecoveryPriorityID)
ON D2.RoleID = F.RoleID
GROUP BY D2.RecoveryManager, D1.AgrarianIndustry, D3.MostlyCritical
HAVING D3.AgrarianIndustry = "Dairy Farms"

The report (OLAP query output) can be utilized by the BCM recovery coordinator to monitor the effectiveness of the current BCM policies in the agrarian sector. If the recovery timeframe requirements for the mostly critical business functions are not fulfilled, then a repetition of the exercise can be implemented, else BCM modifications can be also determined and applied in collaboration with the recovery team members. The following BPMN diagram illustrates the decision-making process which is supported by the above-mentioned OLAP query (Fig. 6).

Model 2: The Risks/Hazards Multidimensional Model

The specific model relies on dataset2 (Table 2). The model is composed of the following derived dimensions (Fig. 7):

- Dimension 1: Agricultural Operational Unit.

The specific dimension is not described in this sub-section since it is described in detail in model 1.

- Dimension 2: Hazards.

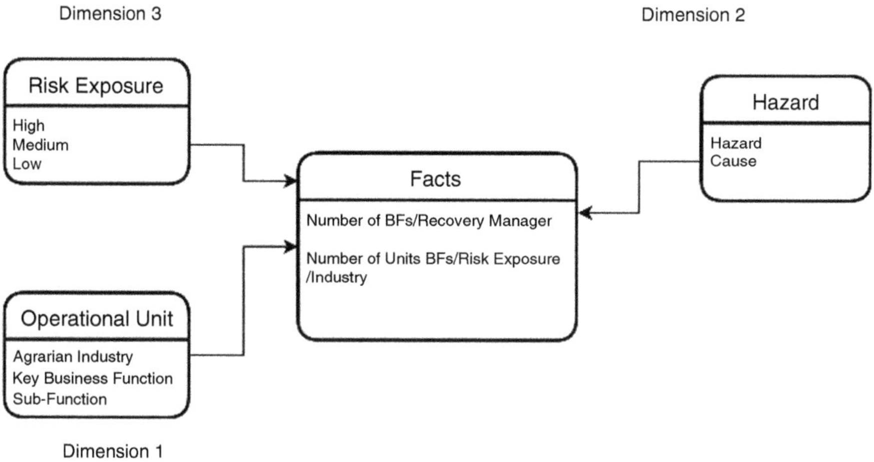

Fig. 8 The risks/hazards star schema

The specific dimension involves two (2) levels of hierarchy, namely the *Hazards* and the *Causes*. Each *hazard (Level 1 in hierarchy)* is related to one or more *causes (Level 2)*. As a consequence, a one-to-many relationship between the two levels of hierarchy exists.

- Dimension 3: Risk Exposure (Level).

The risk exposure dimension includes three (3) different categories which share the same information granularity level, namely *low, medium,* and *high (risk exposure level)* of the key business functions affected by the hazard or the cause.

Similar to the first model, the second multidimensional model can be also represented in the form of a data warehouse star schema (Fig. 8).

In this case, an additional OLAP report can be used as the input descriptive data for effective business continuity management decision making in both the selected agrarian industries.

Example: a BCM Coordinator needs to modify the risk mitigation measures when hazards influence three key business functions or more, the cause is flu pandemic, and the risk exposure is medium. The specific decision relies on an OLAP report which can be generated after the execution of the following query:

```
SELECT COUNT(F. Number of Business Functions), D2.Cause, D3.Medium
FROM Hazards D2
INNER JOIN.
(Operational Unit D1 INNER JOIN (Risk Exposure D3 INNER JOIN Facts F ON
   D3.RiskExposureID = F.RiskExposureID)
ON D1.OperationalUnitID = F.OperationalUnitID)
ON D2.HazardID = F.HazardID
GROUP BY D2.Cause, D3.Medium
HAVING D2.Cause = "Flu Pandemic"
```

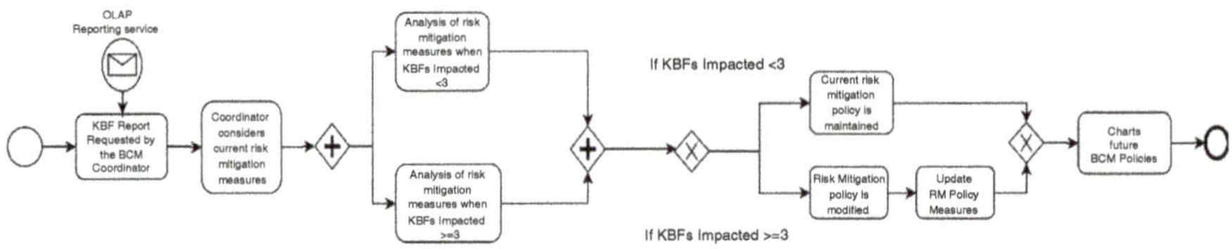

Fig. 9 The BPMN illustration regarding the risk/hazards policy decision-making process

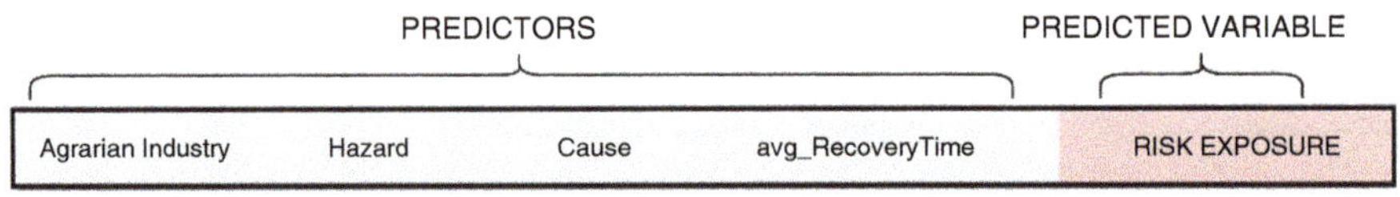

Fig. 10 The variables (columns) included in the exported dataset (candidate classifier)

The decision-making process regarding the risk mitigation measures which can be determined by the BCM coordinator can be illustrated as follows (Fig. 9).

4.2 *Machine Learning Predictive Analytics Based on the Proposed Multidimensional Schemas*

The first part of the results section has been devoted to explaining the value of the currently proposed multidimensional models in terms of descriptive analytics. The examples of the analytical reports provide detailed information of current and past measurements. However, a complete business intelligence multidimensional solution is also expected to host large data volumes and export big datasets which can be used as input for predictive decision making. In this case we refer to predictive analytics based on machine learning techniques. Thus, the second part of the results section includes a representative example based on the full dataset2 (Table 2).

The dataset2 has been imported into the R-package for data mining purposes. The standard steps for data preprocessing have been followed, including *importing data, and splitting of dataset2 into training and testing sets*. The present dataset is intended for classification, else prediction of a class (or category) of a labeled variable. In the example, the predicted labeled variable is the *risk exposure*. The utilized dataset is also composed of the input variables (predictors) which are: *agrarian industry, hazard, cause (type), KBFsImpacted, average_recovery_time (AVG_RT)* (Fig. 10).

The data can be also exported by the multidimensional model2 when adding the *recovery time* column in the table that represents the entity type operational unit. Alternatively, a new dimension can be added in the model, namely recovery priority, where a column that maps the recovery priority level with the proposed recovery time by BCM experts can be added and included in the final report.

For the needs of the current contribution no in-depth analysis of the data mining classification procedure is implemented. The current example is exclusively used for demonstration and has a rather instructive character for agricultural business continuity experts and agronomists who are interested in utilizing advanced data-based decision-making tools and benefit from the value of big datasets in order to discover new decision-making rules based on the existing data. Therefore, a brief delineation of the machine learning activities in the R package is provided.

Data Preprocessing

As it has been mentioned in the previous paragraph, the data preprocessing task involves importing the dataset into the R package and splitting into training and testing sets. The dataset2 is composed of 55 records and six variables. The determined splitting ratio (training set/testing set) is 0.75. The libraries required for splitting the initial dataset is the *caTools*. The following R script is provided for basic data preprocessing in the R package.

```
#importing the dataset
dataset = read.csv('agro.csv')
View(dataset)
library(caTools)
#ensuring reproduction of the same results and splitting the dataset
set.seed(123)
split = sample.split(dataset$RiskExposure,SplitRatio = 0.75)
training_set = subset(dataset, split==TRUE)
testing_set = subset(dataset, split==FALSE)
```

Risk Exposure Classification Based on Decision Tree Induction

For generating and plotting the decision tree from the imported dataset, the classification and regression trees (CART) algorithm has been selected and the library *rpart* is also loaded in the R package. The script for inferring the decision tree (Fig. 11) is also provided:

```
#Decision Tree (dtclassifier).
library(rpart).
#Library for plotting the inferred decision tree.
library(rpart.plot).
#Predicting the risk exposure level based on all the involved input variables.
dtclassifier = rpart(formula = RiskExposure~., data = dataset).
rpart.plot(dtclassifier).
```

From the above illustrated decision tree, the following prediction rule can be inferred:

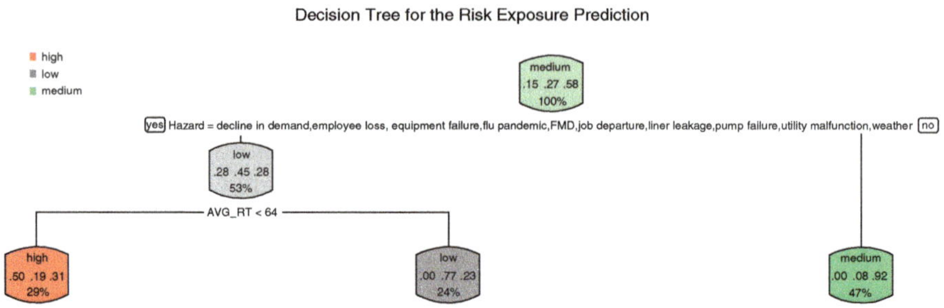

Fig. 11 The decision tree for the Risk Exposure Classifier for key business functions in both agrarian sectors

Table 3 The standard form of the confusion matrix for the binary classification

	Predicted Values: 0	Predicted Values: 1
Observed (actual) Values: 0	TN (true negative)	FP (false positive)
Observed (actual) Values: 1	FN (false Negative)	TP (true positive)

IF Hazard = decline in demand, employee loss, equipment failure, FMD, job departure, liner leakage, pump failure, utility malfunction, weather **THEN** Risk Exposure = 'Low'
IF Average Recovery Time < 64 Hours
THEN Risk Exposure = 'High'
ELSE Risk Exposure = Low
END IF
ELSE Risk Exposure = 'Medium'
END IF

For evaluating the accuracy of the new classifier considering the test data, the confusion matrix [42, 43] is used as a standard evaluation metric for decision tree classifiers. The confusion matrix illustrates the overall true positive, true negative, false positive, and false negative scores by comparing the predictions of the testing dataset (predicted data) with the available data (observed data). When referring to binary classification, the confusion matrix has the form illustrated in the following table (Table 3). The following formula provides the accuracy of the inferred classifier:

$$\text{Accuracy} = \frac{\text{TP} + \text{TN}}{\text{TP} + \text{TN} + \text{FP} + \text{FN}}$$

The R scripts which are used for measuring the predictive accuracy of the classifier based on the testing data when the predicted *risk exposure level* labeled variable is known is provided by the following script:

```
#predicting the accuracy of the decision-making classifier with the test data
pred1 = predict(dtclassifier, type = "class", newdata = testing_set)
#executing the prediction result
pred1
#inferring the confusion matrix
cm1 = table(testing_set$RiskExposure, pred1)
cm1
#computing the business continuity risk exposure classification accuracy
accuracy_RISK_CART <- sum(diag(cm1))/sum(cm1)
#displaying the accuracy result
accuracy_RISK_CART
```

The results of the script are also projected:
Confusion matrix result:

```
high low medium
high 2 0 0
low 2 2 0
medium 2 1 5
accuracy: 0.6428571
```

Base on the projected confusion matrix results, the accuracy of the inducted classifier is 64.28%.

Boosting the *Risk Exposure* classifier's Predictive Power with the 10-Fold Cross-Validation and the Random Forest Techniques

In order to ameliorate the performance of the initial decision tree classifier, increase the robustness of its predictions and achieve more realistic classification results, advanced classification approaches have been also used for the purposes of the present research. More precisely, the 10-fold cross-validation (CV) and the random forests ensemble technique have been also checked in terms of their accuracy. In the R software package, the additional required libraries for the 10-fold CV are the *e1071* and the *caret*. For the random forest, the *caTools* and the *randomForest* packages have been also loaded. The demanded scripts are below illustrated.

In brief, the 10-fold CV is based on 10 different training data/testing data combinations and the overall accuracy of all the final classifier is reported. In the case of the random forest, a significant number of decision trees in inferred (in our case 500 trees have been produced) and the overall accuracy is again based on the average performance of all these classifiers. Again, the evaluation metrics used is the confusion matrix for each classification approach.

```
#10-folds cv
library(e1071)
require(e1071)
library(caret)
```

```
require(caret)
numFolds <– trainControl(method = "repeatedcv", number = 10, repeats = 3)
cpGrid <– expand.grid(.cp = seq(0.01, 0.2, 0.01))
train(RiskExposure~., data = training_set, method = "rpart", trControl = numFolds,
  tuneGrid = cpGrid)
r1cv < –rpart(RiskExposure~., data = training_set, method = "class", cp = 0.13)
#RandomForest for n trees
library(randomForest)
ClassifierRF_RISK = randomForest(RiskExposure~., data = training_set,
  ntree = 500, mtry = 3, importance = TRUE)
ClassifierRF_RISK
plot(ClassifierRF_RISK)
```

The confusion matrix results including the classifiers' accuracy levels are also projected.

```
Confusion matrix for the 10Folds CV.
high low medium.
high 1 0 1
low 0 2 2
medium 2 2 4
accuracy10FCV: 0.5
Confusion matrix for the random forest
high low medium
high 1 0 1
low 0 1 3
medium 2 2 4
accuracy_RISK_rf: 0.4285714
```

The random forest classifier has been proved as the less accurate (42.85%). However, the predictive power of the random forest classifier cannot be easily judged since they perform much better than simple decision tree algorithms in very large datasets when the records and the feature data (involved variables) are more than the available data considered in the present case study. Another way to depict the performance of the random forest classifier is to use the out-of-bag (OOB) error plot. The OOB estimate is used to compute the prediction error of random forests. The OOB estimate is based on observations which are not included in the bootstrap sample. The current OOB error plot shows that when the number of trees is increased the random forest prediction error is reduced and gradually stabilized (Fig. 12).

Finally, the summarized performance of all the tested risk exposure classifiers is also included in the form of a simple and comparative table (Table 4). Based on the achieved results and taking into consideration the limited available dataset, it can be concluded that the agronomists in collaboration with specialized business continuity consultants in the agrarian sector can rely on the predictive power of the suggested classifiers which stem from the proposed multidimensional data model.

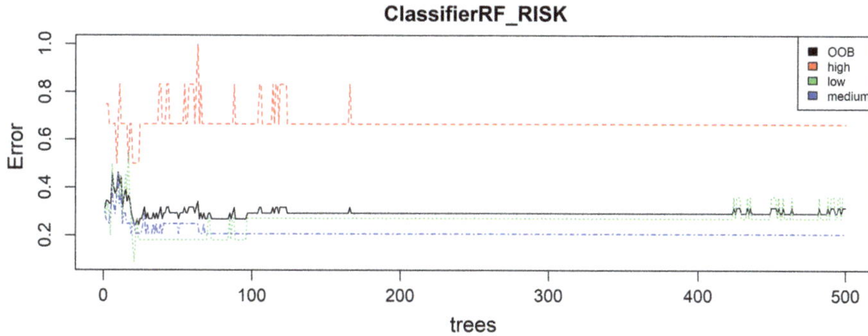

Fig. 12 The business continuity risk exposure classification error plot based on the random forest technique

Table 4 The summarized accuracy results of the three classifiers

Classifier	Accuracy
Decision tree (CART)	64.28%
10Folds CV	50%
Random Forest	42.85%

5 Discussion

The currently utilized data concern the BCM policy for two selected agrarian industries. However, the year that the specific guidelines are documented is not specified. It is assumed that the BCM policies are updated on a regular basis. The advantage of the proposed multidimensional schemas inherits the time-variant (historical data storage) and non-volatile (no changes to historical data are permitted) characteristics of the data warehouse. In the occasion of any future modifications, a time dimension can be added in the proposed multidimensional models. As a consequence, the agronomists or agricultural BCM experts will retain the advantage to retrieve historical agricultural BCM data.

Additionally, based on the available literature regarding business continuity management in agriculture, it can be concluded that the proposal of a unified BCM multidimensional data model for all the agricultural industries and sectors is an extremely hard and demanding task due to the peculiarities of each industry. BCM policies vary not only from one industry to another, but also from one agricultural enterprise to another. As a result, the proposed contribution serves exclusively as a basic or pilot multidimensional proposal including dimensions that exist in all agricultural industries. Starting from this point, agricultural BCM experts and agronomists can be motivated to extend the proposed schemas and create models which meet their own BCM requirements.

Furthermore, the currently proposed multidimensional model is based on the simple star schema of a typical data warehouse solution. The schema was selected contrary to the snowflake schema due to the reduced complexity of the queries utilized as aggregate OLAP operations which facilitate crucial agricultural BCM decisions. However, the other possible schemas are not excluded as alternative data modeling solutions.

Moreover, the data mining classification results which have been inferred can help BCM decision makers in the agrarian sector utilize predictive analytics for formulating future BCM policies in their agrarian industry or test the current policies they apply based on the proposed multidimensional schema. However, it is recommended to gather more data in order to ensure higher predictive accuracy and classification robustness in the domain they serve. For researchers in the agricultural business continuity domains, it is highly recommended to rely also on interviews with selected agricultural companies in order to enrich the dataset and achieve better machine learning prediction models.

Finally, the selection of the appropriate classification algorithm is an issue of major importance. The currently investigated classifiers toward the risk exposure prediction for key agrarian business functions indicate that it is highly beneficial to rely on the OLAP reports which can be exported from the suggested multidimensional data warehouse solutions. However, additional and equally important classification techniques such as gradient boosting, k-nearest neighbors (k-NN), and support vector machine (SVM) can be tested for their accuracy and increase the value of the suggested models. The search of the ideal classifier has been considered to be a task outside the scope of the present chapter. The conducted machine learning classification techniques have a rather instructive character regarding the predictive possibilities of the suggested multidimensional schemas. Exploring the data in order to find the best fitting model requires more data, additional crucial data mining activities such as further data preprocessing, feature selection, and application of the above-mentioned classification techniques.

6 Conclusions

In the agricultural business continuity management domain, the incorporation of business intelligence multidimensional database solutions is a crucial issue. The impact of unexpected business interruptions maybe highly severe in many circumstances. Sophisticated business intelligence solutions may help organizations and enterprises mitigate or avoid the negative consequences from disastrous events or crisis incidents considering that such tools facilitate decisions regarding the *"prevention of serious problems that could develop if no action was taken"* [44].

The present chapter focused on the proposal of data-driven business continuity management decision-making policies in the agricultural sector. The conception of the strategy finds its roots in applying a mixed business intelligence and machine learning approach based on available business continuity management data in two specific agrarian industries, that is the dairy farms and the cattle feed yards. Two datasets which have been constructed based on the available BCM reports have been used for designing two corresponding multidimensional data warehouse conceptual schemas from which OLAP descriptive reports can be exported. Moreover, strong business process modeling tools, namely BPMN, have been used to demonstrate the possibility of agricultural BCM policy formulation based on the OLAP descriptive data. The suggested multidimensional schemas enable decisions for recovery prioritization and risk mitigation policies for key business functions in the two selected agrarian industries.

Additionally, due to the fact that a complete business intelligence solution includes also knowledge-based rules that facilitate the predictive decision--making possibilities provided by the reported data, a machine learning classification approach based on the decision tree induction is also included as an instructive tool for decision making. The classification of the risk exposure (high, medium, low) of key business function in the selected industries is performed with the CART algorithm. The classifier has been developed in the R software package and has been further boosted with the 10-fold cross-validation and the random forest techniques. In general, the predictive possibilities of the investigated classifiers have been proved satisfactory (dt: 64.28%, 10folds cv: 50%, and random forest: 42.85%) taking into account the limited dataset used in the current research.

Future research directions are related to the incorporation of larger data volumes and more data features in order to improve the performance of the current multidimensional schemas with respect to agricultural data-based business continuity decision making. Moreover, more advanced classification techniques will be used and investigated for their accuracy. However, it is believed that the present models and the demonstrated decision-making examples can inspire BCM domain experts, researchers, agronomists, and agribusiness staff to consider the contribution as the starting point for a multidimensional data-driven decision-making approach in their domain with the assistance of both the predictive and the descriptive analytics.

References

1. Kaul, M., Storey, V. C., & Woo, K. (2017). A framework for managing complexity in information systems. *J Database Manage, 28*(1), 31–42. https://doi.org/10.4018/JDM.2017010103
2. Koks, E., Pant, R., Thacker, S., & Hall, J. W. (2019). Understanding business disruption and economic losses due to electricity failures and flooding. *Int J Disaster Risk Sci, 10*, 421–438. https://doi.org/10.1007/s13753-019-00236-y
3. IEEE Risk Insurance. Available at: https://www.ieee.org/about/volunteers/risk-insurance/business-continuity-management.html, Accessed on Sep. 13, 2019.
4. Engemann, K. J. A., & Henderson, D. M. (2012). *Business continuity and risk management: Essentials for an organizational resilience*. Rothstein Associates.
5. Lytos, A., Lagkas, T., Sarigiannidis, P., Zervakis, M., & Livanos, G. (2020). Towards smart farming: Systems, frameworks and exploitation of multiple sources. *Computer Netowrks, 172*(2020).
6. Miller, H. E., & Engemann, K. J. (2012). Using analytical methods in Bussiness continuity planning. In K. J. Engemann, A. M. Gil-Lafuente, & J. M. Merigó (Eds.), *Modeling and*

simulation in engineering, economics and management. MS (2012). Lecture notes in business information processing, Vol. 115 (pp. 2–12). Springer.
7. Caelli, W. J., Kwok, L. F., & Longley, D. (2010). A business continuity management simulator. In K. Rannenberg, V. Varadharajan, & V. C. Weber (Eds.), *Security and privacy – Silver linings in the cloud. SEC 2010. IFIP advances in information and communication technology, vol 330* (pp. 9–18). Springer.
8. Fasolis, E., Vassalos, V., & Kokkinaki, A. (2013). Designing and developing a business continuity plan based on collective intelligence. In C. Douligeris, N. Polemi, A. Karantjias, & W. Lamersdorf (Eds.), *Collaborative, trusted and privacy – Aware e/m-services, I3E 2013, IFIP advances in information and communication technology* (Vol. 399, pp. 278–285). Springer.
9. Šimonová, S., & Šprync, O. (2011). Proactive IT/IS monitoring for business continuity planning, E M Ekon. *Manag., 14*, 57–65.
10. Bialas, a. (2012). Computer support in business continuity and information security management. In A. Kapczyński, E. Tkacz, & M. Rostanski (Eds.), *Internet – Technical developments and applications 2, advances in intelligence and soft computing, vol. 118* (pp. 155–169). Springer – Verlag.
11. International Organization for Standardizations (ISO). (2019). Security and resilience — Business continuity management systems — Requirements. Available at: https://www.iso.org/obp/ui/#iso:std:iso:22301:ed-2:v1:en (Accessed 10 August 2020).
12. Gartner. (2017). Magic Quadrant for Disaster Recovery as a Service. Available at: https://www.gartner.com/doc/3746618/magic-quadrant-disaster-recovery-service, 2017 (Accessed 13 January 2019).
13. Urbanec, J., & Urbancova, H. (2014). Adoption of business continuity management standards in Czech organizations. *Scientia Agriculturae Bohemica, 45*(1), 66–74. https://doi.org/10.7160/sab.2014.450109
14. The Panhandle Regional Planning Commission (PRPC). Texas Dairy Farm Business Continuity Guide. Available at: http://www.theprpc.org/Programs/RegionalServices/Dairy-Industry/prpc_dairy/bcdoc/Texas%20Dairy%20Farm%20BCG%20Template.docx. (Accessed 20 April 2020).
15. The Panhandle Regional Planning Commission (PRPC). Texas Cattle Feed Yard Business Continuity Guide. Available at: http://theprpc.org/Programs/RegionalServices/Cattle-Feedyard-Industry/PRPC_CattleFeed/bcdoc/Texas%20Cattle%20Feedyard%20BCG%20Supplement.docx. Accessed 17 March 2020.
16. BPMN. (2011). Business Process Model and Notation (BPMN), Version 2.0, OMG Document Number: formal/2011-01-03, Standard document. http://www.omg.org/spec/BPMN/2.0. Accessed 20 May 2020.
17. Rausch, P., Sheta, A. F., & Ayesh, A. (Eds.). (2013). *Business intelligence and performance management: Theory, systems and industrial applications*. Springer.
18. Vaisman, A., & Zimanyi, E. (2014). Data warehouse systems: Design and implementation. In *Data centric systems and applications*. Springer-Verlag.
19. Starr, R. (2003). Enterprise Resilience: Managing Risk in the Networked Economy. Available at: http://www.scrlc.com/articles/sbreader_Enterprise_Resilience.pdf. Accessed 8 July 2020. p.66.
20. Snedaker, S. (2007). *Business continuity and disaster recovery planning for IT professionals*. Elsevier.
21. Karmas, A., Tzotsos, A., & Karantzalos, K. (2016). Geospatial big data for environmental and agricultural applications. In S. Yu & S. Guo (Eds.), *Big data concepts, theories, and applications*. Springer. https://doi.org/10.1007/978-3-319-27763-9_10
22. Coble, K. H., Mishra, A. K., Ferrell, S., & Griffin, T. (2018). Big data in agriculture: A challenge for the future. *Applied Economic Perspectives and Policy, 40*, 79–96.
23. Food and Agriculture Organization of the United Nations. (2020). Available at: http://www.fao.org/2019-ncov/q-and-a/en/. Accessed 21 May 2020.

24. Tyrychtr, J., Brozek, J., & Vostrovsky, V. (2015). Multidimensional modelling from open data for precision agriculture. In *Enterprise and organizational Modeling and simulation - 11th international workshop, EOMAS 2015, June, 08 2015* (pp. 141–152). Springer.
25. Tyrychtr, J., Ulman, M., & Vostrovský, V. (2015). Evaluation of the state of the business intelligence among small Czech farms. *Agricultural Economics – Czech, 61*, 63–71.
26. Wang, Z., Li, C., & Liu, W. (2011). A regression analysis of the effect of natural disaster on agriculture. In *IEEE Proc. ICBMEI '11* (pp. 369–372). Guangzhou.
27. Kimball, R., & Ross, M. (2010). *The Kimball Group reader. Relentlessly practical tools for data warehousing and business intelligence*. Wiley.
28. Rainardi, V. (2008). Multidimensional Database. In *Building a data warehouse*. Apress. https://doi.org/10.1007/978-1-4302-0528-9_12
29. Caniupán, M., Bravo, L., & Hurtado, C. A. (2012). Repairing inconsistent dimensions in data warehouses. *Data & Knowledge Engineering, 79–80*(2012), 17–39. https://doi.org/10.1016/j.datak.2012.04.002
30. Pedrycz, W., Al-Hmouz, R., Morfeq, A., & Balamash, A. S. (2014). Building granular fuzzy decision support systems. *Knowl. Based Syst., 58*(2014), 3–10. https://doi.org/10.1016/j.knosys.2013.07.022
31. Grossmann, W., & Rinderle-Ma, S. (2015). *Fundamentals of business intelligence*. Springer-Verlag.
32. Podaras, A. (2017). Conceptual model of the BIA data warehouse. In R. Pergl, R. Lock, E. Babkin, & M. Molhanec (Eds.), *Enterprise and organizational Modeling and simulation, EOMAS 2017, lecture notes in business information processing, vol 298* (pp. 125–133). Springer.
33. Nielsen, M., Plotkin, G., & Winskel, G. (1981). Petri nets, event structures and domains, part I. *Theoretical Computer Science, 13*(1), 85–108.
34. Milani, F. (2019). Business pocesses. In F. Milani (Ed.), *Digital business anaysis*. Springer International Publishing. https://doi.org/10.1007/978-3-030-05719-0
35. Grąbczewski, K. (2014). *Meta-learning in decision tree induction*. Springer International Publishing.
36. Breiman, L., Friedman, J. H., Olshen, R. A., & Stone, C. J. (1984). *Classification and regression trees*. Chapman and Hall.
37. Pinem, A. F. A., & Setiawan, E. B. (2015). Implementation of classification and regression tree (CART) and fuzzy logic algorithm for intrusion detection system. In *Proc. IEEE 3rd international conference on information and communication technology (ICoICT)* (pp. 266–271). Nusa Dua, Bali.
38. R Core Team. (2014). *R: A language and environment for statistical computing. R Foundation for Statistical Computing*. Available at: http://www.R-project.org. Accessed 20 March 2019.
39. Kärkkäinen, T. (2014). On cross-validation for MLP model evaluation. In P. Fränti, G. Brown, M. Loog, F. Escolano, & M. Pelillo (Eds.), *Structural, syntactic, and statistical pattern recognition. S+SSPR 2014. Lecture notes in computer science, vol 8621*. Springer. https://doi.org/10.1007/978-3-662-44415-3_30
40. Jeong, J. H., Resop, J. P., Mueller, N. D., Fleisher, D. H., Yun, K., Butler, E. E., et al. (2016). Random forests for global and regional crop yield predictions. *PLoS One, 11*(6), e0156571. https://doi.org/10.1371/journal.pone.0156571
41. Gibson, D. (2011). *Mitigating risk with a business impact analysis, in managing risks in information systems* (2nd ed.). Jones & Bartlett Learning.
42. Podaras, A. (2018). Measuring the accuracy levels regarding the dual business function criticality classifier. *IEEE Access, 6*, 41598–41606. https://doi.org/10.1109/ACCESS.2018.2860976
43. Yang, F., Watson, P., Koukoula, M., & Anagnostou, E. N. (2020). Enhancing weather-related power outage prediction by event severity classification. *IEEE Access, 8*, 60029–60042. https://doi.org/10.1109/ACCESS.2020.2983159
44. Golab, L., Johnson, T., & Shkapenyuk, V. (2012). Scalable scheduling of updates in streaming data warehouses. *IEEE T. Knowl. Data En., 24*(2012), 1092–1105. https://doi.org/10.1109/TKDE.2011.45

Soybean Price Trend Forecast Using Deep Learning Techniques Based on Prices and Text Sentiments

Roberto F. Silva, Angel F. M. Paula, Gustavo M. Mostaço, Anna H. R. Costa, and Carlos E. Cugnasca

1 Introduction

The soybeans value chain was responsible for producing 352 million tonnes in the 2017–2018 worldwide harvesting season [1]. In addition to human consumption and its use in various products as raw material, soy is also essential for animal production [2]. For this reason, its value chain can be considered as one of the most important agricultural chains worldwide.

It is essential to increase the predictability of the soybean prices, both in the long and short term, to improve the revenues and decision making of the agents in this value chain. Nevertheless, the vast majority of literature about price prediction focus on stocks, such as the important works by Kara et al. [3], Goçken et al. [4], Di Persio and Honchar [5], and Weng et al. [6]. Although these works focus on different markets, stocks, and countries, the techniques used by them can be adapted and applied to forecast prices of agricultural products.

There are two main types of predictions considering price time series, as illustrated in Fig. 1: (i) price prediction, which can be treated as a regression problem, such as in Nasira and Hemageetha [7]; and (ii) price trend prediction, which can be treated as a classification problem, such as in Kara et al. [3]. In the first type, the main interest is on predicting the value for the next period. In the second type, the main goal is to predict if the price will go up or down in the next period. In this chapter, the focus is on the second type, as research indicates it can provide better information for decision making [8].

R. F. Silva (✉) · A. F. M. Paula · G. M. Mostaço · A. H. R. Costa · C. E. Cugnasca
Department of Computer Engineering and Digital Systems, Escola Politécnica da Universidade de São Paulo (USP), São Paulo, Brazil
e-mail: gmostaco@usp.br; anna.reali@usp.br; carlos.cugnasca@usp.br

D. D. Bochtis et al. (eds.), *Information and Communication Technologies for Agriculture—Theme II: Data*, Springer Optimization and Its Applications 183,
https://doi.org/10.1007/978-3-030-84148-5_10

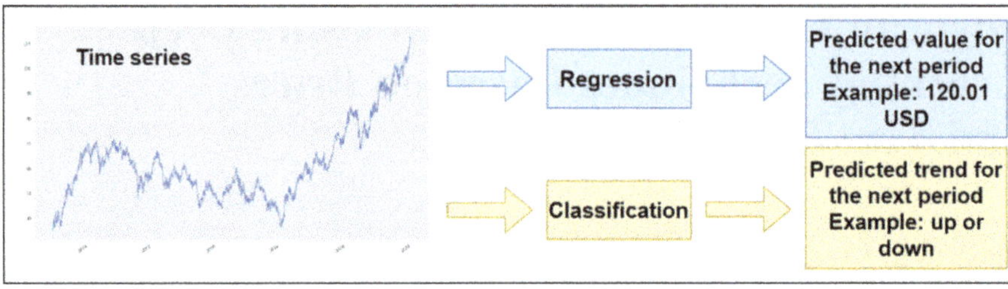

Fig. 1 Types of predictions considering time series

The main techniques used can be divided into the following categories: (i) traditional statistical and econometric models, such as seasonal decomposition, Autoregressive Integrated Moving Average (ARIMA), and Generalized Autoregressive Conditional Heteroskedasticity (GARCH) models, as well as their variations [9, 10]; (ii) traditional machine learning techniques, such as support vector machines (SVM) [11], random forests, among others; and (iii) deep learning techniques, such as multi-layer perceptrons (MLP), convolutional neural networks (CNN), and long short-term neural networks (LSTM) [12–15].

Deep learning techniques are considered state-of-the-art on price and price trend predictions [16, 17], so these techniques were implemented and analyzed in this chapter. Thorough reviews on the use of machine learning techniques for price prediction are available in the literature [5, 6].

An essential finding for forecasting trends and stock prices was that market sentiment could improve forecasting in certain cases [6, 14]. Research by Nassirtoussi et al. [14] and Medhat et al. [18] provide very in-depth reviews on the use of natural language processing (NLP) techniques to extract sentiments from news, forums, and social media to improve price prediction. In summary, it was observed that deep learning models, such as CNN using word embeddings, can provide the best results, and that daily frequency is the most used for forecasting prices and sentiments.

Nevertheless, few papers address the predictability of agricultural product prices. Notable works are from Kanchymalay et al. [11], Xiong et al. [19], Khamis and Abdullah [20], Weng et al. [21], Wang and Gao [22], Xiong et al. [23], and Kantanantha et al. [24]. Section 2 contains a thorough analysis of the main results of these important works.

Most of the studies that analyze time series in the agricultural domain tend to focus on environmental variables. Among these, the most common ones are related to climate or to specific variables that can affect the quantity and quality of production [25]. It is also important to note that most articles in this field focus on high value-added products, such as meat, fruits, and grapes used for wine production. Examples are the article by Kamilaris et al. [25], which contains an extensive review of big data analysis for different agricultural supply chains, and by Verdouw et al. [26], which proposes a model for designing processes for the fruit industries.

In this chapter, the focus is on soybean price trend prediction due to its production volume and importance to the global economy, but the results can be extended to other agricultural commodities with few modifications. The main objective of this chapter is to implement and evaluate deep learning models for soybean price trend prediction in Brazil, considering both the price time series and the analysis of news sentiments. The final trend prediction was treated as a classification problem and the evaluation metric was the F1-score. This metric can be defined as the harmonic mean of precision and recall.[1]

A series of soybean prices[2] from 2006 to 2019 was used to derive the price trend signal, and a dataset of 357 relevant agricultural finance news in Portuguese was used to derive the sentiment signal [28]. The following architectures were evaluated to predict the price trend: (i) one with only the price trend forecasting model (LSTM); (ii) one that uses only the sentiment analysis models (MLP and CNN); and (iii) one that uses an ensemble model. An ensemble is a model that combines the results of two or more separate models to generate a new result. In this chapter, the results of two models were used to generate a trend prediction. One of the models generates a price signal, and the other generates a news sentiment signal. Figure 2 illustrates the methodology used in this chapter.

Then, the main hyperparameters of all models and their impacts were evaluated, as well as the use of different sentiment lexicons or dictionaries for the sentiment analysis. The main research questions of this work are: *"(i) which model results in the best forecast of the soybean price trend in Brazil?"* and *"(ii) does the use of sentiment analysis of news headlines result in better forecasts for the daily trend in soybean prices in Brazil?"*

To the best of our knowledge, this research is one of the few that considers both price and sentiment to make predictions on the agricultural domain, using deep learning techniques that are considered state-of-the-art in the literature. It is also one of the few attempts to conduct this kind of analysis in Portuguese news headlines, as the vast majority of sentiment analysis research is done in English.

The remainder of this chapter is divided into the following sections: Sect. 2 comprises related works, considering the forecasting of agricultural commodity prices and the deep learning model applied in forecasting price trends and sentiment analysis; Sect. 3 describes the methodology used and its related steps; Sect. 4 reports the main results obtained; Sect. 5 discusses implementation aspects, benefits for forecasting agricultural prices and how the models can be adapted for other products; Finally, Sect. 6 presents the conclusions, the observed limitations, and highlights future work.

---

[1] For an in-depth analysis of the precision, recall, and F1 score metrics, the reader can refer to Goutte and Gaussier [27]. For implementation purposes, the reader can refer to the Scikit Learn documentation: https://scikit-learn.org/stable/modules/generated/sklearn.metrics.f1_score.html

[2] https://www.cepea.esalq.usp.br/br

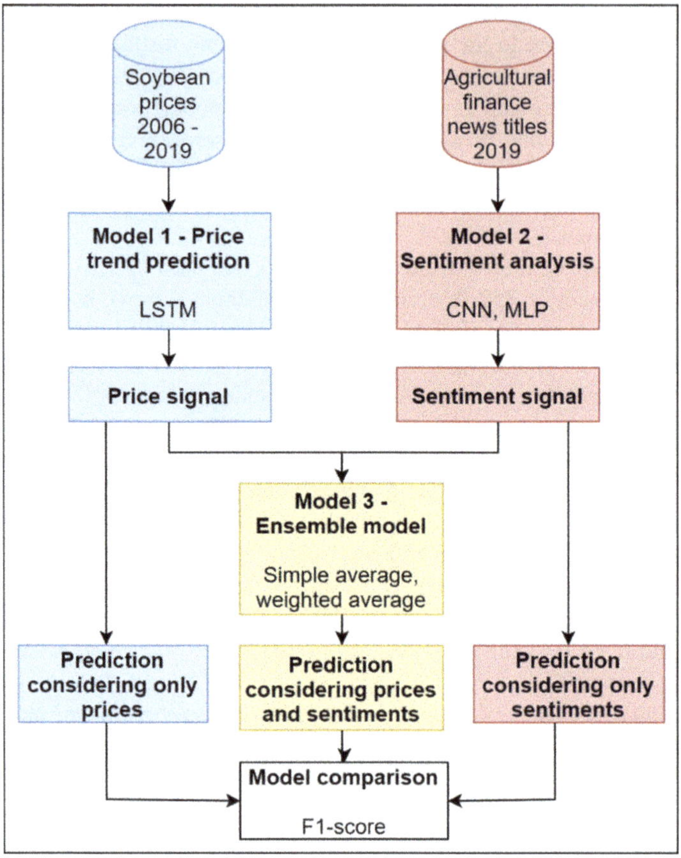

Fig. 2 The methodology used in this chapter. Legend: left (blue color): price trend prediction model; right (red color): sentiment analysis model; center (yellow color): ensemble model, considering the price and sentiment signals; bottom (white color): final model comparison

2 Related Work

This section contains the main concepts and works related to the current chapter, considering the price prediction of agricultural commodities and the use of deep learning for forecasting time series and sentiment analysis of texts.

2.1 Price Prediction of Agricultural Commodities

A value chain is a system of interlinked processes, each adding value to the product of service [29]. Agricultural value chains are the ones that encompass the production, processing, and distribution of agricultural products. One of the most important agricultural commodities is soybean due to its worldwide consumption and use as raw material for different products.

Increasing the predictability of soybean prices, both in the long and short terms is essential to improve the revenue of agents in this value chain. In the long term, better predictability would improve decision making on many aspects, such as how much of the farm area should be used for planting soybeans. In the short term, it could also help with improving revenue and on deciding how much hedging should be acquired. Speculation on the prices is also an important market aspect that can be improved by better short-term predictions [11].

The short-term prediction is the primary concern of this chapter. Nevertheless, the methodology used can be adapted to evaluate different models and agricultural products for short and long-term predictions. Among the factors that can be analyzed for long term predictions are: production costs, insurance costs, and futures contract prices.

One important aspect considered by the agents in this value chain is the amount of soybean to be produced, stored, and distributed during the year. As this is an agricultural commodity, the expected future price is a very important factor affecting the farmers' decision to plant it or not [24]. To estimate the future price, several factors must be considered. Some of them are: expected costs of the main inputs, such as oil, fertilizers, seeds, labor, and others; expected prices of the products at the harvesting time; costs related to insurance and hedging using future contracts; among others [21].

Soybean future contracts are traded on the Chicago Board of Trade (CBOT), but the Dalian Commodity Exchange (DCE) plays a prominent role in the global soybean price prediction [30], and this affects markets worldwide. Several major companies in the industry, which use soybean as a primary product or as an input, are also listed and traded on stock exchanges worldwide. It is possible to use the knowledge and insights from stock market prediction to predict agricultural commodities prices such as soybean.

Most papers in the literature, such as the one by Nasira and Hemageetha [7], focus on regression problems, denoting a lack of research on price trend prediction. For this reason, the daily price trend prediction for soybean was the objective adopted in this chapter.

Notwithstanding, there are few works on price and price trend forecasting for agricultural commodities, mainly because of: (i) the greater difficulty in collecting data when compared to other stock markets; (ii) the greater specificity, since not all countries have well-developed agricultural markets or agricultural production; and (iii) the effect of seasonality and world storage levels [31]. Czudaj [32] has an interesting work on volatility and momentum applied to the trade of agricultural products' future contracts.

A group of techniques that have been used for those tasks is machine learning.[3] These involve developing models used to analyze and recognize patterns on observed data, predicting and evaluating different variables or features [34]. Several papers, such as Nasira and Hemageetha [7], Kara et al. [3], Chong et al. [16], and Di Persio and Honchar [5], among many others, use a group of techniques of machine learning denominated deep learning. The deep learning techniques, which consist mainly of artificial neural networks[4] with different architectures and numbers of hidden layers, help on identifying hidden patterns, create and select features, and are considered state-of-the-art in many fields [16, 17].

Nevertheless, few works consider the impact of the market sentiment on future prices besides the price time series. This is also observed in stock price prediction. It is important to note that the adoption of new NLP techniques such as word embeddings has increased the number of researches related to sentiment analysis for market prediction [14].

Based on the works by Nassirtoussi et al. [14], Ferreira et al. [36], and Medhat et al. [18], it is possible to observe that there are several difficulties on implement sentiment analysis for market prediction, such as: the lack of labeled open datasets; the need to implement complex models to evaluate better and predict the sentiments; and the hardware requirements for training the models. In the next section, the use of deep learning for price trend prediction will be further explored, and some of the main techniques and papers using them will be discussed.

2.2 *Deep Learning for Price Trend Prediction*

Most works in the literature that deal with price trends, such as Kara et al. [3] and Goçken et al. [4], consider that prices may go up or down from one period to the next one. Therefore, the objective is to predict if the trend for the next period is upward or downward.

Deep learning can be useful for those tasks, as the datasets can be easily labeled using the price time series. Also, artificial neural networks can learn nonlinear functions due to the activation functions of the neurons, besides specific architectures such as convolutional and recurrent layers [37, 38].

Nevertheless, this comes at a cost: these models tend to demand a considerable amount of training data to identify the patterns behind them [39, 40]. Since the datasets used in this work are labeled, it uses deep learning to try to understand the patterns behind the data.

[3]For an in-depth review of the main machine learning models applied to price prediction, the readers are referred to the work by Henrique et al. [33].

[4]For an in-depth review of the components of artificial neural networks, the readers are referred to the work by Lipton et al. [35].

Traditional artificial neural networks do not incorporate past predictions into the current prediction, especially if the autocorrelation occurs with a higher lag. Therefore, those methods tend to lose important information when receiving a series with autocorrelation between periods, such as a time series. LSTM[5] is a method proposed by Hochreiter and Schmidhuber [43] that aims to solve this problem by incorporating values from the past into the current learning and prediction. It was the first recurrent neural network (RNN) proposed.

It is used on several domains, from NLP and chatbots to time series analysis, and is considered by many researchers as state of the art on deep learning for price predictions. Some price prediction works that use this method are the ones by Nelson et al. [44], Wang and Gao [22], Di Persio and Honchar [5], among others.

It is possible to use it similarly to the traditional econometric models, considering a specific time window length, which is chosen based on the characteristics of the problem and the dataset. Weng et al. [21] evaluated several methods, including ARIMA and RNN, to predict the price of horticultural products. They concluded that the RNN provided better results for predicting product price trends on several frequencies: daily, weekly, and monthly.

Kanchymalay et al. [11] researched the forecast of crude palm oil prices, considering the prices of several vegetable oils and their relations. The authors evaluated several econometric and machine learning models, such as the multi-layer perceptron (MLP), support vector regression, and the Holt-Winter exponential smoothing technique. They observed that the support vector regression method provided the best accuracy in predicting the prices. Nevertheless, they did not evaluate the use of LSTMs, which could provide better results than those models.

Kantanantha et al. [24] studied two prediction models: the first was used to predict crop yield yearly, and the second was used to predict crop prices. This is an important work for long term planning decisions, such as which crop to plant and when to harvest it to improve farmers' revenues. Their price prediction model provided better results than the ARIMA model.

Khamis and Abdullah [20] analyzed the use of two different artificial neural networks for predicting future wheat price: an MLP and a nonlinear autoregressive model with exogenous inputs. The authors considered as inputs the prices of wheat, oats, barley, and soybeans, and observed that the second model provided a better prediction. Nevertheless, they did not consider the options of using LSTMs or ensemble models to try to improve their predictions.

Xiong et al. [19] explored the task of interval forecasting for different agricultural commodity prices, using several models: vector error correction (VEC); multi-output support vector regression (MSVR); SVR; an ensemble of VEC and MSVR;

---

[5]For an in-depth description of the LSTM model and its components, we refer the reader to the works by Yu et al. [41], Gers et al. [42], Schäfer and Zimmermann [38], and Lipton et al. [35].

an ensemble of ARIMA and MSVR; and an ensemble of ARIMA and MLP. Unlike the traditional price prediction task, in which the objective is to predict a point that indicates the future price in the future, the interval forecasting aims to predict a range of values for the future prices. Their main results showed that the ensemble of VEC and MSVR and the ensemble of ARIMA and MSVR provided the best performance in terms of the mean absolute error percentage metric.

Xiong et al. [23] proposed a model to forecast short, medium, and long-term vegetable prices, combining the seasonal trend decomposition method and the extreme machine learning machines model. First, it decomposes the original price time series into trend, seasonal, and residual components. Then, it applies the extreme learning machine model to forecast both the trend and residual components. The seasonal component is considered as the average of the specific period in the seasonal cycle. Lastly, the three predictions are joined to form the final price prediction.

Xiong et al. [23] compared their proposed model's results with the following models: seasonal ARIMA (SARIMA), time-delay neural network, SVR, and SARIMA with a Kalman filter. Their results showed that the proposed model obtained better results than all the other models, considering two quality metrics: mean absolute scaled error and symmetric mean absolute percentage error. Nevertheless, their work did not explore the possible impacts of market sentiment on the products' prices.

Wang and Gao [22] used an LSTM to predict high and low-price trends for soybean future. They used a traditional artificial neural network as a baseline and concluded that the LSTM had better and satisfactory results. Khamis and Abdullah [20] also used a traditional artificial neural network compared with a nonlinear autoregressive model with exogenous inputs network to predict the price of wheat. They concluded that the autoregressive model provided a better prediction. Nevertheless, the works from Wang and Gao [22] and Khamis and Abdullah [20] did not explore additional variables that could indicate significant price changes, such as market sentiment.

Several studies point to the use of sentiment to improve the quality of price predictions, whether for regression or classification. It is observed that, in addition to the historical price series, market sentiment over time also influences prices, especially in the short term [14, 45]. For intraday and daily prices, considering market sentiment could lead to gains in terms of predictability and trading [14, 45, 46].

In the next section, the use of deep learning for sentiment analysis will be further explored, and the main topics, techniques, and some of the most relevant papers using them will be discussed.

2.3 Deep Learning for Text Sentiment Analysis

Sentiment analysis[6] can be defined as an area of NLP that is concerned with the analysis and extraction of sentiments, emotions, and/or subjectivity from texts in natural language [47, 48]. It has several uses, but the main ones are predicting opinions based on customer analysis of products and customer reviews, and predicting stock market value [14, 49, 50].

The textual data used for market sentiment analysis can have several sources: financial news [13, 31], financial news headlines [14, 45], messages from social networks and forums [48], among others [51]. For an in-depth review of the use of sentiment analysis in finance, the reader can refer to Nassirtoussi et al. [14].

Nassirtoussi et al. [14] and Xing et al. [47] discuss not only the results, which in the majority are positive, but also the models used, the main workflow, and an increasing trend of using model ensembles for market prediction [6, 12]. These consider both price or price trend prediction models and sentiment analysis models to make predictions. Therefore, in this research, the focus was on predicting daily soybean prices using both a model for price trend prediction and sentiment analysis model.

In this chapter, the main concern is on using the hybrid approach for sentiment analysis, which considers the use of machine learning techniques and sentiment lexicons or dictionaries [14, 18, 52]. The deep learning model used, in the case of the present work an MLP or a CNN, will extract linguistic features from the text and use the word embedding and the dictionary's knowledge to predict the sentiment of the news title and learn from the error between its prediction and the target value.

Deep learning was indicated by Nassirtoussi et al. [14], among others [12, 13, 53], as a set of methods that has great applicability in the tasks of sentiment analysis, especially with the adoption of the word embeddings technique [36]. This technique, proposed by Pennington et al. [54], improves the quality of the sentiment analysis by maintaining the proximity between words of a model trained on large text corpora. Among the deep learning models, the CNN[7] is the state-of-the-art in sentiment analysis for forecasting market sentiments, with baseline models such as the bag of words [15].

This chapter follows the growing trend of exploring sentiment analysis using a regression instead of a classification machine learning task, as there are indications that the use of few classes leads to suboptimal results [8]. This is the case of works on sentiment analysis that focus on classification, normally using negative, positive, and neutral as classes.

[6]For an in-depth description of the different models used for sentiment analysis and applications on different domains, the reader is referred to the work by Medhat et al. [18]. For an extensive review of sentiment analysis for market prediction, we refer the reader to the work by Nassirtoussi et al. [14].

[7]For an in-depth description of the CNN model and its components, we refer the reader to the works by Zhang and Wallace [15] and LeCun and Bengio [55].

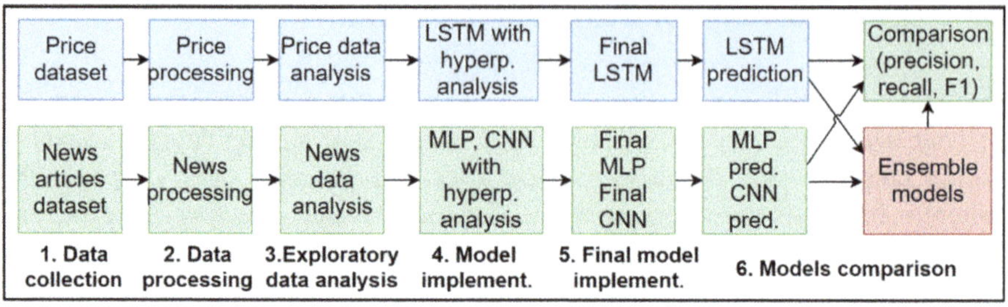

Fig. 3 Steps of the methodology

As in the case of price trend forecasting, few studies use deep learning to forecast sentiment in agricultural markets. For this reason, works that implement sentiment analysis in this domain may provide interesting results for practitioners and researchers.

Lastly, most works use news and social media messages in English only [14, 45, 47]. There is also a lack of dictionaries and works focused on the Portuguese language, such as the work by Neuenschwander et al. [56]. For this reason, an additional contribution of the current chapter is to analyze news and use dictionaries and an embedding in Portuguese instead of English. In the next section, the methodology and its steps will be described, as well as the models, libraries, software, and hardware used in this work.

3 Methodology

The methodology adopted for this work consisted of six main steps, which are illustrated in Fig. 3.

These steps were separated into two groups to be better described: a) Price time series, comprising activities related to time series analysis and price trend prediction; and b) Text sentiment analysis, comprising activities related to sentiment analysis and sentiment prediction. As an exception, in step six, there is a third group of activities related to the implementation of the ensemble model. The steps of the methodology were:

1. *Data collection*

   (a) *Price time series*: the soybean index for the Paranaguá harbor, an important soybean price indicator in Brazil, was downloaded from its official website.[8] The data had a daily frequency and encompassed the period from 03/13/2006 to 12/30/2019. Although there were prices both in Reais and Dollars, only the

[8] https://www.cepea.esalq.usp.br/br/indicador/soja.aspx

price in Dollars was considered. This is because, by using the price in Dollars, it is possible to eliminate the impact of currency exchange rates;
   (b) *Sentiment analysis*: the dataset containing 357 relevant agricultural finance news in Portuguese from Marto et al. [28] was used. The following attributes were considered: ID, news title, source, and sentiment. The sentiment is a value between 0 (very negative) and 5 (very positive), with 3 being neutral. This column was used as a target for the models. For an in-depth description of the dataset and how it was developed, please refer to Marto et al. [28].

2. *Data processing*

   (a) *Price time series*: the data was analyzed to detect outliers and missing data using the following packages: Scikit-Learn,[9] Pandas,[10] Seaborn,[11] and Matplotlib.[12] An additional feature, called daily percent change, was created based on the economic and econometrics literature. It was used as inputs for the price trend prediction models, instead of the pure prices. This is because the time series of daily percentage variation is stationary, while the pure price series is not. The target column (next day's price trend) was also created;
   (b) *Sentiment analysis*: the data were processed using the following packages: Scikit-Learn, Pandas, Seaborn, Matplotlib, nltk,[13] and spacy,[14] for its use on the deep learning models implemented. This consisted of the following activities: tokenization, lemmatization, elimination of numbers and company names, and elimination of stopwords.

3. *Exploratory data analysis*

   (a) *Price time series*: the time series were analyzed using the Pandas, Matplotlib, Seaborn, Scikit-Learn, and Statsmodels[15] packages. The main analyses were related to: (i) trends; (ii) seasonality; (iii) autocorrelation; and (iv) general statistics, including mean, median, high and low prices, as well as the distribution of daily percentage change;
   (b) *Sentiment analysis*: Marto et al. [28] contains an in-depth exploratory analysis of this dataset. Additionally, the distribution of sentiments throughout the dataset and on the training and test subsets was evaluated.

4. *Model implementation and hyperparameters analysis*

   (a) *Price time series*: the model implemented was the LSTM, using the daily percentage change as inputs. For comparing the models, the period from 2006

[9] https://scikit-learn.org/
[10] https://pandas.pydata.org/
[11] https://seaborn.pydata.org/
[12] https://matplotlib.org/
[13] https://www.nltk.org/
[14] https://spacy.io/
[15] https://www.statsmodels.org/

to 2017 was separated as the training subset, and the year of 2018 was used as the validation subset. The model was implemented using the TensorFlow[16] library and contained one LSTM layer. The following hyperparameters were analyzed: (i) number of time steps of the sliding window: 2, 5, 10, and 20; and (ii) number of neurons of the hidden layer: 10, 20, 50, 100, and 200. The loss function used was the binary cross-entropy, as the objective of the model was to predict the trend direction, and the optimizer was the Adam optimizer. The hyperparameters that showed the best result through the analysis were chosen to implement the final model;

(b) *Sentiment analysis*: following the work by Ferreira et al. [36], two models were implemented, MLP and CNN, using a Portuguese version of the GloVe 300 dimensions embedding [54, 57]. They were implemented using the keras[17] library running on top of TensorFlow and trained for 30 epochs. The CNN had one convolutional layer. One month (57 news) was separated as the test subset (and used on step 5), while the rest of the news was used for training through a 3x cross-validation. In both models, the following hyperparameters were analyzed: (i) dictionary, considering no dictionary, Sentilex [58], containing a vocabulary of 82,348 words, OpLexicon [59, 60], containing a vocabulary of 32.192 words, and WordNetAffectBR [61] containing a vocabulary of 291 words; (ii) batch size of 32 and 64; (iii) number of neurons of the dense layers of 50, 100, and 150; (iv) dropout rate, considering no dropout, 0.3, 0.4, and 0.5; and (v) number of hidden layers of 1, 2, 3, 5, and 10. In CNN case, two additional hyperparameters were evaluated: (i) number of filters: 128 and 256; and (ii) size of filters: 2 and 4. The loss function used was the mean squared error (MSE), a traditional function used for training artificial neural networks, and the optimizer used was the Adam optimizer. The objective with those models was to correctly identify the target sentiment value of the specific news title and to re-adjust its weights during the training. The hyperparameters that showed the best result through the analysis were chosen to implement the final model.

5. *Final models' implementation*

(a) *Price time series*: the final models were built using the best hyperparameters identified on step 4-a). They were trained using the training and validation subsets and tested using the final test subset, which consisted of April 2019. This was done for all models evaluated, because the limiting factor was the amount of news in successive days, and April is the only month that contains news for all days, on average with two news headlines per day. The trained models were used both to make predictions on the test set and to provide the price signal for the ensemble model (item c next);

[16] https://www.tensorflow.org/

[17] http://keras.io/

(b) *Sentiment analysis*: the final models were built using the best hyperparameters identified in step 4-b) for the MLP and CNN models. They were trained using the news train subset, containing 278 news headlines, and used both to make a prediction for April 2019 and to provide the sentiment signal for the ensemble model (item c next). The prediction was made using the following workflow: (i) make a sentiment prediction for the current day using each news title in the day; (ii) calculate the average considering all the sentiments of the current day; (iii) make the prediction for the current day's sentiment; and (iv) compare it to the last day's prediction; if it was higher, it signaled an increase, if it was lower, it signaled a decrease. This last signal was considered the trend prediction for the models that used only the sentiment, and as the sentiment signal for the ensemble model.

6. *Models comparison*: ensemble model was implemented combining both the price and the sentiment signals to make a final prediction, which was then compared with the real price trend. Two scenarios were considered: (i) both signals have the same weight and, in the case that they have opposite directions, the direction of the price signal was considered; and (ii) varying weights applied for the sentiment signal, from 0.1 to 0.9, by 0.1 increments. The main objective was to evaluate the weight assigned for each signal to improve the overall trend prediction. Lastly, the final models (LSTM for price trend prediction; MLP and CNN for sentiment prediction; and the ensemble model) were compared, considering the F1-score classification metric. To better visualize the results, charts related to the MLP and CNN model training and heatmaps of the confusion matrices were used. The model that presented the best score for this metric was considered the best model.

The implementation was done using Python with Google Colab[18] in a notebook with the following technical specifications: Intel(R) Core(TM) i7-8700 K CPU @ 3.70GHz, a GeForce GTX 1080 Ti GPU, and 4 × 16 GB DDR4 2133 MHz of RAM.

4 Results

This section contains the description and analysis of the results obtained from the implementation of the different models. The analysis on the next subsections will be described in the following order: (i) exploratory analysis of the dataset; (ii) results of the hyperparameters analysis for each model; and (iii) selection of the best hyperparameters for each model. The last subsection contains an analysis of the results of the predictions of the final trained models on the test subset.

[18] https://colab.research.google.com/

Fig. 4 Line plot of the prices dataset, in USD

4.1 Models Considering Only Prices

The average price in the period was 24.80 USD per soybean bag, with a standard deviation of 5.75 USD. The minimum value on the period, 12.40 USD, was observed in 2006. The maximum value was 45.32 USD and was observed in 2012. Figure 4 illustrates the whole dataset.

The Augmented Dickey-Fuller Test[19] was used to evaluate if the series was non-stationary,[20] as the prices on the period showed an increasing trend. The results of the test confirmed that, indeed, the series was non-stationary. Therefore, it was challenging to model its behavior. For this reason, the daily percentage changes were calculated (similar to the calculation of daily returns for analyzing market stocks).

Figure 5 illustrates the daily percentage change, and the use of the Augmented Dickey-Fuller Test results for this time series indicated that it is stationary. Nevertheless, it is important to note that there is high volatility in 2012, 2013, and 2014. Figure 6 shows the frequency distribution of this time series, indicating that, although its mean is 0.026 and its standard deviation is 1.439, it presents several outliers.

The outliers with more than two standard deviations from the mean were removed to reduce the noise on the neural network's training phase. Several other statistical analyses were conducted, such as the analysis of the autocorrelation and partial

[19]The Augmented Dickey-Fuller Test is a statistical test used to evaluate if a time series is stationary. For implementation details, we refer the reader to the Statsmodels library tutorial (https://www.statsmodels.org/dev/generated/statsmodels.tsa.stattools.adfuller.html)

[20]A time series is considered stationary when its statistical properties do not change over time [62]. For an in-depth description of the different types of stationarity, we refer the reader to the work by Witt et al. [62].

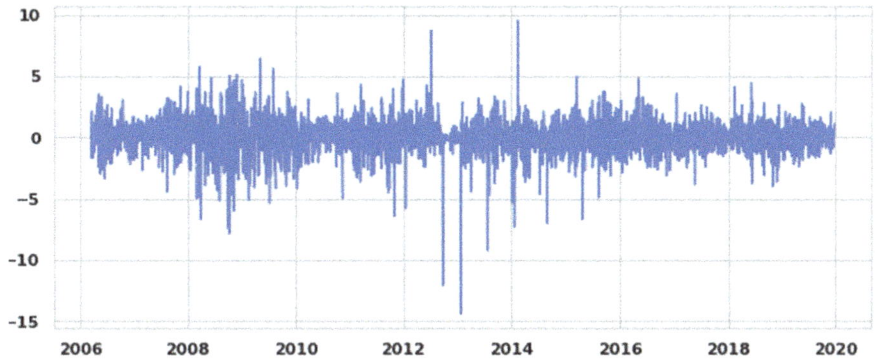

Fig. 5 Plot of the daily percentage changes time series

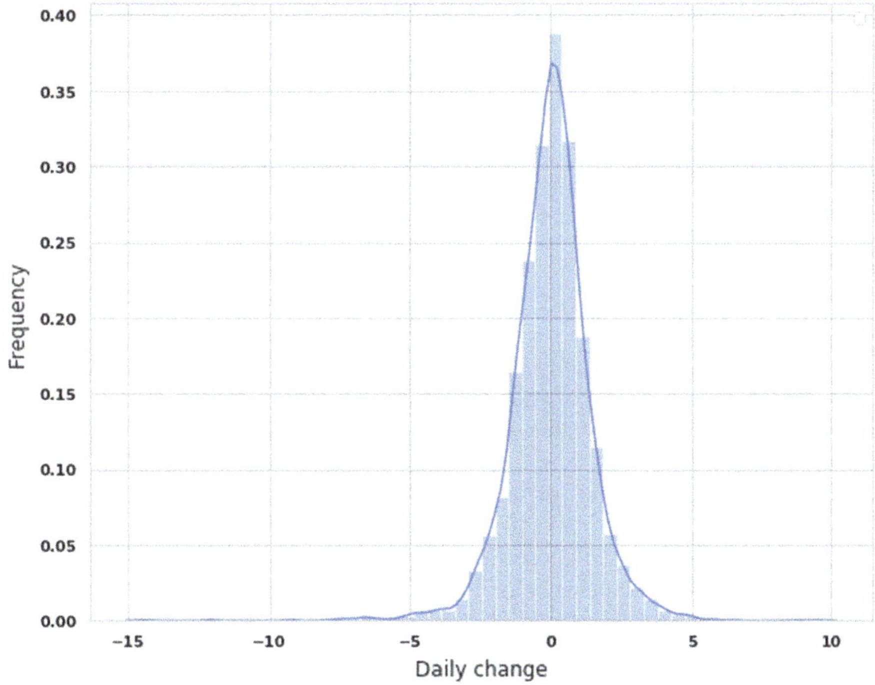

Fig. 6 Frequency distribution of the daily percentage changes time series

autocorrelation functions,[21] both indicating that the daily percentage change and the prices presented an autocorrelation with the immediately previous period, as was expected. Nevertheless, the daily percentage change was chosen as the input for the

[21] Autocorrelation and partial autocorrelation functions evaluate the correlation between a time series and a lagged version of itself for different time lags. The main difference between them is that the first considers both direct and indirect effects of autocorrelation, while the second considers only direct effects. For more information on these functions, we refer the reader to the work of Hamilton and Watts [63]. For implementation details, we refer the reader to the Statsmodels library tutorial (https://www.statsmodels.org/devel/generated/statsmodels.tsa.stattools.acf.html)

Table 1 Results of the hyperparameters analysis for the LSTM model on the validation set, considering the MSE

Hyperparameters	Values	MSE on the validation subset
Time window length	2	No model converged
	5	**0.737**
	10	0.753
	20	0.786
Number of neurons (LSTM layer)	10	0.846
	20	0.739
	50	**0.716**
	100	0.741
	200	0.762

The best value for each hyperparameter is in bold

LSTM, as it contained less noise than the price itself, reflecting the method used in several econometric studies.

First, the number of neurons in the LSTM layer and the length of the time window were chosen as the main hyperparameters for evaluation due to their impacts on the results of this type of model. The values chosen for the number of neurons were: 10, 20, 50, 100, and 200. The values chosen for the time window lengths were: 2, 5, 10, and 20. The implementation considered the MSE loss function, the Adam optimizer, and 60 epochs for training each model. The lower the MSE, the better is the model prediction.

It is important to note that, in Sect. 4.3, the models trained with the best hyperparameter values were used to predict the price trends of the test subset (price trends of April/2019), and the F1-score metric was used to evaluate the quality of those predictions. Table 1 contains the results of this analysis, considering the average MSE of the validation subset of all models implemented with each hyperparameter value.

Based on: (i) the analysis of the MSE of each model; (ii) the analysis of the hyperparameters; (iii) the convergence of the model (as several models did not converge); and (iv) on predictions made on the validation set (mainly to check if the behavior of the predictions behaved like a random walk model[22]), four models were chosen for the final analysis (Sect. 4.3).

It was important that the models analyzed did not show a random behavior, a random walk or predicted a constant value. The models chosen were: (i) with a time window of 5 and 50 neurons (T5N50); (ii) with a time window of 5 and 200 neurons (T5N200); (iii) with a time window of 10 and 50 neurons (T10N50); and (iv) with a time window of 20 and 200 neurons (T20N200).

[22]The random walk model considers that the series history is not important for predicting the next period, only the value of the last period plus a white noise error. For a thorough analysis of the random walk hypothesis for stock market prices, we refer the reader to the work by Fama [64].

Table 2 Results of the implementation of the four LSTM models, considering the MSE

Model	MSE on the validation subset
T5N50	0.786
T5N200	**0.721**
T10N50	0.803
T20N200	0.839

The best result is in bold

Among the results on Table 1, it is important to observe two findings: (i) some time-window lengths presented a faster convergence than others, while no model with a time-window of two converged; and (ii) the difference in MSE between choosing the best value for the number of neurons (50) vs. choosing the worst value (10) was of around 15%. Nevertheless, both results can be highly influenced by the dataset, so more experiments are needed to evaluate those findings.

Also, it was observed that the number of epochs the models took to converge was more related to the time window length than to the number of neurons in the LSTM layer. This may have happened because shorter time windows could be capturing more noise than the longer time windows, but more experiments are needed, especially with larger datasets, to better evaluate this hypothesis.

Table 2 illustrates the results for the training of the four chosen models, considering the number of epochs that best suited that model. This was defined by evaluating each model's learning curve. The best performing model (T5N200) presented an MSE that was 14% lower than the worst performing model (T20N200). Nevertheless, as the dataset may contain considerable noise (as cited before), it is important to evaluate the trend predictions of all models in the model comparison (Sect. 4.3).

In the next subsection, the models that consider only the sentiment analysis and prediction are developed and analyzed. They generated both predictions and the sentiment signal, which were used by the ensemble model.

4.2 Models Considering Only Text Sentiments

The dataset used for the sentiment analysis contained 357 agricultural finance news headlines in Portuguese, each with a sentiment label that ranged from very negative to very positive. It is important to note that: (i) each news title was labeled by three different researchers; (ii) the sentiment label used was the simple average of the values of the labels given by the researchers; (iii) 50% of the news in the dataset were positive, and only five of them contained a very negative label; (iv) an analysis of the labels in the whole dataset have not shown autocorrelation between the sentiment labels, aggregated by day or week; and (v) the dataset contained, after removing for stopwords, 678 relevant words. For a further description of the development of this dataset and its characteristics, please refer to Marto et al. [28].

Table 3 Results of the hyperparameters analysis for the MLP model on the validation set for each lexicon, considering the MSE

Hyperparameters	Values	No dictionary	OpLexicon	SentiLex	WordnetAffectBR
Batch size	**32**	0.578	0.568	**0.565**	0.568
	64	0.878	0.834	**0.832**	0.833
Number of neurons (hidden layer)	**50**	0.655	0.645	**0.642**	**0.642**
	100	0.732	0.732	**0.718**	0.719
	150	0.769	0.749	**0.744**	0.749
Dropout rate	**No drop.**	0.621	0.610	**0.607**	0.611
	0.3	0.755	0.736	**0.735**	**0.735**
	0.4	0.919	0.900	0.897	**0.895**
	0.5	1.020	1.001	1.001	**0.994**
Number of hidden layers	**1**	0.671	0.660	**0.653**	0.656
	2	0.824	0.811	**0.807**	**0.807**
	3	0.960	**0.924**	0.925	0.925
	5	1.097	1.083	1.083	**1.082**
	10	0.972	**0.956**	0.966	0.963

The best value for each hyperparameter and each line are in bold

First, the hyperparameters for both models were analyzed through implementation using a three-fold cross-validation.[23] A function similar to grid-search was used to generate and train models containing all the possible combinations between hyperparameters.

The MSE loss function of the neural networks was considered for this evaluation, mainly for two reasons: (i) it is widespread in the literature; and (ii) it is also used in the models implemented in Sect. 4.1.

Table 3 contains the results of this analysis for the MLP, considering the average MSE of the validation set of all models implemented with each hyperparameter value. The final hyperparameters chosen were: the batch size of 32; 50 neurons in the hidden layer; 1 hidden layer; the SentiLex dictionary; and no dropouts on the model.

Among the results in Table 3, it is important to observe three major findings: (i) using any dictionary is better than using no dictionary; (ii) the SentiLex dictionary provided the best results; and (iii) less complex hyperparameter values tended to result in a lower MSE. The first aspect was also observed by Ferreira et al. [36]. Therefore, these results are aligned with the literature.

Nevertheless, more experiments with larger and more diverse agricultural news datasets are needed, especially related to the observation that less complex hyperparameter values resulted in a lower MSE. These results can be attributed to the use of a well-balanced dataset and the fact that the lexicons used contained the words with sentiments similar to the targets in the training subset.

[23] Cross-validation is a technique used to improve the pattern detection of deep learning models, especially on small datasets, by separating the training subset into several splits. For a thorough analysis of this concept, we refer the reader to the works by Prechelt [65] and Hu et al. [66].

Table 4 Results of the implementation of the MLP model with the best hyperparameters on the validation set for each lexicon, considering the MSE

Metric	No dictionary	OpLexicon	SentiLex	WordnetAffectBR
MSE	0.631	0.621	**0.619**	0.622

The best value is in bold

Table 5 Ranking of hyperparameters for the MLP, considering the difference between best and worst values for MSE of the validation subset

Hyperparameters	Value	Difference
Number of hidden layers	1	47.32%
Dropout rate	No dropout	42.01%
Batch size	32	37.15%
Number of neurons (hidden layer)	50	17.30%
Dictionary	SentiLex	2.22%

Table 4 illustrates the results for the training of the MLP using the best values for each hyperparameter. The use of these values showed a significant improvement in reducing the MSE compared to other values for those hyperparameters. The use of the SentiLex dictionary resulted in the lowest MSE.

Table 5 illustrates the raking of the hyperparameters analyzed and the differences between implementations using the best and the worst values for each hyperparameter. An interesting finding is that the number of hidden layers was the most important hyperparameter since using 1 hidden layer resulted in an MSE 47.32% lower than using five hidden layers. This can be explained by the characteristics of the dataset, as seen before.

The second most important hyperparameter was the dropout rate, since using no dropout rate resulted in an MSE 42.01% lower than using a 0.5 dropout rate. One of the main reasons that explain this behavior is that, as the dataset was small, by increasing the dropout rate, the difficulty of identifying the correct patterns in the data was also increased. According to Zhang and Wallace [15], high dropout rates tend to cause this behavior. Nevertheless, more experiments are needed to evaluate if this behavior would persist on a larger dataset.

The batch size was also an important hyperparameter for the MLP. Using a batch size of 32 reduced the MSE by 37.15% compared to using a batch size of 64. This result depended on several factors and was more difficult to evaluate than the other hyperparameters. However, there was evidence that this may be influenced by the dataset and the word embedding used.

Using 50 neurons in the hidden layer instead of 150 led to a reduction of the MSE of 17.30%. As observed before, that may be explained by the relatively balanced dataset. Lastly, an important finding was that using a dictionary improves the results versus not using one, but this was a marginal gain. For the MLP, this resulted in a reduction of only 2.22% on the MSE. This may be highly affected by the word embedding quality since a worse word embedding would probably present better results when a dictionary is added.

Table 6 Results of the hyperparameters analysis for the CNN model on the validation subset for each lexicon, considering the MSE

Hyperparameters	Values	No dictionary	OpLexicon	SentiLex	WordnetAffectBR
Batch size	**32**	0.380	0.372	0.370	**0.369**
	64	0.541	0.532	0.534	**0.529**
Number of neurons (hidden layer)	**50**	0.383	0.373	0.373	**0.371**
	100	0.412	0.538	**0.400**	0.403
	150	0.405	**0.401**	0.402	**0.401**
Dropout rate	**No**	0.389	0.381	**0.378**	0.381
	drop.	0.398	0.388	0.389	**0.387**
	0.3	0.428	0.420	0.420	**0.416**
	0.4	0.473	0.465	0.463	**0.457**
	0.5				
Number of hidden layers	**1**	0.380	0.372	0.371	**0.370**
	2	0.424	0.415	**0.414**	0.417
	3	0.462	0.454	0.455	**0.450**
	5	0.512	0.511	**0.509**	0.511
	10	0.476	**0.465**	**0.465**	0.468
Filters size	2	0.395	0.383	**0.382**	0.384
	4	0.387	0.382	0.380	**0.379**
Number of filters	128	0.396	0.387	0.388	**0.386**
	256	0.386	0.377	**0.375**	0.377

The best value for each hyperparameter and each line is in bold

Nevertheless, with the word embedding used, the dictionaries had a very low impact on the results. An additional factor that could increase the impact of using dictionaries is the use of specific sentiment dictionaries for the agricultural domain. To the best of our knowledge, no such dictionaries exist in this domain.

Table 6 contains the results for the CNN implementations analysis, considering the MSE of the validation set, and the average result for each hyperparameter value. The final hyperparameters chosen were: the batch size of 32; 50 neurons in the hidden layer; 1 hidden layer; the SentiLex and the WordnetAffectBR dictionaries (as both had the same final results); no dropouts on the model; a filter size of 4; and 256 filters.

Similarly to the MLP, using any dictionary was better than using no dictionary. Nevertheless, both the SentiLex and the WordnetAffectBR presented similar results. In general, all the results were better than the MLP, and the best values for the hyperparameters used in both models were the same. Additionally, CNN had two hyperparameters that were not present in the MLP: filter size and the number of filters. The results observed were aligned with the literature.

Table 7 illustrates the results for the training of CNN using the best values for each hyperparameter. The use of these values showed a significant improvement in reducing the MSE compared to other values for those hyperparameters. The use of the SentiLex or the WordnetAffectBR dictionaries resulted in the lowest MSE.

Table 8 illustrates the ranking of the hyperparameters analyzed and the differences between implementations using the best and the worst values for each

Table 7 Results of the implementation of the CNN model with the best hyperparameters on the validation set for each lexicon, considering the MSE

Metric	No dictionary	OpLexicon	SentiLex	WordnetAffectBR
MSE	0.384	0.376	**0.375**	**0.375**

The best values are in bold

Table 8 Ranking of the different hyperparameters for the CNN, considering the MSE of the validation subset

Hyperparameters	Best value	Difference between best and worst values
Dropout rate	No dropout	55.36%
Batch size	32	51.30%
Number of hidden layers	1	46.63%
Number of neurons (hidden layer)	50	25.30%
Size filters	4	4.09%
Dictionary	SentiLex or WordnetAffectBR	2.34%
Number filters	256	0.29%

hyperparameter. In contrast with the MLP, the dropout rate was the most important hyperparameter. Using no dropout reduced the MSE by around 55% compared to using a dropout of 0.5.

The reasons for these results are similar to the ones already commented for the MLP. Notwithstanding, the order is different for the CNN, with the dropout rate, the batch size, and the number of hidden layers being the most important ones, at the same time that the size of filters, the use of dictionaries, and the number of filters brought only marginal gains, of less than 5%.

Concerning the dictionaries, the explanation is similar to what was observed for the MLP. For the size and number of filters, this is also aligned with the results from the MLP, in which the word embedding quality and the quality of the dataset are more important than those hyperparameters. However, more experiments should be conducted to evaluate the impact of other word embeddings and dimensions to understand this observation better.

The best CNN model presented results for the MSE that was around 39% lower than the MLP, showing that this model was better to identify patterns in the data. This result is in line with the literature, which considers CNNs and LSTMs as superior models for sentiment analysis in comparison to MLPs.

Figures 7, 8 and 9 illustrate the learning curves of the MLP, CNN with SentiLex and CNN with WordnetAffectBR models, respectively, with the best hyperparameters for each. Based on the behavior of the curves in those charts, it is possible to observe that: (i)) the models had improved their identification of patterns throughout the training period considerably; (ii) the three models converged without leading to overfitting on epoch 30; and (iii) the CNN models achieved a lower MSE,

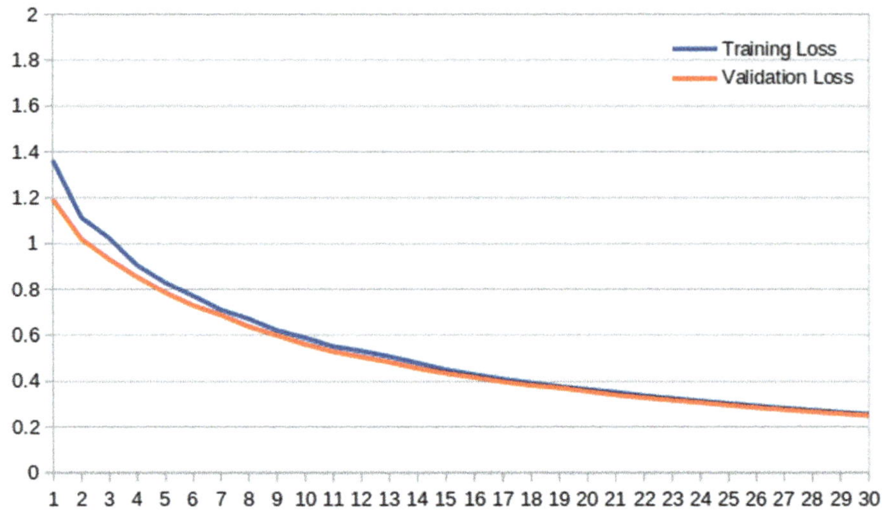

Fig. 7 Learning curve for the MLP model, considering the MSE as the loss function

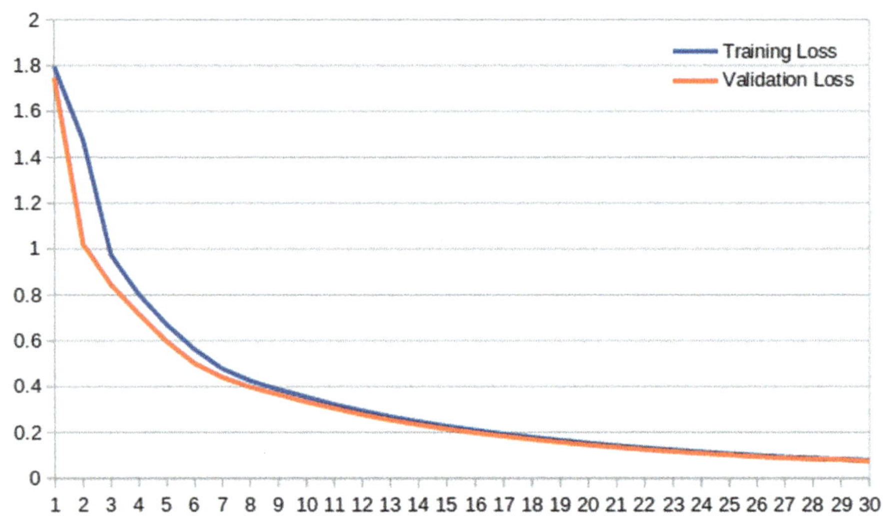

Fig. 8 Learning curve for the CNN model with the SentiLex dictionary, considering the MSE as the loss function

in comparison with the MLP. Therefore, it is possible to infer that all models were suitable for sentiment predictions and that the CNNs were better candidates for the ensemble model.

In the next subsection, an ensemble model is developed, based on the price and the sentiment signals. As the LSTM presented the best results for price prediction, this model was used to provide the price signal. Regarding the sentiment signal, this

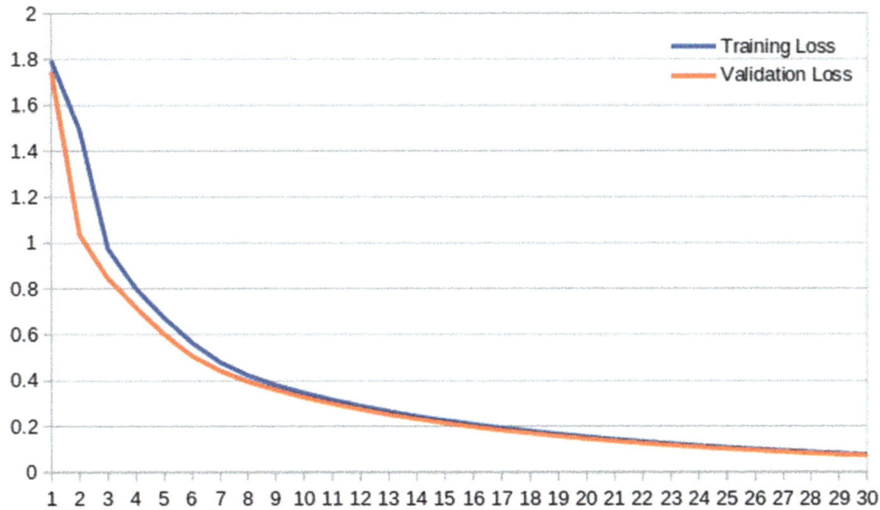

Fig. 9 Learning curve for the CNN model with the WordnetAffectBR dictionary, considering the MSE as the loss function

was provided by the CNN, since it presented considerably better results when compared to the MLP, with a reduction of around 39% on its MSE in comparison with the best MLP model implemented.

4.3 *Ensemble Model Considering Prices and Text Sentiments*

The main goal of the models implemented in this subsection was to use both the price trend signal (originated from the best of the four models from Sect. 4.1) and the sentiment signal (originated from the best of the three models in Sect. 4.2) to generate a new trend prediction. This was compared with the target prediction, which was generated from the price time series.

First, the four price trend prediction models were evaluated on the test subset, based on the F1-scores of their predictions. The target was the trend for the next day, and it had two options: up or down. In the case the price or the trend predicted remained flat, which is unusual, it was considered as a continuation of the last period's trend. It is important to observe that the test subset encompassed 20 days of the month of April 2019. Table 9 contains the results of those metrics for each model.

It is possible to observe that, although the results were similar, the model that obtained the best results was the T5N200 (time window length of 5, 200 neurons on the LSTM layer), and the worst model was the T5N50. Therefore, the price signal that was considered on the ensemble model came from the T5N200 model.

Table 9 F1-score for the predictions of each price trend model

Model	F1-score
T5N50	0.48
T5N200	**0.60**
T10N50	0.58
T2N200	0.55

The best value is in bold

Table 10 F1-score for the predictions of each sentiment analysis model

Model	F1-score
MLP	0.54
CNN with SentiLex	0.45
CNN with WordnetAffectBR	**0.55**

The best value is in bold

Table 11 F1-score for the predictions of each scenario of the ensemble model

Scenario	F1-score
Scenario 1: 0.10 weight for sentiment and 0.90 weight for price	**0.60**
Scenario 2: 0.50 weight for sentiment and 0.50 weight for price (simple average)	**0.60**
Scenario 3: 0.60 weight for sentiment and 0.40 weight for price	0.55

The best values are in bold

In the case of the sentiment signal, a similar approach was used: generate one signal per day for each day on the test subset, based on the news from that day, and then compare this with the real trend (the same target used for the analysis of the price trend prediction models). It is important to note that, on average, each day had around three news, resulting in a subset of 57 news.

The sentiment signal generated was an average of the predicted sentiments of all news headlines on that day. If the result was less 0.5, it was considered negative. If it was higher than 0.5, it was considered positive. If it was exactly 0.5, it was considered the same value as the past period's predicted trend. The metric considered was also the F1-score.

Table 10 contains the results for each model. It is possible to observe that the CNN with SentiLex was the worst model, while the CNN with WordnetAffectBR was the best one. For this reason, the sentiment signal used on the ensemble model was provided by the CNN with WordnetAffectBR. Even though the MLP also presented similar results, it was not considered for generating the sentiment signal as it showed the worst results when training and validating the model using the MSE (Sect. 4.2).

Lastly, the ensemble model was developed, considering the following scenarios: a simple average of the price and sentiment signals, and the use of different weights for each signal, from 0.1 to 0.9, with increments of 0.1. Table 11 shows the results of those scenarios. Not all scenarios are shown because there was no difference between the scenarios with 0.2, 0.3, and 0.4 weights for the sentiment signal, and between the scenarios with 0.7, 0.8, and 0.9 weights.

Table 12 F1-score for the best models in each category

Inputs for prediction	Model	F1-score
Only price signal	**T5N200**	**0.60**
Only sentiment signal	CNN with WordnetAffectBR	0.55
Price and sentiment signals	**Ensemble 1: 0.10 weight for sentiment and 0.90 weight for price**	**0.60**
Price and sentiment signals	**Ensemble 2: 0.50 weight for sentiment and 0.50 weight for price**	**0.60**

The best values are in bold

The most interesting finding is that the use of too much weight on the sentiment signal, considering the datasets used and the models implemented, leads to a worse trend prediction. This can be observed in the results of scenario 3. The results of scenarios 1 and 2 were the same, which may indicate that considering a small weight of sentiment can lead to better results. To better visualize and contextualize all the results, Table 12 contains a summary of the F1-score for the best models considered in each of the situations: only price signals, only sentiment signals, and an ensemble.

It is possible to infer, based on the results of Table 12, that using only the sentiment signal leads to worse results than using only the price signal or using an ensemble of price and sentiment signals. More experiments, especially with a larger dataset, are essential for a better assessment of this hypothesis. As for the ensemble models and the model considering only the price signal, no differences were observed. Nevertheless, in Sect. 5, some of the possible impacts of using one type of model in relation to the other are explored.

5 Discussion

This section contains the discussions regarding the main impacts of the implemented models and the importance of using deep learning techniques for agricultural products price prediction.

5.1 Benefits of Deep Learning for Agricultural Price Prediction

In general, the results observed in this chapter for the price trend prediction models (Sect. 4.1) are aligned with the literature. The exploration of hyperparameters allowed the identification of which ones cause the most impact on implementing LTSM for price trends prediction, which can be of interest to both practitioners and researchers.

It would be interesting to see if the same analysis applied to different datasets (especially the ones related to market indexes and stocks) would lead to similar results. In general, market indexes of developed countries with an efficient market (such as is the case of SP500 in the USA market), tend to have an increasing trend over the long term. For the dataset used in this work, the total period is too short to detect a similar trend, so it is possible that the models' behavior could change considerably on more mature and efficient markets with larger datasets.

Notwithstanding, the use of deep learning models for markets that are not so efficient, such as the agricultural commodities markets, could lead to important findings in relation to the traditional econometric models, such as ARIMA, SARIMAX, GARCH, and others. In future work, predictions of those models will be compared to the ones obtained in this chapter to better evaluate if there are possible tangible benefits in terms of increasing trend predictability for agricultural commodities prices.

Another interesting observation that should be better explored, with larger and more diverse datasets (especially the ones with intraday prices and several news headlines per day) is if the use of ensemble models can improve the trend prediction in comparison to those using only the price signal. It is expected that this will happen in the short term (daily or intraday predictions), although it was not verified in the experiments. For the long term, it is expected that the news sentiment signal will have a smaller effect, as news is quickly incorporated on the prices, especially in efficient markets.

One reason that may explain the results obtained is that the news dataset, besides being small, had few data points with very positive or very negative sentiments. Therefore, the quantity of news that could significantly influence the price in the short term was low. For this reason, it would be interesting to test the same models in a dataset that had such extremes (both positive and negative), to evaluate if the ensemble model would provide a faster adaptation to the expressed sentiment. If so, it could improve decision making and the use of resources.

Essentially, for the agricultural value chains, it could have several benefits, for example: (i) helping farmers to decide when to sell or hedge their production; (ii) helping the different agents in the value chain to predict prices of raw materials and finished products; and (iii) improving decision making for warehousing and distribution, especially for commodity traders. It could also provide more data for agents that speculate on future commodity prices.

Some of the main advantages of using deep learning models in this context are: (i) the possibility to incorporate different features that cannot be considered using the traditional econometric models; (ii) automatic feature generation, detecting trends that would demand manual labor using other techniques; and (iii) the possibility of exploring not only prediction but also the impacts of taking certain decisions, using techniques such as deep reinforcement learning.

5.2 Adaptation and Uses for Other Products

The methodology used in this work can be adapted to several other products and value chains, whether agricultural or not. Two of the main requirements are: (i) availability of a high-quality dataset for the sentiment analysis, containing news, news headlines, or messages from social media; and (ii) availability of a high-quality dataset for the time series analysis and price prediction. Ideally, both datasets should encompass periods of high and low trends.

One difficulty related to the models' implementation in this work was their diverse nature. CNN has a different architecture than MLP and LSTM, and thus, each has its own important set of hyperparameters and requires a different procedure to input the dataset. It is essential to understand these characteristics to be able to implement the models successfully. The exploration of these different models, when forecasting trends in agricultural commodity prices, may motivate new explorations by other researchers.

Although neglected in many cases, one important factor is to conduct a thorough exploratory data analysis before the development and fitting of the deep learning models. It is essential to analyze if the time series has, among other attributes: (i) stationarity; (ii) seasonality; (iii) a long term trend; (iv) the presence of cycles; (v) outliers which may impact on the models' pattern identification; (vi) balance among its data points and targets; (vii) autocorrelation, considering different time lags; and (viii) additional features available that may be incorporated into the dataset.

Lastly, one important aspect worth mentioning is that several other models can be explored, including supervised and unsupervised machine learning. They could be used before implementing the price and sentiment prediction models (such as the use of clustering algorithms to find hidden patterns in feature data), substituting those models, or after them (such as reinforcement or deep reinforcement learning for decision making and trading). These should be explored in future work.

6 Conclusions

Improving the forecast of agricultural commodity prices trends has several benefits, such as improving decision making and revenue for the value chain agents and helping farmers decide when to sell or hedge their production. Deep learning techniques can be used to improve predictability in comparison to traditional models. This work explored the use of deep learning techniques for: (i) generating a price signal from the soybeans price in Brazil, (ii) generating a sentiment signal from agricultural finance news, and then (iii) predicting the price trend.

For the price signal generation, the LSTM model was implemented, evaluating two important hyperparameters. For the news sentiment signal generation, an MLP model and a CNN model were implemented after analyzing several important hyperparameters. Lastly, the best model predictions for each of these two categories (prediction based only on prices and prediction based only on news sentiments) were chosen to be part of an ensemble model, which generated a final prediction.

The three models that provided the best results were: (i) an LSTM model to produce a price signal; (ii) an ensemble model composed of a simple averaging of the LSTM and the CNN models with the WordnetAffectBR dictionary; and (iii) an ensemble model composed of a weighted averaging of the LSTM model (90% of the weight) and the CNN model with the WordnetAffectBR dictionary (10% of the weight). Among the main findings, it is important to highlight: (i) some hyperparameters have a great influence on the models' results for all categories; (ii) the use of dictionaries did not improve the sentiment prediction considerably as it was expected based on the literature; (iii) using too much weight on the sentiment signal could lead to a worse trend prediction in the ensemble model; and (iv) the CNN model provided a better sentiment signal than the MLP model.

The main limitations that affected this work were: (i) lack of open datasets on agricultural finance and market news, especially those labeled; (ii) lack of sentiment dictionaries that are specific for the agricultural domain, especially in Portuguese; (iii) short duration of the price time series, in comparison with other time series; and (iv) lack of open datasets for intraday prices of agricultural commodities.

The future work that is envisaged is related to: (i) improving and enlarging the news dataset; (ii) implementing unsupervised learning models in an attempt to identify patterns on the news datasets; (iii) implementing deep reinforcement learning models to improve decision making; and (iv) evaluate also the traditional econometric models, such as ARIMA, SARIMA, SARMAX, and GARCH.

Acknowledgments This work was supported by Itaú Unibanco S.A. through the Itaú Scholarship Program, at the Centro de Ciência de Dados (C^2D), Universidade de São Paulo, Brazil, and by the National Council for Scientific and Technological Development (CNPq Grants numbers 307027/2017-1, 425860/2016-7 and 167227/2017-3).

References

1. FAO. Food and Agriculture Organization of the United Nations. (2020). *FAOSTAT database*. FAO. Retrieved Januarey 31, 2020 from http://www.fao.org/faostat/en/#data/QC.
2. Baber, J. R., Wickersham, T. A., Sawyer, J. E., Freetly, H. C., Brown-Brandl, T. M., & Hales, K. E. (2020). Effects of diet type on nutrient utilization and energy balance in drylot heifers. *Journal of Animal Science, 98*(1), 1–8.
3. Kara, Y., Boyacioglu, M. A., & Baykan, Ö. K. (2011). Predicting direction of stock price index movement using artificial neural networks and support vector machines: The sample of the Istanbul stock exchange. *Expert Systems with Applications, 38*(5), 5311–5319.
4. Goçken, M., Özçalici, M., Boru, A., & Dosdogru, A. T. (2016). Integrating metaheuristics and artificial neural networks for improved stock price prediction. *Expert Systems with Applications, 44*, 320–331.
5. Di Persio, L., & Honchar, O. (2016). Artificial neural networks architectures for stock price prediction: Comparisons and applications. *International Journal of Circuits, Systems and Signal Processing, 10*, 403–413.

6. Weng, B., Lu, L., Wang, X., Megahed, F. M., & Martinez, W. (2018). Predicting short-term stock prices using ensemble methods and online data sources. *Expert Systems with Applications, 112*, 258–273.
7. Nasira, G. M., & Hemageetha, N. (2012). Vegetable price prediction using data mining classification technique. In: *International conference on pattern recognition, informatics and medical engineering (PRIME-2012)*. IEEE. pp. 99–102.
8. Bollen, J., Mao, H., & Zeng, X. (2011). Twitter mood predicts the stock market. *Journal of Computational Science, 2*(1), 1–8.
9. Adebiyi, A. A., Adewumi, A. O., & Ayo, C. K. (2014). Stock price prediction using the ARIMA model. In: 2014 UKSim-AMSS 16th international conference on computer modeelling and simulation. IEEE. pp. 106–112.
10. Rotela, J. R. P., Salomon, F. L. R., & Pamplona, E. O. (2014). ARIMA: An applied time series forecasting model for the Bovespa stock index. *Applied Mathematics, 5*(21), 3383–3391.
11. Kanchymalay, K., Salim, N., Sukprasert, A., Krishnan, R., & Hashim, U. R. A. (2017). Multivariate time series forecasting of crude palm oil price using machine learning techniques. *IOP Conference Series: Materials Science and Engineering, 226*(012117), 1–9.
12. Ghosal, D., Bhatnagar, S., Akhtar, S., Ekbal, A., & Bhattacharyya, P. (2017). IITP at SemEval-2017 task 5: An ensemble of deep learning and feature based models for financial sentiment analysis. In: *Proceedings of the 11th international workshop on semantic evaluation (SemEval-2017)*, ACL, Canada, pp. 899–903.
13. Mansar, Y., Gatti, L., Ferradans, S., Guerini, M., & Staiano, J. (2017). Fortia-FBK at SemEval-2017 task 5: Bullish or bearish? Inferring sentiment towards brands from financial news headlines. In: *Proceedings of the 11th international workshop on semantic evaluation (SemEval-2017)*, ACL, Canada, pp.1–6.
14. Nassirtoussi, A. K., Aghabozorgi, S., Wah, T. Y., & Ngo, D. C. L. (2014). Text mining for market prediction: A systematic review. *Expert Systems with Applications, 41*(16), 7653–7670.
15. Zhang, Y., & Wallace, B. C. (2017). A sensitivity analysis of (and practitioners' guide to) convolutional neural networks for sentence classification. In: *Proceedings of the 8th international joint conference on natural language processing*, AFNLP, Taiwan, pp.253–263.
16. Chong, E., Han, C., & Park, F. C. (2017). Deep learning networks for stock market analysis and prediction: Methodology, data representations, and case studies. *Expert Systems with Applications, 83*, 187–205.
17. Jiang, Z., Liu, C., Hendricks, N. P., Ganapathysubramanian, B., Hayes, D. J., & Sarkar, S. (2018). Predicting county level corn yields using deep long short term memory models. arXiv preprint arXiv:1805.12044, pp.1–26.
18. Medhat, W., Hassan, A., & Korashy, H. (2014). Sentiment analysis algorithms and applications: A survey. *Ain Shams Engineering Journal, 5*(4), 1093–1113.
19. Xiong, T., Li, C., Bao, Y., Hu, Z., & Zhang, L. (2015). A combination method for interval forecasting of agricultural commodity futures prices. *Knowledge-Based Systems, 77*, 92–102.
20. Khamis, A., & Abdullah, S. N. S. B. (2014). Forecasting wheat price using backpropagation and NARX neural network. *The International Journal of Engineering and Science, 3*(11), 19–26.
21. Weng, Y., Wang, X., Hua, J., Wang, H., Kang, M., & Wang, F. Y. (2019). Forecasting horticultural products price using ARIMA model and neural network based on a large-scale data set collected by web crawler. *IEEE Transactions on Computational Social Systems, 6*(3), 547–553.
22. Wang, C., & Gao, Q. (2018). High and low prices prediction of soybean futures with LSTM neural network. In: *2018 IEEE 9th international conference on software engineering and service science (ICSESS)*. IEEE. pp.140–143.
23. Xiong, T., Li, C., & Bao, Y. (2018). Seasonal forecasting of agricultural commodity price using a hybrid STL and ELM method: Evidence from the vegetable market in China. *Neurocomputing, 275*, 2831–2844.

24. Kantanantha, N., Serban, N., & Griffin, P. (2010). Yield and price forecasting for stochastic crop decision planning. *Journal of Agricultural, Biological, and Environmental Statistics, 15*(3), 362–380.
25. Kamilaris, A., Kartakoullis, A., & Prenafeta-boldú, F. X. (2017). A review on the practice of big data analysis in agriculture. *Computers and Electronics in Agriculture, 143*, 23–37.
26. Verdouw, C. N., et al. (2010). Process modelling in demand-driven supply chains: A reference model for the fruit industry. *Computers and Electronics in Agriculture, 73*(2), 174–187.
27. Goutte, C., & Gaussier, E. (2005). A probabilistic interpretation of precision, recall and F-score, with implication for evaluation. In *European conference on information retrieval, ECIR* (pp. 345–359). Springer.
28. Marto, T., Silva, R. F., Paula, A. M., Costa, A. H. R., & Cugnasca, C. E. (2019). Development of a labelled dataset of Brazilian agricultural market news titles for sentiment analysis. In: *XII Congresso Brasileiro de Agroinformática (SBIAGRO)*, Brazil, pp.528–537.
29. Wolfert, S., Ge, L., Verdouw, C., & Bogaardt, M. J. (2017). Big data in smart farming–a review. *Agricultural Systems, 153*, 69–80.
30. Han, L., Liang, R., & Tang, K. E. (2013). Cross-market soybean futures price discovery: Does the Dalian commodity exchange affect the Chicago Board of Trade? *Quantitative Finance, 13*(4), 613–626.
31. Hamadi, H., Bassil, C., & Nehme, T. (2017). News surprises and volatility spillover among agricultural commodities: The case of corn, wheat, soybean and soybean oil. *Research in International Business and Finance, 41*, 148–157.
32. Czudaj, R. L. (2020). The role of uncertainty on agricultural futures markets momentum trading and volatility. *Studies in Nonlinear Dynamics & Econometrics,* 24(3), 20180054.
33. Henrique, B. M., Sobreiro, V. A., & Kimura, H. (2019). Literature review: Machine learning techniques applied to financial market prediction. *Expert Systems with Applications, 124*, 226–251.
34. Ghahramani, Z. (2003). Unsupervised learning. In: *Summer School on Machine Learning*. Springer: Berlin, 72–112.
35. Lipton, Z. C., Berkowitz, J., & Elkan, C. (2015). A critical review of recurrent neural networks for sequence learning. arXiv preprint arXiv:1506.00019, pp. 1–38.
36. Ferreira, T., Paiva, F., Silva, R., Paula, A., Costa, A., & Cugnasca, C. (2019). Assessing reegression-based sentiment analysis techniques in financial texts. In: *Proceedings of the XVI National Meeting on Artificial and Computational Intelligence*. SBC. pp.729–740.
37. Bahrammirzaee, A. (2010). A comparative survey of artificial intelligence applications in finance: Artificial neural networks, expert system and hybrid intelligent systems. *Neural Computing and Applications, 19*(8), 1165–1195.
38. Schäfer, A. M., & Zimmermann, H. G. (2007). Recurrent neural networks are universal approximators. *International Journal of Neural Systems, 17*(4), 253–263.
39. Pasini, A. (2015). Artificial neural networks for small dataset analysis. *Journal of Thoracic Disease, 7*(5), 953–960.
40. Shaikhina, T., & Khovanova, N. A. (2017). Handling limited datasets with neural networks in medical applications: A small-data approach. *Artificial Intelligence in Medicine, 75*, 51–63.
41. Yu, Y., Si, X., Hu, C., & Zhang, J. (2019). A review of recurrent neural networks: LSTM cells and network architectures. *Neural Computation, 31*(7), 1235–1270.
42. Gers, F. A., Schmidhuber, J., & Cummins, F. (1999). Continual prediction using LSTM with forget gates. In *Neural Nets WIRN Vietri-99* (pp. 133–138). Springer.
43. Hochreiter, S., & Schmidhuber, J. (1997). Long short-term memory. *Neural Computation, 9*(8), 1735–1780.
44. Nelson, D. M., Pereira, A. C., & Oliveira, R. A. (2017). Stock market's price movement prediction with LSTM neural networks. In: *Proceedings of the 2017 international joint conference on neural networks (IJCNN).* IEEE, USA, pp.1419–1426.

45. Feuerriegel, S., & Gordon, J. (2019). News-based forecasts of macroeconomic indicators: A semantic path model for interpretable predictions. *European Journal of Operational Research, 272*(1), 162–175.
46. Joseph, K., & Garcia, P. (2018). Intraday market effects in electronic soybean futures market during non-trading and trading hour announcements. *Applied Economics, 50*(11), 1188–1202.
47. Xing, F. Z., Cambria, E., & Welsch, R. E. (2018). Intelligent asset allocation via market sentiment views. *IEEE Computational Intelligence Magazine, 13*(4), 25–34.
48. Zhang, W., & Skiena, S. (2010). Trading strategies to exploit blog and news sentiment. In: *Fourth international AAAI conference on weblogs and social media*, AAAI, USA, pp.375–378.
49. Mukherjee, S., & Bhattacharyya, P. (2012). Feature specific sentiment analysis for product reviews. In *International conference on intelligent text processing and computational linguistics* (pp. 475–487). Springer.
50. Wang, H., Can, D., Kazemzadeh, A., Bar, F., & Narayanan, S. (2012). A system for real-time twitter sentiment analysis of 2012 U.S. presidential election cycle. In: *Proceedings of the ACL 2012 system demonstrations*, ACL, Jeju Island, Korea. pp.115–120.
51. Ren, R., Wu, D. D., & Liu, T. (2018). Forecasting stock market movement direction using sentiment analysis and support vector machine. *IEEE Systems Journal, 99*, 1–11.
52. Ruder, S., Ghaffari, P., & Breslin, J. G. (2016). A hierarchical model of reviews for aspect-based sentiment analysis. In: *Proceedings of the 2016 Conference on empirical methods in natural language processing*, Austin, ACL, Texas, USA, pp. 999–1005.
53. Zhang, L., Wang, S., & Liu, B. (2018). Deep learning for sentiment analysis: A survey. *Wiley Interdisciplinary Reviews: Data Mining and Knowledge Discovery, 8*(4), 1–25.
54. Pennington, J., Socher, R., & Manning, C. D. (2014). Glove: Global vectors for word representation. In: *Proceedings of the 2014 conference on empirical methods in natural language processing (EMNLP)*, ACL, Qatar, pp.1532–1543.
55. LeCun, Y., & Bengio, Y. (1995). Convolutional networks for images, speech, and time series. *The handbook of brain theory and neural networks, 3361*(10), 1–14.
56. Neuenschwander, B., Pereira, A., Meira Jr, W., & Barbosa, D. (2014). Sentiment analysis for streams of web data: A case study of Brazilian financial markets. In: *Proceedings of the 20th Brazilian symposium on multimedia and the web*, ACM, Brazil, pp.167–170.
57. Hartmann, N., Fonseca, E., Shulby, C., Treviso, M., Rodrigues, J., & Aluisio, S. (2017). Portuguese woerd embeddings: Evaluating on word analogies and natural language tasks. arXiv preprint arXiv:1708.06025, pp.1–7.
58. Silva, M.J., Carvalho, P., & Sarmento, L. (2012). Building a sentiment lexicon for social judgement mining. In: *Lecture notes in computer science (LNCS)/lecture notes in artificial intelligence (LNAI), international conference on computational processing of Portuguese (PROPOR)*, Coimbra, Portugal, pp. 218–228.
59. Souza, M., & Vieira, R. (2011). Construction of a portuguese opinion lexicon from multiple resources. In: *Proceedings of the 8th Brazilian symposium in information and human language technology*, Brazil, pp. 59–66.
60. Souza, M., & Vieira, R. (2012). Sentiment analysis on twitter data for Portuguese language. In: *Proceedings of the 10th international conference computational processing of the Portuguese language*, Springer, Berlin, Heidelberg, pp. 241–247.
61. Pasqualotti, P. R., & Vieira, R. (2008). WordnetAffectBR: uma base lexical de palavras de emoções para a língua portuguesa. *RENOTE-Revista Novas Tecnologias na Educação, 6*(1), 1–10.
62. Witt, A., Kurths, J., & Pikovsky, A. (1998). Testing stationarity in time series. *Physical Review E, 58*(2), 1800–1810.

63. Hamilton, D. C., & Watts, D. G. (1978). Interpreting partial autocorrelation functions of seasonal time series models. *Biometrika, 65*(1), 135–140.
64. Fama, E. F. (1995). Random walks in stock market prices. *Financial Analysts Journal, 51*(1), 75–80.
65. Prechelt, L. (1998). Automatic early stopping using cross validation: Quantifying the criteria. *Neural Networks, 11*(4), 761–767.
66. Hu, M. Y., Zhang, G., Jiang, C. X., & Patuwo, B. E. (1999). A cross-validation analysis of neural network out-of-sample performance in exchange rate forecasting. *Decision Sciences, 30*(1), 197–216.

Use of Unsupervised Machine Learning for Agricultural Supply Chain Data Labeling

Roberto F. Silva, Gustavo M. Mostaço, Fernando Xavier, Antonio M. Saraiva, and Carlos E. Cugnasca

1 Introduction

A supply chain (SC) can be defined as a network of companies responsible for fulfilling customer segments [1]. Agricultural SCs involve the production, processing, and distribution of agricultural products, such as grains, fruits, flowers, vegetables, and animal products. A SC can be divided into links or stages characterized by the groups of companies that conduct specific types of activities, like farms, processing industries, logistics service providers, distribution centers, retailers, among others [1]. Each company present in an SC is denominated as an agent. SCs generate significant amount of data that can be characterized as having a large volume and variety, originated from different sources (several data flows), processed on different systems, and stored on different databases and formats [2].

In general, this data is not standardized when exchanged among links and agents in the same link in the SC. Therefore, interoperability problems are usually observed when different agents are involved [3]. These problems tend to increase due to the coexistence of technologies on the different stages of the data lifecycle, such as: data generation, collection, storage, and processing, resulting in data in different formats and resolutions [3]. Nevertheless, the complexity and time needed to standardize the data generated by all agents are prohibitive.

In this sense, methods that permit a better understanding and use of the data generated on agricultural SCs are welcome. For that, it is essential to divide it into categories. The three main categories, related to the systems that produce those data, are: (i) product identification and traceability; (ii) environmental monitoring;

R. F. Silva (✉) · G. M. Mostaço · F. Xavier · A. M. Saraiva · C. E. Cugnasca
Department of Computer Engineering and Digital Systems, Escola Politécnica da Universidade de São Paulo (USP), São Paulo, Brazil
e-mail: gmostaco@usp.br; fxavier@usp.br; saraiva@usp.br; carlos.cugnasca@usp.br

D. D. Bochtis et al. (eds.), *Information and Communication Technologies for Agriculture—Theme II: Data*, Springer Optimization and Its Applications 183,
https://doi.org/10.1007/978-3-030-84148-5_11

and (iii) processes monitoring [1, 4–6]. Systems in the categories (i) and (ii) produce mainly structured data, while systems in the category (iii) produce mainly non-structured data [1, 4–6].

With the increasing adoption of the Internet of Things (IoT) technologies on agricultural SCs [7], the quantity of data that is generated and its associated problems tend to increase. Also, lack of interoperability, long periods of implementation, and lack of communication are some of the problems of adopting these technologies without proper standardization and coordination among agents [8].

The main objective of this chapter is to evaluate if unsupervised machine learning techniques can be successfully used to identify clusters and generate labels for a non-standardized unlabeled agricultural SC dataset (Fig. 1). This would contribute to reducing interoperability problems that can be observed in real-life scenarios. This is one of the few attempts to address this problem using unsupervised machine learning techniques to the best of our knowledge.

In this chapter, two techniques are used to cluster agricultural SC data and generate labels related to the systems that originated them. These are k-means [9, 10] and Self Organizing Maps (SOM) [11]. Principal Component Analysis (PCA) was also used for dimensionality reduction.

The principal research question for this chapter was: "Can unsupervised machine learning techniques contribute to agricultural SC data labeling?" This was evaluated through the results of the implementation of four models: (i) k-means; (ii) SOM; (iii) PCA with two features and k-means; and (iv) PCA with three features and k-means. A dendrogram, created with the Agglomerative Hierarchical Clustering technique [12, 13], was also used to understand the data structure better. Figure 2 illustrates the basic methodological elements of the chapter.

This chapter is divided into six sections: Sect. 1 contains the introduction, research motivation, the main concepts related to agricultural supply chains, and the description of the objective and research question; Sect. 2 contains a description of unsupervised learning, as well as examples used in agriculture; Sect. 3 contains the description of the methodology used and its related steps; Sect. 4 contains the report of the main results obtained; Sect. 5 contains the discussion concerning the results, the implementation aspects and the difficulties regarding the models; and Sect. 6 contains the conclusions, the observed limitations and plans for future work.

2 Unsupervised Machine Learning in Agriculture

Unsupervised machine learning techniques can be described as a group of machine learning techniques that focus on identifying structure, patterns, and clusters on unlabeled datasets [14]. For this reason, most of these methods focus on evaluating the relations between individual points and features in a dataset, as well as their distributions. Besides data labeling and pattern identification, unsupervised machine learning techniques can also be used as an alternative for dimensionality reduction [14]. For a thorough review of machine learning techniques applied in

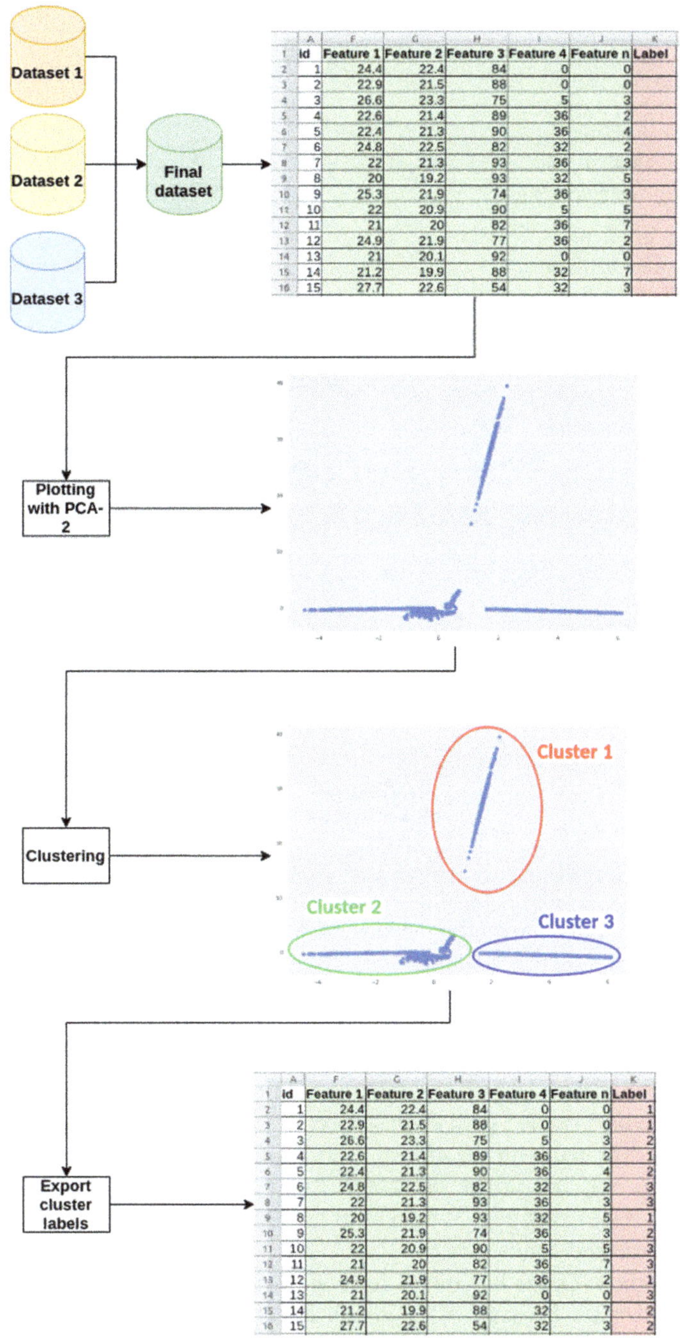

Fig. 1 Generic workflow for using clustering models for generating dataset labels. Legend: the first row illustrates the data fusion process and the fact that, in this chapter, the data contains no labels; the second row illustrates the use of a Principle Components Analysis (PCA) with two components to visualize the data in two dimensions; the third row illustrates the use of the clustering model and a plot of its results, considering the PCA with two components; and the last row illustrates the process of exporting the cluster groups as labels for the original dataset, labeling the data

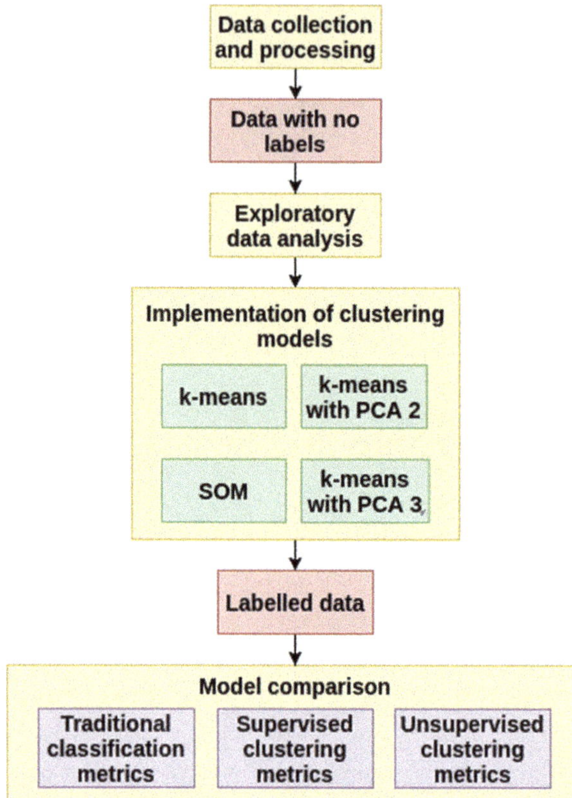

Fig. 2 The methodology used in this chapter. Legend: in yellow, the main steps of the methodology used, from data collection to comparing the final models implemented; in red, the datasets that were built; in green (middle of the figure), the clustering models that were implemented; and in purple (bottom of the figure), the three types of metrics used for comparing the models

the agricultural domain, including unsupervised learning techniques, refer to Liakos et al. [15].

The k-means is a technique, dated 50 years ago, that is still being used due to its good results and fast execution time [9]. Figure 3 contains an example of the use of the k-means clustering model. Figure 3a represents the two-dimensional input data for the k-means method, which is composed of three major steps: (i) the initialization creates seed points that are used as cluster centers (Fig. 3b); (ii) then all points in the dataset are associated with the closest clusters, according to a predefined distance measure (Fig. 3c); and (iii) repeat step (ii) until no more points change between clusters (Fig. 3d, e); then Fig. 3f represents the final clustering situation obtained by the k-means method convergence.

Although the k-means is a robust method, it can lead to local optima instead of global optimization because of its initialization. The k-means++ algorithm was proposed to reduce the impacts associated with this problem [16]. The k-means++ algorithm differs from the standard k-means because of its randomized seeding technique, in which the initial cluster centers are chosen in a specific way to avoid local optima. It has a considerably lower processing time compared to the use of

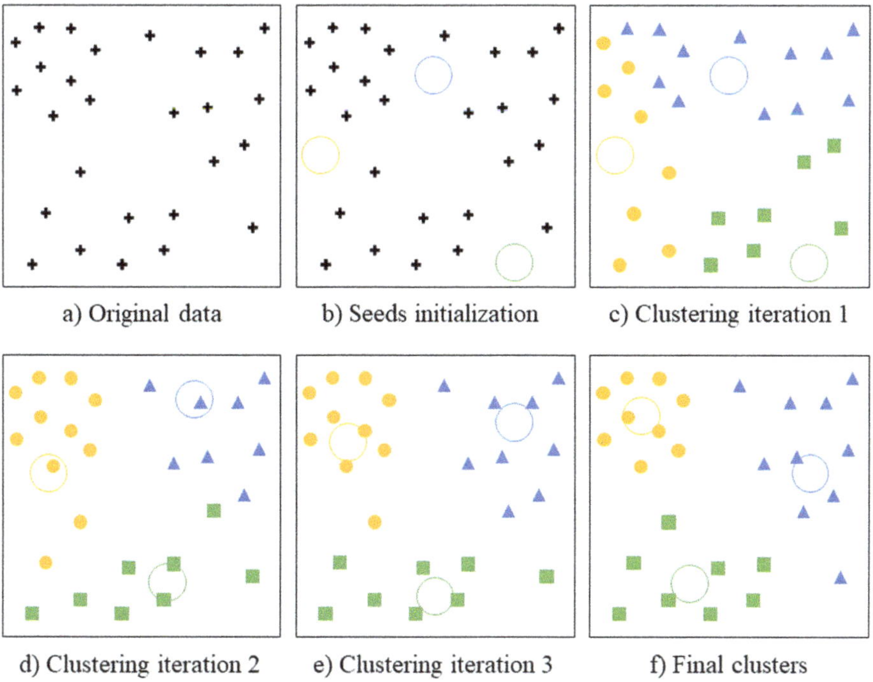

Fig. 3 Example of the k-means algorithm in two-dimensional data. Legend: (**a**) Original input data; (**b**) Initialization with tree seed points selected as cluster centers; (**c**) First iteration with the assignment of data points to clusters; (**d**) Second and (**e**) Third iterations with cluster centers and point labels updating; (**f**) Final clustering situation obtained by the k-means method convergence

multiple random initializations and can achieve better accuracy. For those reasons, for the k-means model, it was used in the present chapter.

A comparison between six different unsupervised machine learning models on precision agricultural data developed by Mehta et al. [17] concluded that the k-means model provided the best results on the Food and Agriculture Organization (FAO) soybean dataset, both in terms of time taken for the completion and of cluster instances. The methods considered were k-means, Density-Based Spatial Clustering of Applications with Noise (DBSCAN), Ordering Points To Identify the Clustering Structure (OPTICS), agglomerative hierarchical clustering, divisive hierarchical clustering, and cobweb.[1]

The DBSCAN [18] is a clustering model that aggregates points close together and treats points in low-density regions as outliers [17]. Therefore, it is a method that considers the density of the data points in different regions. The OPTICS [19] is a

[1] For a thorough description of the steps of each clustering model, we refer the reader to the work by Mehta et al. [17].

clustering model developed to overcome problems of the DBSCAN when dealing with data with variable density [17].

The agglomerative hierarchical clustering [13] is a widely used form of hierarchical clustering. It is considered a bottom-up approach, as it starts with each data point as an individual cluster and aggregates the clusters one step at a time until it achieves a single cluster [17]. The divisive hierarchical clustering [20] is the inverse approach, starting with a single cluster containing all data points and further dividing it until it achieves one cluster for each data point [17].

The cobweb [21] is based on a different clustering model, denominated conceptual clustering. According to Mehta et al. [17] and Fisher [21], the cobweb model uses two main factors to form the clusters: (i) the structure of the data points (as the other models discussed); and (ii) the conceptual description of the classes or data points (which is not considered on the previous models).

Jain [9] and Steinley [10] present a thorough analysis of the k-means model, its variations, and its applications. In the agricultural domain, there are several important implementations of the k-means model. Ramesh et al. [22] propose a parallelization method for the k-means models applied to large agricultural databases. A similar implementation can be used for supply chain data, as these tend to generate millions of data points.

Another important model for unsupervised machine learning is self-organizing maps (SOM). It is a neural network model created in 1982 [11] to identify data clusters and reduce dimensionality in the dataset. It also generates a projection of the n-dimensional data into a 2D map [23]. This is called a cluster map. Unlike the k-means model, it preserves the data topology, allowing the identification of hidden patterns that may not be observed in other models. This is especially observed on datasets with a considerable number of features that may have a connection between them. It can also be used for cluster analysis and labeling generation, but it is not nearly as used as the k-means algorithm. It is implemented in this chapter and evaluated in comparison to the k-means model. Figure 4 summarizes the training algorithm for the SOM model.

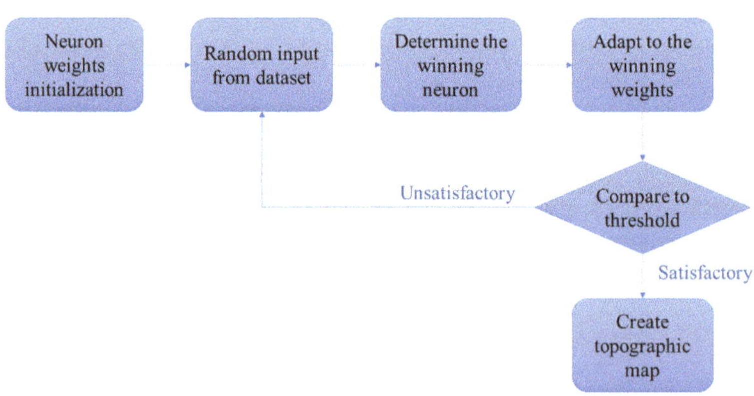

Fig. 4 Simplified view of the training algorithm for the SOM model

There is one neuron for each input node in the SOM model, each having a vector of weights associated. First, the initialization creates a randomized set of weights for them. The training process then involves several training iteration steps until the model can learn the items in the dataset. There will be one winning neuron for each pattern to be recognized in the dataset during these iterations, meaning that there will be one vector of weights for each pattern. Next, this winning neuron will have its weights adjusted to present a stronger response to a similar input in the future, minimizing the error in this response. Because each pattern is represented by a different set of neurons, their ability to recognize one specific pattern will increase. The result of the training process is an ordered neuron location and a meaningful coordinate system for the input features so that a topographic map can be created for the input patterns considering the coordinates of the associated winning neurons for each one of these patterns [11, 23, 24].

An interesting implementation of SOM for environmental purposes can be observed in the research of Jeong et al. [25]. They used SOM to identify patterns in a heterogeneous dataset containing economic, social, and environmental data on a South Korea river basin. The model resulted in four clusters, which allowed for a better understanding of which features were more important for understanding the conservation and urbanization impacts of different river basin areas. A thorough analysis of the impacts of the features was conducted, and the clusters were applied to the river map. The authors concluded that the SOM provided important results for their analysis and can be used on environmental problems such as the characterization of rivers on a basin level.

An example of another unsupervised machine learning technique that maintains the structure of the data is hierarchical clustering. An interesting implementation of this model in the agricultural domain was the article of Ruß and Kruse [26], in which it was used for management zone delineation in precision agriculture, considering several features with spatial correlation. This is essential for defining the ways to better manage the farm processes in a specific area.

The agglomerative hierarchical clustering model is more computationally intensive than the k-means model [17]. Nevertheless, as the SOM contains a significantly higher number of parameters compared to those models [27], it is the most computationally intensive of the three models implemented in this chapter.

In this chapter, the agglomerative hierarchical clustering model is used to generate a dendrogram that is used as an input to evaluate the number of clusters on the data. The next section of this chapter contains a description of the methodology used. It contains information related to its steps, the models implemented, codes, and the main assumptions.

3 Methodology

Figure 5 illustrates the methodology adopted in this chapter which is divided into seven main steps.

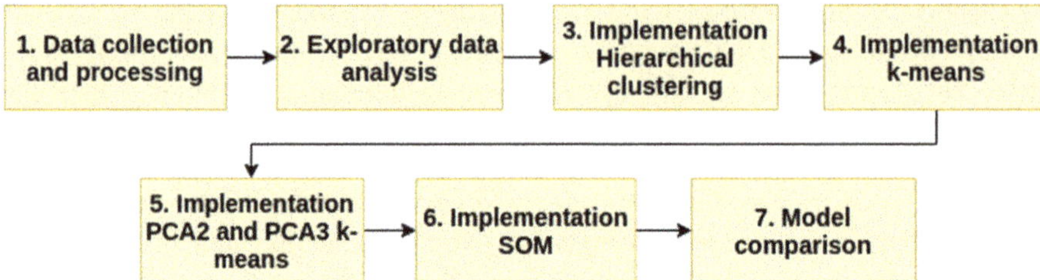

Fig. 5 Steps of the methodology

1. *Data collection and processing* using Python on the Jupyter IDE[2] and LibreOffice Calc.[3] It was composed by five main activities: (i) data collection from different data sources; (ii) data preprocessing, with unique identification of rows, separation of target column from the data body, dealing with missing data (with zeros and interpolations), and.csv exporting; (iii) data processing,[4] elimination of errors and outliers; (iv) data fusion,[5] identification of common features on the datasets, merging of those features, solving missing data problems and selection of final features to enter the model; and (v) data normalization,[6] using the standard scaler[7] for PCA, k-means and hierarchical clustering algorithms, and the MinMax scaler for the SOM. The Scikit-Learn[8] package was also used to perform the scaling;

[2]https://jupyter.org/

[3]https://www.libreoffice.org/discover/calc/

[4]Data preprocessing and processing techniques focus on eliminating faulty data, dealing with missing values, and preparing the data to be used by the knowledge extraction and pattern recognition models. For an in-depth review of data processing and preprocessing techniques, the reader is referred to the work by Van den Broeck et al. [28]. If the reader is interested in processing time series data, please refer to the work by Wang and Wang [29]. If the reader is interested in processing natural language data, please refer to the work by Sun et al. [30].

[5]Data fusion can be described as the combination of multiple datasets or data sources to improve the quality of the information [31]. It is important to note that these datasets may contain different features or variables. For an in-depth review of the different techniques used for data fusion, the reader is referred to the work by Castanedo [31].

[6]Data normalization is an essential step for most machine learning models, especially for artificial neural networks. According to Singh and Singh [32], it can be defined as transforming the features to a specific range. Two of its main advantages are: allowing for faster model training, and avoiding problems due to features with very distinct value ranges. For more information on the use of data normalization for classification, please refer to the work by Singh and Singh [32].

[7]For implementation aspects of the standard scaler technique, the reader is referred to the scikit-learn library documentation: https://scikit-learn.org/stable/modules/generated/sklearn.preprocessing.StandardScaler.html

[8]For implementation aspects of the MinMax scaler technique, the reader is referred to the scikit-learn library documentation: https://scikit-learn.org/stable/modules/generated/sklearn.preprocessing.MinMaxScaler.html

2. *Exploratory data analysis* using the Pandas, Matplotlib, and Seaborn packages, aiming to understand the data better. This was an essential step as the dataset was composed of seven unrelated datasets from different sources;
3. *Implementation of the Agglomerative variation of the Hierarchical Clustering model*[9] using Scikit-Learn, considering four variations related to the linkage: ward, complete, single, and average. A dendrogram[10] was also generated to evaluate better the number of clusters based on the data structure;
4. *Implementation of the k-means++ model*[11] using Scikit-Learn, evaluating the following hyperparameters: tolerance (0.001, 0.01, and 0.1), number of initializations (100, 300, and 500), and maximum number of iterations (100, 300, and 500). The elbow method[12] was also implemented to evaluate better the number of clusters based on the data structure;
5. *Implementation of PCA2 and PCA3 k-means++*. The principal components analysis (PCA) method[13] is an important dimensionality reduction technique that generates new features composed of the existing features. These are called components. In this chapter, two variations of this technique were considered: PCA-2 (generating two components that capture maximum information from the dataset) and PCA-3 (generating three components that capture maximum information from the dataset). The elbow method was also implemented to better evaluate the number of clusters based on the data structure. The PCA was used to process the data before using the k-means model. The latter was performed using Scikit-Learn, evaluating the following hyperparameters: tolerance (0.001, 0.01, and 0.1), number of initializations (100, 300, and 500), and maximum number of iterations (100, 300, and 500);

---

[9]The Agglomerative is a commonly used variation of the hierarchical clustering models. Initially, each data point is considered a cluster. The algorithm then joins neighbor clusters, using a linkage method to evaluate the dissimilarity between them and to choose which clusters should be aggregated [12]. A dendrogram can then be used to visualize the results of each step of the algorithm.

[10]For an in-depth description of the different methods that are commonly used to estimate the optimal number of clusters for hierarchical clustering models, please refer to the work by Zambelli [33].

[11]For a detailed analysis of the k-means model, the reader is referred to the works by Jain [9] and Steinley [10]. For implementation aspects of the k-means model, the reader is referred to the scikit-learn library tutorials on its implementation: https://scikit-learn.org/stable/modules/generated/sklearn.cluster.KMeans.html and https://scikit-learn.org/stable/auto_examples/cluster/plot_kmeans_digits.html#sphx-glr-auto-examples-cluster-plot-kmeans-digits-py

[12]For a detailed description of the elbow method, as well as other methods for determining the optimal number of clusters for applying the k-means model, please refer to the works by Kodinariya and Makwana [34] and Zambelli [33].

[13]For a detailed analysis of the PCA method, as well as examples of uses, please refer to the work by Jolliffe and Cadima [35]. For implementation aspects related to PCA, the reader is referred to the scikit-learn library tutorials on its implementation: https://scikit-learn.org/stable/modules/generated/sklearn.decomposition.PCA.html and https://scikit-learn.org/stable/auto_examples/decomposition/plot_pca_iris.html#sphx-glr-auto-examples-decomposition-plot-pca-iris-py

6. *Implementation of the SOM model*[14] using MiniSom[15] [38], with the triangle and Gaussian neighborhood functions and evaluating the following parameters: learning rate (0.1, 0.2 and 0.5), sigma (2, 3, 5 and 10), and the number of rounds (100, 500, 1000, and 5000). Maps of the implementations were generated to evaluate the clustering quality better;
7. *Model comparison* using Scikit-Learn, divided into three parts: (i) traditional supervised learning metrics[16]: confusion matrix, precision, recall, and F1-score for each cluster; (ii) supervised clustering metrics[17]: adjusted rand index, mutual info, homogeneity, completeness, v-measure, and Fowlkes-Mallows index; and (iii) unsupervised clustering metrics: silhouette, Davies-Bouldin index, and Calinski-Harabasz index.

While the supervised metrics focus on measuring how well the predicted labels match the correct labels, the unsupervised ones focus on evaluating clustering quality, mainly through intra and inter-cluster distances.

Some simplifying assumptions were made due to two main factors: (i) the lack of practical research on clustering agricultural SC data; and (ii) the absence of open and standardized datasets that could be readily used in this chapter. These assumptions were:

1. The predicted labels and clustering groups were related to the three main systems described in Sect. 1: (i) product identification and traceability; (ii) environmental monitoring; and (iii) processes. In future work, the prediction of SC links could also be explored.
2. The considered SC was composed of four links, representing a generic agro-industrial SC: farm, industry, transportation, and retailer. Each of those links could contain one or multiple agents.
3. The dataset used for clustering was formed by the fusion of seven open datasets, described in Table 1. Each one was an independent dataset and represented partial data generated by different links in the SC and as the most common information architecture present on it. The selection of the datasets was based on a thorough

[14] For a detailed analysis of the SOM model, the reader is referred to the works by Kohonen [11], Yin [36], and Liu and Weisberg [37]. For implementation aspects of the SOM model, the reader is referred to the Minisom library tutorials on its Github repository: https://github.com/JustGlowing/minisom

[15] https://github.com/JustGlowing/minisom

[16] For implementation aspects as well as descriptions of the supervised learning metrics used in this chapter, please refer to the scikit-learn library documentation: https://scikit-learn.org/stable/modules/model_evaluation.html#classification-metrics

[17] For implementation aspects as well as descriptions of the supervised and unsupervised clustering metrics used in this chapter, please refer to the scikit-learn library documentation on clustering performance evaluation: https://scikit-learn.org/stable/modules/clustering.html#clustering-evaluation

Table 1 Open datasets selected to form the final dataset

Dataset	SC link and system	Description	Total number of features
01	Farm – 2	Environmental data collected[a]	9 (6 common, 3 unique)
02	Farm – 3	Herbicide application per plot[b]	10 (5 common, 5 unique)
03	Farm – 3	Plant growth per plot[6]	12 (5 common, 7 unique)
04	Farm – 3	Productivity per plot[6]	6 (5 common, 1 unique)
05	Transportation – 1	Transportation – farm to industry[c]	9 (6 common, 3 unique)
06	Industry – 2	Environmental data – industry[d]	15 (8 common, 7 unique)
07	Industry – 3	Operational data – industry[e]	12 (5 common, 7 unique)

[a]INMET - Data from Rio Verde, GO – http://www.inmet.gov.br
[b]Scribner et al. [39]; Silva et al. [40]; Omer et al. [41]; Agridat – https://github.com/kwstat/agridat
[c]Cashew Truck Arrivals – https://www.kaggle.com/extralime/cashew-truck-arrivals
[d]Wsn-indfeat-dataset – https://github.com/apanouso/wsn-indfeat-dataset
[e]Vega shrink-wrapper degradation – https://www.kaggle.com/inIT-OWL/vega-shrinkwrapper-runtofailure-data

search for open databases on Kaggle,[18] Datahub,[19] and Github.[20] Datasets features were analyzed during the data fusion activity (in step 1) and the exploratory data analysis (in step 2).

4. The final dataset represented the situation of agro-industrial SCs with information sharing among links, in which agents send their data to a shared database in the cloud, without standardization. If no information is shared among links, a data collection step (conducted on step 1) is needed. Therefore, the methodology proposed in this chapter can be used for both SCs that have a central database and SCs that have only local databases.
5. For the traditional supervised learning metrics, the F1-score[21] was considered the most important one, as it is a harmonic mean of precision and recall, leading to better overall results. The best models were chosen for the supervised and unsupervised clustering metrics, considering the same weight for all the metrics. Therefore, the model that accumulated the best results on most metrics was considered the best one.

Finally, the implementation was done using a Google Colab[22] TPU with 2 Intel (R) Xeon (R) CPUs with 2.30GHz and 13GB RAM. The number of data points per

[18]https://www.kaggle.com

[19]https://datahub.io

[20]https://github.com

[21]For an in-depth analysis of classification metrics, including the F1 score, we refer the reader to the work by Goutte and Gaussier [42]. For implementation purposes, the reader can refer to the Scikit Learn documentation: https://scikit-learn.org/stable/modules/generated/sklearn.metrics.f1_score.html

[22]https://colab.research.google.com

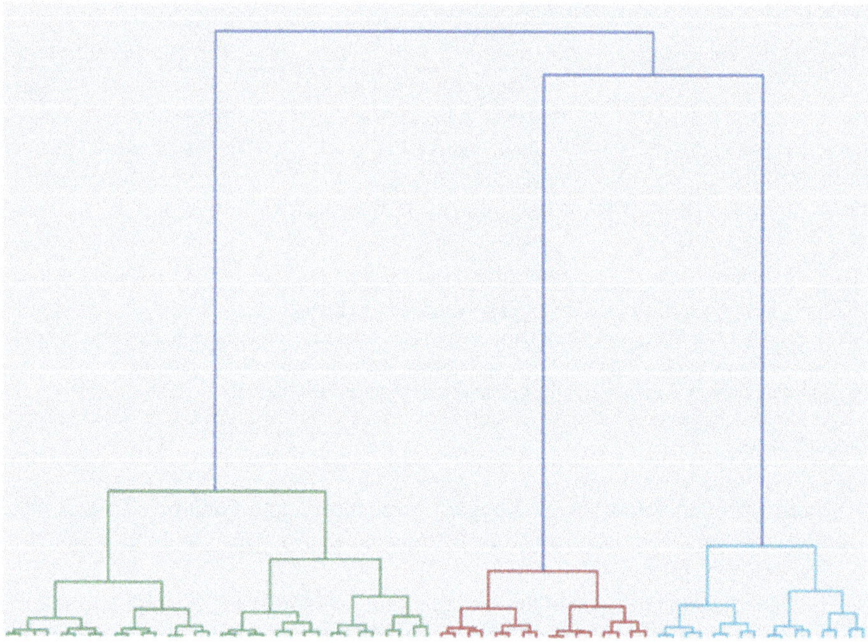

Fig. 6 Hierarchical clustering dendrogram

cluster were: Cluster 1 (System 1) = 20,527; Cluster 2 (System 2) = 15,153; and Cluster 3 (System 3) = 671. The code is available on an open Github repository.[23]

4 Results

This section contains the description and analysis of the results obtained from the implementation of the different models. Since the results of the Hierarchical Clustering models and the k-means implementations were the same, these were excluded from the analysis conducted in this chapter.

Two observations are important: (i) the dendrogram, illustrated in Fig. 6, pointed to the selection of three well-defined clusters, based on the data structure; and (ii) the implementations can differ considerably if the linkage method used is not the Ward method, showing considerably worse results. The Ward method is a commonly used linkage method for the agglomerative hierarchical clustering models [43]. Therefore, the Ward method is the most recommended one, due to the structure of the dataset. Nevertheless, it is important to observe that this result could be different on datasets with a different data structure. Therefore, it is essential to conduct a hyperparameter

[23] https://github.com/rfsilva1/unsupervised-learning-agriculture

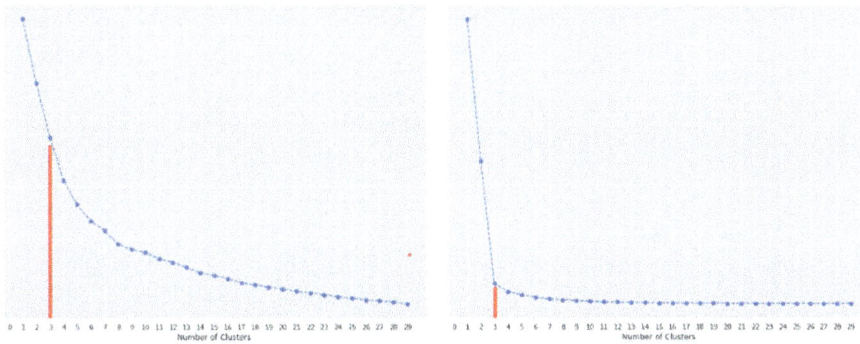

Fig. 7 Elbow method for the k-means (left) and PCA2 + k-means (right) implementations. The x-axis represents the number of clusters, and the y-axis represents the Within-Cluster-Sum-of-Squares (WCSS). The red line marks WCSS for $k = 3$ (three clusters). The main objective of the k-means model is to cluster the data into k clusters, minimizing the WCSS, which can be defined as the squared difference between each point in a cluster and the cluster mean [44]. The k is an input for the model

analysis to evaluate the different linkage methods and their impacts on the clustering results.

The analysis in this section is described in the following order: first, the results of the hyperparameter analysis for each model are discussed; then, the results of the traditional supervised learning, the supervised clustering, and the unsupervised clustering metrics are described, considering the implementation of the final models (the ones with the best results on the hyperparameter analysis). Lastly, the model that best fits the problem in this chapter is chosen.

The results from the hyperparameter analysis (tolerance, number of initializations, and maximum number of iterations) for the k-means++ algorithm implementation, were equal. Therefore, it is possible to infer that, for this dataset, the hyperparameters analyzed and their range, the algorithm can find the best result considering three clusters. This is also true for the PCA-2 and PCA-3 implementations. The main difference observed among those three models is related to the unsupervised clustering metrics.

Another interesting point to observe is the difference between results from the k-means, PCA-2 + k-means, and PCA-3 + k-means implementations. Those can be observed in Fig. 7, which illustrates the elbow method applied for the k-means and PCA2 + k-means. In the case of k-means, this method points out to a number between six and seven clusters, while for PCA-2 + k-means it indicates exactly three clusters, as was expected by analyzing the hierarchical clustering dendrogram. Nevertheless, an analysis of the metrics is also needed to choose between models.

The analysis of the hyperparameters for the SOM model (described in Sect. 3) pointed out that the neighborhood function and the sigma were the most important ones. The learning rate had less impact than those two hyperparameters. The number of iterations was the least important hyperparameter. One possible explanation for

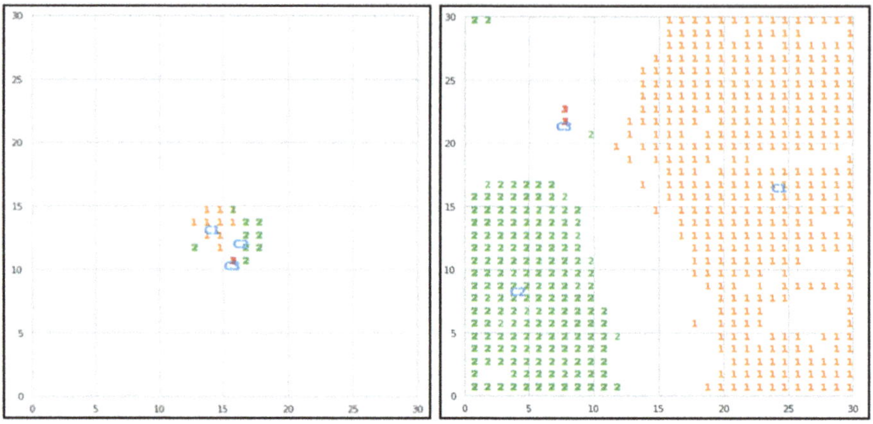

Fig. 8 Cluster map for the SOM algorithm, considering the triangle (left) and Gaussian (right) neighborhood functions

this behavior is that the algorithm using the Gaussian neighborhood function better approximates the number of clusters and the data structure than the triangle function.

This can be seen in Fig. 8, which illustrates the cluster map[24] for the best models (from the hyperparameters analysis) for both functions. It is possible to observe that the triangle function barely separates the clusters, which may induce errors in further observations. On the other hand, the cluster map for the Gaussian function presented distinct clusters, suggesting that this neighborhood function can describe the structure of the dataset better.

The final hyperparameters chosen for the k-means, PCA-2 + k-means, and PCA-3 + k-means were: 3 clusters, k-means++ algorithm, 10 initializations, 300 maximum iterations, and tolerance of 0.0001. The k-means++ algorithm was used for all the k-means implementations. The final hyperparameters, chosen for the SOM model were: Gaussian neighborhood function, a learning rate of 0.5, a sigma of 10, and 1000 iterations.

Table 2 contains the results of the traditional supervised learning metrics for the k-means, PCA2 + k-means, PCA3 + k-means, and SOM implementations. It is important to observe that, since the results for the first three models were equal, they are presented only once. One factor that may explain this result is that the PCA analysis components reflect specific features close to each other, which are also the ones used by the k-means to separate the clusters.

It is possible to observe that, even though all models showed the same F1-score for Cluster 1, there were differences for the other clusters. The SOM model presented the best F1-score for all clusters, resulting in a higher weighted average F1-score

[24] Cluster maps are used with the SOM model to graphically illustrate the clusters' spatial relations [24]. These maps can be drawn using different formats. For a thorough exploration of different cluster maps, as well as best practices, please refer to the work by Samsonova et al. [24].

Table 2 Results of the traditional supervised learning metrics for all models

Models	Metrics	Cluster 1	Cluster 2	Cluster 3	Weighted avg.
k-means PCA2 + k-means PCA3 + k-means	Precision	0.86	1.00	0.00	0.90
	Recall	1.00	0.81	0.00	0.90
	F1-score	0.93	0.90	0.00	0.90
SOM	Precision	1.00	1.00	0.12	0.98
	Recall	0.86	0.86	1.00	0.84
	F1-score	**0.93**	**0.92**	**0.22**	**0.91**

The best results for each cluster and metric are in bold

(0.91). A higher weighted average precision (0.98) is also presented. Another important fact is that, although the SOM model predicted the data points from Cluster 3 better (resulting in an F1-score of 0.22 vs. 0.00 for the k-means implementations), it was still a very low score. One possible explanation for that, was due to a low number of observations in this specific cluster, leading to a data imbalance problem,[25] and also that this cluster had fewer features than the other two.

Table 3 contains the results of the supervised clustering metrics for the k-means, PCA2 + k-means, PCA3 + k-means, and SOM implementations. As with the traditional supervised learning metrics, the k-means and PCA + k-means implementations resulted equal, and therefore, are presented only once.

Like the traditional supervised learning metrics, the clustering metrics also need the true labels of the data to evaluate the quality of the clustering implementation. The SOM model also showed the best results for this group of metrics, obtaining the best results for five of the six evaluated metrics. Most of these metrics evaluate different aspects of the implemented clustering technique, but they are mainly related to the evaluation of: (i) correct association of data points from the same label with the right cluster; (ii) similarity between clusters; and (iii) the relation between precision and recall for each cluster.

Table 4 contains the results of the unsupervised clustering metrics for the k-means, PCA2 + k-means, PCA3 + k-means, and SOM implementations. Unlike the previous metrics, these have shown considerable differences between the models implemented. It is also important to observe that these measures are related to the variations of intra and inter-cluster distances.

The PCA-2 + k-means model presented the best results for the unsupervised clustering metrics, except for the Calinski-Harabasz metric, which was considerably worse than the other models. The SOM model presented the worst results for the Davies-Bouldin index and intermediary values for the Calinski-Harabasz and Silhouette metrics.

From the analysis of all the considered metrics in the three groups, it is possible to conclude that, for the current dataset, the SOM model with the Gaussian function

[25]For an in-depth exploration of the data imbalance problem, refer to the work by Kotsiantis et al. [45].

Table 3 Results of the supervised clustering metrics for all models

Metrics	k-means PCA2 + k-means PCA3 + k-means	SOM
Adjusted Rand	0.682	**0.748**
Mutual info	0.629	**0.697**
Homogeneity	0.595	**0.806**
Completeness	**0.667**	0.614
V-measure	0.629	**0.697**
Fowlkes-mallows	0.848	**0.865**

The best result for each metric is bold

Table 4 Results of the unsupervised clustering metrics for all models

Metrics	k-means	PCA2 + k-means	PCA3 + k-means	SOM
Silhouette	0.429	**0.799**	0.708	0.512
Davies-Bouldin	0.933	**0.219**	0.354	1.393
Calinski-Harabasz	**10,289.228**	226,688.890	40,095.433	36,032.946

The best result for each metric is bold

provided the best results. It had medium to lower results on the unsupervised clustering metrics but was considerably better than the other models on the traditional supervised and supervised clustering metrics.

5 Discussion

This section contains a discussion of the main impacts of the models implemented and the use of unsupervised machine learning techniques for supply chain data. Four main aspects are discussed: (i) the time needed to train each model and how this can impact its use; (ii) the main difficulties in implementing the models and how they can impact real-life use cases; (iii) the main benefits of the implemented models for SC traceability; and (iv) possible applications for other SCs.

5.1 Training Time for Each Model

The first important aspect related to the results obtained in this chapter is related to the time required to train the models and the impact this may have on real-life scenarios. The k-means model is considerably faster than the SOM, which is expected, as any artificial neural network needs to perform more operations and have more components than the k-means model.

In the experiments conducted in this chapter, each iteration of the algorithms took: 0.16 s for k-means, 86.00s for SOM, 0.34 s for PCA-2 + k-means, and 0.33 s for PCA-3 + k-means. Therefore, even though the SOM was chosen as the best model for the current research problem, it may take a considerable amount of time to train on real-life SC datasets, which may have millions of data points (instead of the 36.5 k points present in the dataset used in this chapter).

It is also important to note that this has significant impact on the hyperparameter analysis, especially for the SOM model, as each grid search took more than 2 h 20 m to complete. This is an essential step for every clustering problem, as a thorough search among the hyperparameters' values may better capture the structure of the data and their patterns and avoid local optima.

Nevertheless, it should be noted that these observations are related to the models' training phase. After the model has been trained and evaluated using the specific situation's quality metrics, making predictions is not a time-consuming task. Based on the results obtained in this chapter, it is possible to infer that using both unsupervised and supervised metrics is essential. Although the datasets are not labeled in real-life scenarios, the model can be trained on a partially labeled dataset, and then its results can be extended for the whole dataset.[26]

5.2 *Implementation Difficulties*

The second important aspect is related to the difficulties in implementing the models used in this chapter. The k-means has several implementations on R and Python, the most used languages for machine learning and data science. Therefore, its implementation is straightforward, and the code is already validated by a considerable amount of applications.

Notwithstanding, the SOM model is not implemented on the main machine learning libraries used by practitioners (such as scikit-learn, tensorflow, and keras). Therefore, this model has several implementations, and the data scientist must evaluate and choose the best one among them for the specific use case. Fortunately, the minisom implementation has been validated by more than 30 research works and is well documented [38].[27] Some examples are the articles by Mansha et al. [47], Stoean et al. [48], and Riese, Keller, and Hinz [49]. To the best of our knowledge, there is still no widespread use of SOM in real-life scenarios. In conclusion, as the minisom implementation gets more visibility, it tends to increase the number of implementations in real-life scenarios.

Some of the main reasons that help explain the results obtained in this chapter were: (i) the data imbalance among the different labels (System 1 contained

[26]The strategy of partially labelling the dataset and then using the partially labelled subset to train the model and predict the labels for the whole dataset is a form of semi-supervised learning. For a detailed description of semi-supervised learning models, please refer to the work by Bagherzadeh and Asil [46].

[27]For a full list of papers that used the Minisom implementation of SOM, we refer the reader to its official Github repository: https://github.com/JustGlowing/minisom

only 671 data points, while System 2 contained 20,527 data points, and System 3 contained 15,153 data points); (ii) the heterogeneity of the data itself; and (iii) the considerably small total size of the dataset. Nevertheless, these results can be used as a baseline for other implementations of unsupervised learning on the agricultural SCs domain.

5.3 Benefits for SC Traceability

The third important aspect is related to the benefits of implementing the analyzed models for traceability in agricultural SCs. The results obtained indicate that unsupervised machine learning models can successfully be used for automatic data labeling of agricultural SC data. They can also help identify patterns and clusters in the data and help to understand the relative importance of the different features in those datasets. An unsupervised machine learning model could be used to label the data before it goes to a middleware, where it would be processed and inserted in a cloud database for the whole SC.

This could have direct and indirect impacts on SC management. The direct impacts would be mainly related to traceability, improving the identification of data origins and the system that generated them, and identifying patterns in the data and the features that can contribute to the definition on which cluster a specific data point belongs. Among the indirect impacts, the most significant one would be to provide information for improving the agents' decision-making in the chain.

5.4 Adaptation to Other SCs

The fourth important aspect is related to applying the results obtained in this chapter on other SCs. Even though the SOM model performed better than the k-means in the dataset used, it is not possible to guarantee that this is always the case. Other SCs, with data generated from other systems, could have datasets with a significantly different structure, leading to different results.

Nevertheless, this is an initial analysis for the agricultural domain, so more experiments with different datasets (both from agricultural and from other domains) would help to better understand the use of unsupervised machine learning techniques for data labeling of SC data. The methodology used in this chapter and its code can be replicated on other datasets, further exploring the use of this group of models on SC data.

Lastly, for the implementation of algorithms that demand an adequate number of clusters as inputs, such as the k-means, there are several methods for determining the number of clusters based on the data structure, such as those used in this chapter (dendrogram and elbow method). Nevertheless, as it was observed, they may result in different numbers of desirable clusters, especially when considering the use of PCA. Therefore, previous knowledge on the problem's domain is essential for implementing unsupervised learning techniques that use the number of clusters as an input.

6 Conclusions

Agricultural SCs generate an enormous quantity of heterogeneous data. To improve its use for decision making, it is important to label the data with the information from its originating system. The three most important systems are: (i) product identification and traceability; (ii) environmental monitoring; and (iii) processes monitoring. Nevertheless, it is too costly to manually label the data, as millions of data points are generated in an SC.

In this chapter, a dataset representing a generic agricultural SC was built, and two important unsupervised machine learning models, k-means and SOM, were used to identify patterns on the data and label the data points. The use of PCA was also evaluated. The SOM was the model that presented the best results from the three groups of metrics evaluated (traditional supervised learning, supervised clustering, and unsupervised clustering). It obtained a weighted average for the F1-score of 0.91 and was the only model able to detect the smallest cluster data points. The use of the Gaussian neighborhood function presented the best cluster separation, as evidenced by the cluster map generated.

The primary limitations that impacted this chapter were: (i) lack of open datasets to evaluate agricultural SCs, especially considering the whole chain; and (ii) lack of frameworks for analyzing heterogeneous, non-standardized data generated by the three main systems throughout agricultural SCs. Future work involves: (i) adding more data to the dataset, especially from processes monitoring, to reduce problems of data imbalance; (ii) implementing other clustering models, such as DBSCAN; and (iii) implementing other dimensionality reduction techniques that may improve the clustering quality, such as the t-Distributed Stochastic Neighbor Embedding (t-SNE).

Acknowledgments This work was supported by Itaú Unibanco S.A. through the Itaú Scholarship Program, at the Centro de Ciência de Dados (C^2D), Universidade de São Paulo, Brazil, by the National Council for Scientific and Technological Development (CNPq), and also by the Coordenação de Aperfeiçoamento de Pessoal de Nível Superior - Brazil (CAPES) - Finance Code 001.

References

1. Chopra, S., & Meindl, P. (2013). *Supply chain management: Strategy, planning, and operation* (5th ed., 528pp). Pearson Education.
2. Corella, V. P., Rosalen, R. C., & Simarro, D. M. (2013). SCIF-IRIS framework: A framework to facilitate interoperability in supply chains. *International Journal of Computer Integrated Manufacturing, 26*(1–2), 67–86.
3. Huang, C. C., & Lin, S. H. (2010). Sharing knowledge in a supply chain using the semantic web. *Expert Systems with Applications, 37*(4), 3145–3161.
4. Pang, Z., Chen, Q., Han, W., & Zheng, L. (2015). Value-centric design of the internet-of-things solution for food supply chain: Value creation, sensor portfolio and information fusion. *Information Systems Frontiers, 17*(2), 289–319.
5. Verdouw, C. N., Vucic, N., Sundmaeker, H., & Beulens, A. (2013). Future internet as a driver for virtualization, connectivity and intelligence of agri-food supply chain networks. *International Journal on Food System Dynamics, 4*(4), 261–272.
6. Verdouw, C. N., Wolfert, J., Beulens, A. J. M., & Rialland, A. (2016). Virtualization of food supply chains with the internet of things. *Journal of Food Engineering, 176*, 128–136.
7. Verdouw, C., Sundmaeker, H., Tekinerdogan, B., Conzon, D., & Montanaro, T. (2019). Architecture framework of IoT-based food and farm systems: A multiple case study. *Computers and Electronics in Agriculture, 165*(104939), 1–26.
8. Harris, I., Wang, Y., & Wang, H. (2015). ICT in multimodal transport and technological trends: Unleashing potential for the future. *International Journal of Production Economics, 159*, 88–103.
9. Jain, A. K. (2010). Data clustering: 50 years beyond k-means. *Pattern Recognition Letters, 31*(8), 651–666.
10. Steinley, D. (2006). K-means clustering: A half-century synthesis. *British Journal of Mathematical and Statistical Psychology, 59*(1), 1–34.
11. Kohonen, T. (1982). Self-organized formation of topologically correct feature maps. *Biological Cybernetics, 43*(1), 59–69.
12. Hartigan, J. A. (2000). 'Statistical clustering'. *International Encyclopedia of the Social and Behavioral Science*, Yale University, 15014–15019. https://citeseerx.ist.psu.edu/viewdoc/summary?doi=10.1.1.38.1277
13. Lance, G. N., & Williams, W. T. (1967). A general theory of classificatory sorting strategies: 1. Hierarchical systems. *The Computer Journal, 9*(4), 373–380.
14. Ghahramani, Z. (2003). Unsupervised learning. In: *Summer School on Machine Learning*. Springer: Berlin, 72–112.
15. Liakos, K., Busato, P., Moshou, D., Pearson, S., & Bochtis, D. (2018). Machine learning in agriculture: A review. *Sensors, 18*(8), 1–29.
16. Arthur, D., & Vassilvitskii, S. (2007). K-means++: The advantages of careful seeding. In *Proceedings of the 18th annual ACM-SIAM symposium on discrete algorithms* (pp. 1027–1035). ACM.
17. Mehta, P., Shah, H., Kori, V., Vikani, V., Shukla, S., & Shenoy, M. (2015). Survey of unsupervised machine learning algorithms on precision agricultural data. In *2015 international conference on innovations in information, embedded and communication systems (ICIIECS)* (pp. 1–8). DRDO.
18. Ester, M., Kriegel, H. P., Sander, J., & Xu, X. (1996). A density-based algorithm for discovering clusters in large spatial databases with noise. In: *Proceedings of the 2nd international conference on knowledge discovery and data mining*, AAAI Press, pp. 226–231.
19. Ankerst, M., Breunig, M. M., Kriegel, H. P., & Sander, J. (1999). OPTICS: Ordering points to identify the clustering structure. *ACM SIGMOD Record, 28*(2), 49–60.
20. Gowda, K. C., & Ravi, T. V. (1995). Divisive clustering of symbolic objects using the concepts of both similarity and dissimilarity. *Pattern Recognition, 28*(8), 1277–1282.

21. Fisher, D. H. (1987). Knowledge acquisition via incremental conceptual clustering. *Machine Learning, 2*(2), 139–172.
22. Ramesh, V., Ramar, K., & Babu, S. (2013). Parallel k-means algorithm on agricultural databases. *International Journal of Computer Science Issues (IJCSI), 10*(1), 710–713.
23. Kind, M. C., & Brunner, R. J. (2014). SOMz: Photometric redshift PDFs with self-organizing maps and random atlas. *Monthly Notices of the Royal Astronomical Society, 438*(4), 3409–3421.
24. Samsonova, E. V., Kok, J. N., & IJzerman, A. P. (2006). TreeSOM: Cluster analysis in the self-organizing map. *Neural Networks, 19*(6–7), 935–949.
25. Jeong, K. S., Hong, D. G., Byeon, M. S., Jeong, J. C., Kim, H. G., Kim, D. K., & Joo, G. J. (2010). Stream modification patterns in a river basin: Field survey and self-organizing map (SOM) application. *Ecological Informatics, 5*(4), 293–303.
26. Ruß, G., & Kruse, R. (2011). Exploratory hierarchical clustering for management zone delineation in precision agriculture. In *Industrial conference on data mining* (pp. 161–173). Springer.
27. Mingoti, S. A., & Lima, J. O. (2006). Comparing SOM neural network with fuzzy c-means, k-means and traditional hierarchical clustering algorithms. *European Journal of Operational Research, 174*(3), 1742–1759.
28. Van den Broeck, J., Cunningham, S. A., Eeckels, R., & Herbst, K. (2005). Data cleaning: detecting, diagnosing, and editing data abnormalities. *PLoS Med, 2*(10), e267, pp.966–970.
29. Wang, X., & Wang, C. (2019). Time series data cleaning: A survey. *IEEE Access, 8*, 1866–1881.
30. Sun, W., Cai, Z., Li, Y., Liu, F., Fang, S., & Wang, G. (2018). Data processing and text mining technologies on electronic medical records: A review. *Journal of Healthcare Engineering, 2018*, 1–9.
31. Castanedo, F. (2013). A review of data fusion techniques. *The Scientific World Journal, 2013*, 1–19.
32. Singh, D., & Singh, B. (2019). Investigating the impact of data normalization on classification performance. *Applied Soft Computing, 105524*, 1–23.
33. Zambelli, A. E. (2016). A data-driven approach to estimating the number of clusters in hierarchical clustering. *F1000Research, 5*, 1–11.
34. Kodinariya, T. M., & Makwana, P. R. (2013). Review on determining number of cluster in k-means clustering. *International Journal of Advance Research in Computer Science and Management Studies, 1*(6), 90–95.
35. Jolliffe, I. T., & Cadima, J. (2016). Principal component analysis: A review and recent developments. *Philosophical Transactions of the Royal Society A: Mathematical, Physical and Engineering Sciences, 374*, 1–16.
36. Yin, H. (2008). The self-organizing maps: Background, theories, extensions and applications. In *Computational intelligence: A compendium* (pp. 715–762). Springer.
37. Liu, Y., & Weisberg, R. H. (2011). A review of self-organizing map applications in meteorology and oceanography. *Self-organizing maps: Applications and Novel Algorithm Design*, InTech:Croatia, 253–272. https://books.google.com.br/books?hl=en&lr=&id=k-SgDwAAQBAJ&oi=fnd&pg=PA253
38. Vettigli, G. (2013). *MiniSom: minimalistic and NumPy-based implementation of the Self Organizing Map*. Available at: https://github.com/JustGlowing/minisom/. Accessed on: October 15th, 2020.
39. Scribner, E. A., Battaglin, W. A., Dietze, J. E., & Thurman, E. M. (2003). *'Reconnaissance data for glyphosate, other selected herbicides, their degradation products, and antibiotics in 51 streams in nine Midwestern States', open-file report 03–217* (102pp). U.S. Geological Survey.
40. Silva, A. M. D., Degrande, P. E., Suekane, R., Fernandes, M. G., & Zeviani, W. M. (2012). Impacto de diferentes níveis de desfolha artificial nos estádios fenológicos do algodoeiro. *Revista de Ciências Agrárias, 35*(1), 163–172.

41. Omer, S. O., Abdalla, A. W. H., Mohammed, M. H., & Singh, M. (2015). Bayesian estimation of genotype-by-environment interaction in sorghum variety trials. *Communications in Biometry and Crop Science, 10*, 82–95.
42. Goutte, C., & Gaussier, E. (2005). A probabilistic interpretation of precision, recall and F-score, with implication for evaluation. In *European conference on information retrieval (ECIR)* (pp. 345–359). Springer.
43. Murtagh, F., & Contreras, P. (2012). Algorithms for hierarchical clustering: An overview. *Wiley Interdisciplinary Reviews: Data Mining and Knowledge Discovery, 2*(1), 86–97.
44. Mohamad, I. B., & Usman, D. (2013). Standardization and its effects on k-means clustering algorithm. *Research Journal of Applied Sciences, Engineering and Technology, 6*(17), 3299–3303.
45. Kotsiantis, S., Kanellopoulos, D., & Pintelas, P. (2006). Handling imbalanced datasets: A review. *GESTS International Transactions on Computer Science and Engineering, 30*(1), 25–36.
46. Bagherzadeh, J., & Asil, H. (2019). A review of various semi-supervised learning models with a deep learning and memory approach. *Iran Journal of Computer Science, 2*(2), 65–80.
47. Mansha, S., Babar, Z., Kamiran, F., & Karim, A. (2016). Neural network based association rule mining from uncertain data. In *International conference on neural information processing* (pp. 129–136). Springer.
48. Stoean, C., Stoean, R., Becerra-García, R. A., García-Bermúdez, R., Atencia, M., García-Lagos, F., Velázquez-Pérez, L., & Joya, G. (2019). Unsupervised learning as a complement to convolutional neural network classification in the analysis of saccadic eye movement in spino-cerebellar ataxia type 2. In *International work-conference on artificial neural networks* (pp. 26–37). Springer.
49. Riese, F. M., Keller, S., & Hinz, S. (2020). Supervised and semi-supervised self-organizing maps for regression and classification focusing on hyperspectral data. *Remote Sensing, 12*(1), 1–23.

www.ingramcontent.com/pod-product-compliance
Ingram Content Group UK Ltd.
Pitfield, Milton Keynes, MK11 3LW, UK
UKHW022213051026
15624UKWH00001B/145
* 9 7 8 3 0 3 0 8 4 1 4 7 8 *